The Technology Pork Barrel

The Technology Pork Barrel

Linda R. Cohen
& Roger G. Noll

With Jeffrey S. Banks, Susan A. Edelman,
and William M. Pegram

The Brookings Institution
Washington, D.C.

1775 Massachusetts Avenue, N.W., Washington, D.C. 20036

Library of Congress Cataloging-in-publication data:

The Technology pork barrel / Linda R. Cohen and Roger G. Noll.
p. cm.
Includes bibliographical references and index.
ISBN 0-8157-1508-0 (cloth)—ISBN 0-8157-1507-2 (paper)
1. Research, Industrial—Government policy—United States.
2. Technology and state—United States. 3. Research and development contracts, Government—United States. I. Cohen, Linda R., 1952–
II. Noll, Roger G.
HC110.R4T43 1991 91-8832
338′.064—dc20 CIP

9 8 7 6 5 4 3 2 1

The paper used in this publication meets the minimum requirements of the American National Standard for Information Sciences—Permanence of paper for Printed Library Materials, ANSI Z39.48-1984.

THE BROOKINGS INSTITUTION

The Brookings Institution is an independent organization devoted to nonpartisan research, education, and publication in economics, government, foreign policy, and the social sciences generally. Its principal purposes are to aid in the development of sound public policies and to promote public understanding of issues of national importance.

The Institution was founded on December 8, 1927, to merge the activities of the Institute for Government Research, founded in 1916, the Institute of Economics, founded in 1922, and the Robert Brookings Graduate School of Economics and Government, founded in 1924.

The Board of Trustees is responsible for the general administration of the Institution, while the immediate direction of the policies, program, and staff is vested in the President, assisted by an advisory committee of the officers and staff. The by-laws of the Institution state: "It is the function of the Trustees to make possible the conduct of scientific research, and publication, under the most favorable conditions, and to safeguard the independence of the research staff in the pursuit of their studies and in the publication of the results of such studies. It is not a part of their function to determine, control, or influence the conduct of particular investigations or the conclusions reached."

The President bears final responsibility for the decision to publish a manuscript as a Brookings book. In reaching his judgment on the competence, accuracy, and objectivity of each study, the President is advised by the director of the appropriate research program and weighs the views of a panel of expert outside readers who report to him in confidence on the quality of the work. Publication of a work signifies that it is deemed a competent treatment worthy of public consideration but does not imply endorsement of conclusions or recommendations.

The Institution maintains its position of neutrality on issues of public policy in order to safeguard the intellectual freedom of the staff. Hence interpretations or conclusions in Brookings publications should be understood to be solely those of the authors and should not be attributed to the Institution, to its trustees, officers, or other staff members, or to the organizations that support its research.

Foreword

COMMERCIAL RESEARCH and development is widely recognized as an important source of economic growth, and so quite naturally the U.S. federal government has had a long history of sponsoring selected R&D projects aimed at developing new technology for private use. In the post-1970 era, when the United States has experienced slower productivity growth and an erosion of competitive advantage in several important industries, many policy analysts and public officials have proposed expanding the federal role in commercial R&D. Some argue that the government should focus on "generic" research to build broad technical foundations for industries. Others believe that the government should pursue R&D through large commercial demonstration projects in which the government builds the first production facilities that use new technology.

In this study, Linda R. Cohen and Roger G. Noll, assisted by Jeffrey S. Banks, Susan A. Edelman, and William M. Pegram, examine the histories of six major commercial R&D programs. Going beyond a simple assessment of the economic and political success of these programs, the authors seek to ascertain the factors that lead to successful programs. The core challenge to public officials is not only to find ways to organize and manage R&D programs so they are free from large swings in annual expenditures, but also to be sensitive to the fact that because R&D projects are risky, some are likely to fail and should be canceled or redirected in midstream.

Some successful R&D programs, such as the program to make photovoltaic electricity generation economically attractive, have suffered from highly unstable annual budgets. Some unsuccessful projects, such as the Clinch River Breeder Reactor, have continued to receive support long past the time when the project was clearly destined to fail, because of inflexibilities in project design and political imperatives. The authors conclude that these problems are inherent in long-term, risky investments by the government, and thus constitute an argument for favoring generic research activities rather than large-scale commercialization projects. They also conclude that both generic and commercialization programs would be more likely to succeed if responsibility for the programs were less fragmented among agencies and congressional committees.

Linda R. Cohen is an associate professor of economics at the University of California, Irvine, and Roger G. Noll is Morris M. Doyle Centennial Professor in Public Policy at Stanford University. Both were formerly members of the Brookings staff in the Economic Studies program. Jeffrey S. Banks is associate professor of economics and political science at the University of Rochester; Susan A. Edelman is assistant professor of economics in the Graduate School of Business at Columbia University; and William M. Pegram is a staff economist in the Bureau of Economics at the Federal Trade Commission.

The authors wish to thank their many colleagues who provided criticism and suggestions on drafts of all or part of the book: Gardner Brown, Bruce Cain, Nathaniel Cohen, Edward Denison, Jeffrey Dubin, Burton Edelson, Morris Fiorina, Robert Fri, John Gates, Leland Johnson, John Ledyard, John Logsdon, Mollie Macauley, Mathew McCubbins, Richard Nelson, Lawrence Papay, Paul Portney, Edward Posner, Alice M. Rivlin, Charles L. Schultze, and Dwain Spenser. They also thank Julia Leighton for research assistance and Deborah Bernheim, Laura Dolly, Joy Hansen, Deborah Johnston, Dee Koutris, and Carolyn Lloyd for secretarial support. Finally, the authors specially acknowledge the encouragement of the much-beloved Joseph A. Pechman, who helped shape the original research project leading to this book and whose guidance and support were indispensable in bringing the project to fruition.

Brookings and the authors are grateful to many organizations for providing financial support for the work in preparing this book: the Caltech Environmental Quality Laboratory through a grant from the Exxon Educational Foundation, the Center for Advanced Study in the Behavioral Sciences through a grant from the National Science Foundation, the Hoover Institution, the Stanford Center for Economic Policy Research through grants from the John and Mary Markle Foundation, the Smith Richardson Foundation, the Alfred P. Sloan Foundation, the Andrew W. Mellon Foundation, and the University of Michigan School of Law through its Sunderland Fellowship program.

The views expressed here are the authors' alone and should not be ascribed to the persons or organizations whose assistance is acknowledged or to the trustees, officers, or staff members of the Brookings Institution.

May 1991 Bruce K. MacLaury
Washington, D.C. *President*

Contents

Tables

Figures

The Technology Pork Barrel

Chapter 1

New Technology and National Economic Policy

LINDA R. COHEN & ROGER G. NOLL

AMERICAN public policy has a long history of technological optimism: a belief that "technological growth will . . . expand resources ahead of exponentially increasing demands."[1] The historical economic and military success of the United States forms the basis for America's technological optimism. Despite a few decade-long periods of serious economic setbacks, persistent historical growth in per capita income has caused Americans to reject alternative views of the economy, such as Marxism and Malthusianism, the two most prominent pessimistic theories of long-run economic conditions in industrialized capitalistic societies. Likewise, the success of the United States in all-out wars and its freedom from the devastation of modern warfare contribute to optimism about technological advances in weaponry.

In the public sphere technological optimism leads government officials to respond to a serious national problem by throwing technology at it. Even when science provides no basis for believing that a "technical fix" is feasible, scientists and engineers are called on to invent not only the appropriate technical solution but the fundamental new knowledge that will make it possible. And usually, amid controversy about its feasibility, important figures in the scientific community rally to the challenge and predict their own eventual success. Often, but not always, they are right.

The government's optimism about technology knows neither programmatic, partisan, nor ideological bounds. In national defense, for example, the Reagan administration had the strategic defense initiative (Star Wars), but previous administrations pursued the nuclear airplane, Skybolt, and

1. William Ophuls, *Ecology and the Politics of Scarcity: Prologue to a Political Theory of the Steady State* (San Francisco: Freeman, 1977), p. 116.

the TFX—and atomic weapons, radar, and the Polaris missile. President Ronald Reagan, despite his stated objective of reducing the role of government in the economy, supported programs to develop commercially attractive breeder reactors, nuclear fusion, orbital manufacturing facilities (Space Lab), and rocket planes (the Orient Express). Presidents Richard Nixon, Gerald Ford, and Jimmy Carter responded to the energy crises of the 1970s by injecting billions of federal dollars into research and development (R&D) on virtually every known and suspected technology for producing energy. Indeed, technology was asked not only to carry the burden of weaning the nation from imported oil but also to offset perverse regulatory and macroeconomic policies. Before that, the Kennedy-Johnson administration gave the nation the Apollo program, the supersonic transport (SST), and the war on cancer; the Eisenhower administration created the semiconductor industry and gave the nation Atoms for Peace with nuclear power that would be "too cheap to meter"; and the Roosevelt and Truman administrations financed the birth of the computer industry.

Active federal participation in the development and encouragement of technological innovation has even more distant historical roots. In 1836 Congress appropriated $30,000 to subsidize Samuel Morse's first telegraph, an experimental line from Washington, D.C., to Baltimore.[2] The land-grant colleges were created after the Civil War in part to advance technology in resource-based industries, and in the 1890s the Department of Agriculture supported research that produced the first high-yield hybrid seeds.[3]

These examples illustrate the historical pervasiveness of federal policies to develop new technology. They also illustrate the heterogeneity of the purpose, success, and means of R&D programs.

Programs differ in the extent to which they blend three conceptually distinct purposes: contributing to the general scientific and technological base from which new technologies will be developed, advancing technology in the production and performance of goods and services for which government is the dominant customer (for example, national defense), and developing advanced technologies for producing goods and services primarily for the private sector.

The success of programs in all three categories has been mixed. Failures have been sufficiently common, especially in recent years, to cause federal

2. Gerald W. Brock, *The Telecommunications Industry* (Harvard University Press, 1981), p. 56.

3. R. E. Evenson, "Agriculture," in Richard R. Nelson, ed., *Government and Technological Progress: A Cross-Industry Analysis* (Pergamon, 1982), pp. 233–82.

R&D to be sometimes regarded as another component of the pork barrel. Despite this reality, the list of successes is hardly empty, even if the list is restricted to technologies that have been widely adopted in the private sector. Examples are aircraft, telegraphy, hybrid seeds, computers, semiconductors, and communications satellites.

The federal government has used a wide range of methods to support R&D. At one extreme are formulaic approaches that allow decisionmakers in private organizations to make all the important decisions about the size and scope of research projects. Examples are tax incentives for R&D, sales and price guarantees for the products of new technologies, institutional support for universities (as exemplified by the creation of land-grant colleges), and the Independent Research and Development (IR&D) program for contractors of the Department of Defense. At the other extreme are targeted research, development, and demonstration projects that are managed by government agencies and financed by project-specific appropriations and authorizations. Examples are the space shuttle, the Power Reactor Demonstration Program, and most of the commercial demonstration projects for new energy technology that were undertaken in the late 1970s.

The focus of this book is on one category of programs to support R&D: federal projects to develop new commercial technology for the private sector. The following characteristics pertain to this category of projects:

—When the project is adopted, commercialization is a sufficiently important objective that the program cannot be considered a success unless the private sector adopts the technology at the conclusion of the program.

—The program is financed by direct federal expenditures through the normal budgetary process with review by the Office of Management and Budget and appropriations by Congress.

—Government officials are deeply involved in technical design and management decisions.

—From the beginning, the government is committed to support the development of prototypes and the first commercial demonstration, not just the early research or a component of a broader private research effort.

—Although the advocates of the programs may have complex reasons for favoring them, each program is justified initially on strictly economic criteria: that is, it will reduce costs or enhance performance by a sufficient amount to make it an economically attractive investment, even though the private sector alone is unwilling or unable to undertake the project.

Most federal programs to support R&D do not fit into this category.

In this book we exclude defense programs, even those with important commercial spillovers, as long as the purpose of the program is purely military. For example, the military supported much of the development of computer technology in the late 1940s and 1950s, and important advances in commercial computing arose from projects supported by the Department of Defense; however, prospects for commercial success did not influence the design and operation of the program. Indeed, most people in the computer industry were surprised by the development of a big commercial market.[4]

We also exclude programs to support basic (or, more accurately, generic) research that is not oriented to any specific applications and that has no federal commitment, or even expectation, of support should a potential application emerge. Most of the research supported by the National Science Foundation belongs in this category, as do the high-energy physics program in the Department of Energy and the space science program in the National Aeronautics and Space Administration. Even though the research supported by these programs may contribute to the technological base for future public and private applications, and may be used to justify them, these programs are not managed to produce specific commercial results. Finally, we exclude programs that rely exclusively on indirect forms of federal support for private R&D, such as the R&D tax credit or the use of federal procurement to guarantee a market to the developer of a new technology.

We believe that federal commercial R&D projects are an analytically useful category for three reasons. First, they are theoretically distinct in the kinds of decisionmaking problems and incentives they create for public officials. The economic success of commercial R&D programs depends on acceptance in the market. In the private sector the problems of relating uncertain current research to uncertain future market conditions are demanding and risky enough, but they are even harder for public officials, who usually have less access to information about markets than their counterparts in the private firms. Moreover, the public decisionmaker must consider political forces that would be of little or no concern to a private organization. These range from narrowly distributive characteristics of a program—the geographic and organizational identities of its beneficiaries—to the relationship of the program to broad national issues such as

4. Barbara Goody Katz and Almarin Phillips, "The Computer Industry," in Nelson, ed., *Government and Technological Progress*, pp. 162–232.

economic growth, national prestige, national security, and international technological leadership. For reasons elaborated in chapter 4, government commercialization projects have distinct political ramifications and managerial problems owing to this combination of high performance uncertainty, market-oriented objectives, and the prosaic political constraints of direct expenditure programs.

Second, federal commercial R&D projects constitute a practically relevant category in the debate over the role of the government in affecting the structure of the economy, that is, industrial policy. Proponents of industrial policy believe government should do more than simply maintain a healthy economic environment in which private individuals, through millions of decentralized decisions, determine the rise and fall of industries and geographic locations. Although various industrial policies have been proposed, a common element is active government participation in decisions about the allocation of investment capital and research among industries. A key issue in this debate is whether the government is likely to do more harm than good in deciding which industries and which technologies to support. One influential advocate of industrial policy has put the issue succinctly:

> To be successful, industrial policies require that government participants have some ability to exercise discretion and judgment. These participants have to have some freedom from the geographic constraints that so often are built into American legislation. . . . For example, if Congress insisted that research and development funds or investment banking funds be spent equally in each Congressional district, industrial policies could not possibly work.[5]

In an era when congressional campaigns cost millions of dollars and involve hundreds of volunteers, geography is not the only distributional consideration likely to tempt a legislator. In addition, government agencies and private companies face a different set of management problems. Although advocates of industrial policy have examined private organizational failures in pursuing new commercial technologies, parallel forms of management problems, which must be overcome to pursue a commercial R&D project successfully, can also arise in a public organization.

5. Lester C. Thurow, *The Case for Industrial Policies* (Washington: Center for National Policy, 1984), p. 21.

Third, commercial R&D programs constitute an important part of national R&D. Regardless of how the debate on industrial policy is resolved, history provides convincing evidence that federal programs to develop new technologies for the private sector are unlikely to disappear.

Our purpose in writing this book is to shed light on a core policy question: how do economic and political factors influence the performance of commercial R&D programs, and to what extent are these influences predictable and controllable? Obviously, federal commercial R&D projects are not all failures, nor are they all successes, which raises the question about why performance is variable. One possibility is blind chance: not all highly uncertain leaps into an uncharted technical frontier will be rewarded; however, as technological optimists, we might expect some successes if the direction of the leap is at least informed by expertise. Nevertheless, our theoretical and empirical research convinces us that much of the observed variability in the performance of these projects is systematic and predictable. The extent to which political forces work to the detriment of a program depends on the economics of the industry undertaking the project and receiving its benefits and on the nature of the political issues that give rise to it. Moreover, the factors influencing the success of a project are not only predictable but in part controllable, in the sense that the details of the way the program is set up can affect its expected performance. Unfortunately, some conditions that make a project more likely to succeed—for example, an industry in which competition among firms contributes to underinvestment in R&D—also make it vulnerable to premature cancellation, even if it is on target to technical and economic success.

The first four chapters of the book explain these conclusions in detail and provide their conceptual foundations. At the center of our analytical framework is the underlying economics of a federal commercialization project: the economic rationale for the program. In the rest of this chapter we examine the importance of R&D to the national economy and noneconomic national goals, explaining why the nation's research effort is likely to be a matter of public attention. Chapter 2 deals with two questions: why might the private sector underinvest in R&D, and why might the federal government have valid economic reasons to focus support on particular industries and technologies? In chapter 3 we address the issue of managing R&D commercialization programs, again from the standpoint of economic efficiency. The object is to identify the decision problems faced by an efficient manager and discuss how these problems are solved. Chapter 4 examines the politics of commercialization programs. We identify the external political forces acting on elected and bureaucratic officials as they decide to undertake a commercialization program and then decide subsequently, on the basis of accumulated experience, whether to continue, terminate, or redirect the program. The objective is to identify how and why the performance objectives of a program can come into conflict with other political forces, and how these conflicts are resolved.

The remainder of the book applies the analytical framework of the first four chapters to study the history of specific commercialization programs. Chapter 5 discusses the methods used to apply our conceptual framework to the study of a program. Chapters 6 through 11 contain detailed technical, economic, and political histories of six programs: the supersonic transport, communications satellites, the space shuttle, the breeder reactor, synthetic fuels, and photovoltaics. Each chapter describes the technical problem that the program was designed to solve, evaluates the economic rationale of the program throughout its life, and examines its political support from the president, Congress, and the bureaucracy to detect its political assets and liabilities. In each case the authors seek to identify points in the history of a program when its economic attractiveness diverges from its political appeal and to explain how this conflict was resolved. In the concluding chapter we draw together these findings, plus information taken from other case studies, to make several policy recommendations.

R&D and National Goals

When asking why government pays for private R&D, two kinds of answers may be sought: why *does* government support the activity, and why *should* it do so? The first is about the behavior of public officials: what motivates their actions, what forces influence their decisions, and how are their decisions constrained by the institutions and information available to them? The second is about values: what pursuits of government are legitimate, how can conflicting objectives be comparatively evaluated, and what makes decisions either better or worse? Both groups of questions are complicated, and substantial further progress in social science is necessary before either can be examined and answered comprehensively. Yet both are important, and we will deal with each, beginning with the second.

R&D and Economic Growth

The argument for government support for research and development begins with the observation that R&D is important to the national economy.

Since the pioneering work of Moses Abramovitz and Robert Solow,[6] economists have developed various ways to apportion the sources of economic growth to increased use of productive resources and to improvements in knowledge (for example, technological change). The consensus is that the latter is probably the single most important source of growth in per capita national income. Moreover, societies with high wages can continue to experience high rates of growth only if they are continuously on the edge of the technical frontier. If know-how is roughly the same everywhere, rapid growth in a high-wage society is unlikely to be sustained in competition with a low-wage society.[7] As an empirical matter, the most economically advanced nations tend to be the principal producers and exporters of the most technically sophisticated products.[8]

These conclusions are based on many research studies that apply the economic theory of production to the problems of accounting for increases in national income. For any production unit, whether a company or a nation, output is assumed to be functionally related to the use of inputs, which are conventionally categorized as labor, capital equipment, and natural resources (land, water, air, and so on). The functional relation between inputs and outputs represents the technology of production: the managerial and technical know-how that enables people to engage in transforming inputs into useful products. One can view technological progress as continuing changes in the form of the production function that lead to more output for a given amount of inputs. The research on growth accounting consists of various approaches and refinements to measuring shifts in the production function.

The research by Solow, and later in greater depth by Edward Denison,[9]

6. Moses Abramovitz, "Resource and Output Trends in the U.S. since 1870," *American Economic Review*, vol. 46 (May 1956, *Papers and Proceedings, 1955*), pp. 1–23; and Robert M. Solow, "Technological Change and the Aggregate Production Function," *Review of Economics and Statistics*, vol. 39 (August 1957), pp. 312–20.

7. Richard R. Nelson, "Policies in Support of High Technology Industries," ISPS Working Paper 1011, Yale University, 1984.

8. Raymond Vernon, ed., *The Technology Factor in International Trade* (National Bureau of Economic Research, distributed by Columbia University Press, 1970); and Louis Schorsch, *Federal Support for R&D and Innovation* (Congressional Budget Office, 1984), pp. 34–36.

9. The most important contributions by Edward F. Denison include *The Sources of Economic Growth in the United States and the Alternatives Before Us* (New York: Committee for Economic Development, 1962); *Accounting for United States Economic Growth, 1929–69* (Brookings, 1974); *Accounting for Slower Economic Growth: The United States in the*

estimates the contributions of traditional inputs to total output either through time or between nations and then calculates a "residual" growth in output that is not attributable to increases in these inputs. Solow found that the unexplained residual accounted for approximately 80 percent of the growth in output per worker between 1909 and 1949. So surprising was this finding that it gave rise to additional research that is still continuing, the aim of which is to whittle down the unexplained residual by using more sophisticated statistical techniques, by adding more detailed categories of inputs, and by trying to measure directly some of the important sources of increased technical know-how, such as education and R&D.

The work of Denison is especially important in this regard. Using detailed data from the system of national accounts, Denison breaks down into many categories the sources of differences in growth rates through time and among countries. His studies of growth in the United States examine the contribution of education, reductions in production inefficiencies such as misallocation of inputs, and societal factors such as crime and the extent of regulation. His international comparisons also deal with how much R&D expenditures affect differences in growth rates. Perhaps most important, his meticulous exposition of his methods makes abundantly clear all the points in the analysis where his estimates depend on key technical assumptions. Denison, and others, has therefore been able to undertake studies of the dependence of his results on his technical assumptions.

Denison's basic findings on the sources of growth in total potential national income from 1929 to 1969 are reported in the following table:[10]

Source of growth	*Percent*
Increased labor input	26.7
Increased capital input	14.7
Increased land input	0.0
Economies of scale	10.6
Improved resource allocation	8.8
Miscellaneous	0.3
Education	12.0
Advances in knowledge, etc.	27.0

The last two categories in the table, education and advances in knowledge,

1970s (Brookings, 1979); and, with Jean-Pierre Poullier, *Why Growth Rates Differ: Postwar Experience in Nine Western Countries* (Brookings, 1967).

10. Denison, *Accounting for United States Economic Growth*, p. 128.

pertain most closely to technological progress. Education refers to the years of schooling of the work force. Denison adjusts labor input to account for the fact that an hour of work by a college graduate is more productive than an hour by an uneducated worker. The contribution of education to growth represents the increase in output due to the rising average educational level of the population over time. Part of the earnings effect of educational attainment is related to technological progress. Increased education, among other things, is associated with increased ability to develop or to use new technologies.

The last category, advances in knowledge, is what remains of the residual detected by Solow. The category comprises all sources of growth in output that could not be measured directly. To call the category "advances in knowledge" is somewhat controversial; it is really a catchall for all unmeasured or improperly measured sources of growth. Nevertheless, other elements that would be included in this category are unlikely to have contributed anything like one-fourth of the historical growth in the U.S. economy; hence the consensus has developed that most of this number represents advances in technical know-how.

Denison has also calculated the relative contributions of the same factors to growth in output per worker, which produces the allocation to technological change comparable to Solow's estimate. According to Denison, advances in knowledge account for 46 percent of the growth in output per labor-force participant and 48 percent of the growth in output per employed worker. Education adds another 21 percent. Thus if one begins with Solow's 81 percent residual, Denison's more detailed work cuts the amount of growth per worker that is not explained by traditional production inputs to only 69 percent; additional education, some of which is related to technical progress, reduces the unexplained residual to about half, most of which is plausibly attributable to technological change.

Although most economists accept these conclusions, many believe the attribution of the residual to technological progress is on sufficiently shaky scientific ground to warrant developing a formal model to estimate the process by which advances in knowledge take place.[11] Although doing so can be complex, the basic idea is to see how productivity growth is related to measurable inputs in the process that produces new technology. For example, Frederic Scherer finds that in the postwar era R&D has added to

11. For a good summary of these issues, see Richard R. Nelson, "Research on Productivity Growth and Differences: Dead Ends and New Departures," *Journal of Economic Literature*, vol. 19 (September 1981), pp. 1029–64.

the rate of growth by about 1 percentage point a year, or about half the annual rate of growth in productivity.[12] This result is roughly equal to Denison's unexplained residual; however, Scherer's estimates of the productivity of R&D are higher than other estimates. Denison and Griliches produce very low estimates,[13] but more recent work that takes account of the asset features of R&D and its interindustry spillovers produces results that are much larger than those, though lower than Scherer's.[14] We are convinced that Scherer's general approach is sound; however, because important methodological issues remain to be resolved, the proper conclusion to draw from his and other studies is that a large part of the productivity residual can be explained by R&D activity—at least one-quarter, and probably more. Whatever the precise figure, the clear conclusion is that technological progress is a vital source of economic growth and R&D a vital source of technological progress.

Other Benefits of R&D

Advancing technology is also alleged to provide other benefits, although here the argument is not backed by extensive theoretical and empirical research. Nevertheless, the arguments about other benefits are held passionately by some participants in the debate about technology policy and so can play a role in shaping federal R&D projects. These other benefits are quasi-economic because, though they deal with noneconomic objectives of society, they easily mesh with the superstructure of economic analysis—the rational choice of means to do as well as one can according to some preselected set of objectives. The most important examples pertain to national prestige, national security, international relations, and the domestic political system.

Technological progress within a nation's boundaries, as seen in more detail in the six case studies, is often said to contribute to national prestige and therefore to be something of an end in itself. Thus some Americans arguably derived special pleasure from Neil Armstrong's landing on the moon and the introduction of the Cray I supercomputer because they were "made in America." If so, part of the value of technical progress is

12. Frederic M. Scherer, "Inter-Industry Technology Flows and Productivity Growth," *Review of Economics and Statistics*, vol. 64 (November 1982), pp. 627–34.

13. Denison, *Sources of Economic Growth*; and Zvi Griliches, "Research Expenditures and Growth Accounting," in B. R. Williams, ed., *Science and Technology in Economic Growth* (Wiley and Sons, 1973), pp. 59–83.

14. Denison, *Accounting for Slower Economic Growth*, pp. 122ff.

that it is a consumption good. The presence of a large, successful publications business producing popularized literature about science and technology lends credence to the claim that new technology is more than a means to other ends. This observation is perhaps most obviously true of the space program, one purpose of which is the exploration of the solar system.

The national defense issue amounts to an argument for hedging bets by preempting expertise in rapidly evolving technologies. Because technological progress is unpredictable, research explicitly oriented toward military applications may miss an important opportunity for improved defense. Hence a society that is progressing rapidly across a wide spectrum of technical horizons is more likely to find the unanticipated nuggets of new defense technology. Even if it is not first to find them, it is likely to be positioned to explore quickly a new horizon that others have discovered. This argument is upheld qualitatively by many historical examples. The military did not support the initial breakthroughs in knowledge and technology that made possible such militarily important items as airplanes, nuclear weapons, ballistic missiles, and semiconductors, yet in each case the U.S. military has been at the forefront of their development and application in weapons systems.

A related defense issue is that national security is enhanced if the nation controls potentially destabilizing technologies. For example, preeminence in the commercial use of nuclear energy allows a nation to be influential in controlling the use of nuclear technology and the proliferation of nuclear weapons.

In international relations the role of technology is to make a country less vulnerable (or, stated less tactfully, more powerful) in dealing with other countries, especially for negotiating international agreements in which relative military strength is unimportant. Here technological change becomes a means not only for making improvements in productivity or product quality but also for increasing the feasible options in international relations. Should another nation or group of nations succeed in monopolizing an economic resource (for example, OPEC), technology can limit its ability to exploit the monopoly. Similarly, if a nation develops a unique and widely valuable technology, it can act as something of a monopolist, extracting political concessions as part of a deal to give other nations access to it. Insofar as nations have objectives that are noneconomic (for example, the Arab states and the United States vis-à-vis Israel), technical developments can affect the extent to which any one nation achieves those objectives.

Another objective served by technological progress is the stability and smooth operation of the internal political process. A society oriented toward such progress requires a more educated work force both to develop new technologies and to understand how to implement ideas developed by others. Thus technological progress provides an economic carrot for education that makes individuals willing to stay in school longer and society willing to spend more public funds on education. To the extent that education contributes to decisionmaking capabilities and to the extent that democratic political systems work better if voters and the people they elect are better decisionmakers, technical progress can provide a more smoothly functioning political system.

Technological progress is also said to contribute to overall confidence in the social system. Confidence is an elusive term, but it seems most meaningfully related to the concept of "demoralization costs" developed by Michelman.[15] When people believe they are being threatened by capricious decisions by others, it is argued, they undertake unproductive activities either to protect themselves or to "get even" with the cause of the problem. Presumably in a zero-sum society, where one person's gain is perceived as acquired at another person's expense, government is seen primarily as the force that picks winners and so is held responsible for personal well-being. Such a society is likely to be authoritarian or, if democratic, to be unstable in the sense that each election threatens to change the ruling coalition and the policies adopted to favor it. The fundamental indeterminacy of majority rule in situations of incompatible objectives produces this instability, which in turn generates the defensive actions described by Michelman as well as a general unwillingness to invest in the future because of fear of expropriation.

Technological change leads to a situation in which, for the most part, life is continually getting better for most people. As a result, government policies that foster technological progress and contribute to continuing improvement in personal living standards may avoid the perception that government's role is simply to award favors. In fact, as shown in especially convincing detail by Kieweit, Americans normally hold the government

15. Frank I. Michelman, "Property, Utility, and Fairness: Comments on the Ethical Foundations of 'Just Compensation' Law," *Harvard Law Review*, vol. 80 (April 1967), pp. 1165–1258.

responsible for the general well-being of society, but not as responsible for their personal economic problems.[16]

Undesirable Effects of Technological Change

These arguments about the economic and social value of technical progress constitute the normative component of American technological optimism: that technological change is beneficial. Their validity is controversial on several grounds. Most obviously, they ignore the detrimental effects of advancing technology.

The adoption of new technologies is likely to cause shifts in the composition of national output, at least temporarily disrupting the lives of people whose source of income has been displaced. If the new technology is promoted by government, the losers are hardly likely to absolve government from responsibility for their personal circumstances. Moreover, private decisions to adopt new technologies will undervalue spillover effects. For example, profit-seeking businesses have no incentive to develop new technologies that serve only to reduce pollution, but they do have an incentive to pursue technologies that have small cost advantages but major environmental disadvantages.[17]

With respect to national defense, the *best* outcome of a technological arms race in an otherwise stable world is that *both* sides either will fail to make progress or will simultaneously discover and adopt a new, more powerful weapons system, thereby preserving stability while devoting scarce resources to the effort. If so, technological progress in the military sphere has a negative net value, and its presence is a sort of prisoners' dilemma: all major powers must pursue new technologies (at significant expense) to end up, on balance, at best no better off, and perhaps, if the technology is destabilizing, much worse off.

As for other social goals, technology is but one narrow means to these ends. National prestige, faith in the future, and a well-functioning democracy might be even better served by promoting the arts, recreation, or religion. Even if technology has been beneficial in the past, there is no reason to believe it will continue on that path indefinitely. Opportunities for further technical progress may be diminishing, or the relationship between various noneconomic goals and technology may depend on the

16. D. Roderick Kieweit, *Macroeconomics and Micropolitics: The Electoral Effects of Economic Issues* (University of Chicago Press, 1983).

17. James E. Krier and Clayton P. Gillette, "The Un-Easy Case for Technological Optimism," *Michigan Law Review*, vol. 84 (December 1985), pp. 405–29.

wealth and economic structure of the society. How we will think and behave in a nation with per capita income of $50,000 cannot be reliably forecast by a simple extrapolation from circumstances in which per capita income is $20,000 or less. Technological optimism is, then, at best an atheoretical historical projection.

Implications for Understanding Policy Decisions

Examining the debate about the value of technical progress provides several useful insights for understanding R&D policy. First, most of the components of the argument are unquantifiable and probably not susceptible to resolution by technical analysis. Indeed, it is difficult to imagine how one would go about even assessing their relative importance. The main exception to this generalization is the narrowly economic impact of technical progress. The evidence on the relation between technological progress and economic growth overwhelmingly supports the view that, even counting disruption and environmental effects, technology is responsible for much of the historical improvement in living standards in modern society. Because government is held responsible for the overall performance of the economy, it is likely to be concerned with the technological base of its industries.

Second, the noneconomic issues, whatever their actual significance, are important in the policy domain because they spark passionate debate and motivate politically relevant action. One property of majority-rule voting is "positive responsiveness"—that is, if the preferences of voters shift in favor of a particular policy and if the voting mechanism has an equilibrium outcome, then the equilibrium will shift in the same direction.[18] Although no political institutions actually use simple majority rule, one can reasonably expect them to exhibit positive responsiveness. This means, for example, that if events cause some members of Congress to assign increased importance to one of the noneconomic issues, policy is likely to be affected; hence an analysis of why the government changes R&D policy must consider whether changes in perceptions about the noneconomic variables played a role.

Third, most of the discussion of the role of technological change has considered its advantages and disadvantages as a unified, holistic concept. In fact, technology advances fitfully and unevenly through many small

18. William H. Riker, *Liberalism against Populism: A Confrontation between the Theory of Democracy and the Theory of Social Choice* (San Francisco: Freeman, 1983), pp. 45–46.

changes. Some of these changes are more beneficial, or more prone to important noneconomic consequences, than others are. Thus the most sensible response to the debate about the general value of new technology is to convert it to more focused (and more manageable) debates about specific areas of technological advance. An optimist would probably conclude that it usually makes sense to promote technological change but that exceptions might have to be made in some circumstances. A more pessimistic view would be that the task at hand is to identify the specific circumstances under which it makes sense to promote new technology. In either event, the policy debate becomes focused on individual cases and thus more oriented toward distributive effects. Instead of a ubiquitous phenomenon that more or less benefits everyone, technological progress becomes a force that gives rise to a distributive debate about who will be included, and who excluded, from its benefits.

The preceding analysis explains why a nation's R&D effort is a matter of importance to its citizens, and why R&D policy is likely to be debated in part on a case-by-case basis, but it does not constitute an argument for government programs to support the development of new commercial technologies. The importance of some kinds of R&D in achieving national goals does not imply a need for federal commercial R&D projects any more than the argument that shoes are valuable implies that government should subsidize the shoe industry. Besides showing that R&D is important, the case for federal commercial R&D must be based on an argument that private industry, acting in response to market incentives, will underinvest in R&D.

Chapter 2

Government Support for Commercial R&D

LINDA R. COHEN & ROGER G. NOLL

IF RESEARCH and development is important to attaining national goals, the adequacy of the R&D effort in the private economy is a matter of public concern. And because there are theoretical and empirical grounds for believing that the private sector is prone to underinvest in R&D, the matter of government support for commercial research is also of public concern. This chapter summarizes the principal rationales for active government support for R&D, including programs aimed at developing new commercial techniques for specific industries. It also examines the recent history of government-supported R&D to ascertain the extent to which the United States and other leading industrial powers are involved in commercial R&D.

Many studies have estimated the rate of return to investment in R&D.[1] One important finding is that the profitability of private R&D exceeds that of other investments, usually by a substantial amount. The second finding is that the social returns to private R&D are larger still. Although much of this work does not distinguish between the average return and the incremental return to the last dollar spent, the basic finding remains: the return to additional investment in R&D would be high. Moreover, because the studies generally do not take account of all sources of returns to R&D, such as the benefits to consumers from improvements in product quality, they tend to underestimate the social returns to investment in private R&D.

1. Many are summarized in Congressional Budget Office, *Federal Support for R&D and Innovation* (Washington, April 1984), pp. 27ff.

The Causes of Underinvestment in R&D

If the private sector devotes too little effort to research, why? The answer has two components: not all the returns to investment in R&D can be appropriated by the innovator, and both capital markets and large organizations do not accurately evaluate the kinds of risks involved in pursuing commercial R&D.

The returns to innovation accrue from two sources: a firm achieves lower costs of production, or it introduces a new or better product that can be sold at a profit. To recoup the cost of R&D, an innovator must succeed in charging prices that are higher than average production costs. But other firms, observing that the first has lower costs or a better product, may be able to imitate its technology at far lower cost than the cost of the R&D that made the innovation possible. Successful imitation will drive the price of the product down toward the average cost of production, perhaps leaving an operating margin insufficient to recover the costs of the original R&D. If this occurs, the principal beneficiaries of the innovation become the purchasers of the product, not its producers. In one sense imitative competition is socially desirable because the ensuing price reductions increase sales and consumer benefits. But if firms believe that most of the benefits of innovation will accrue to imitators and customers, they will have less incentive to invest in R&D. The resulting underinvestment in R&D would explain the observed empirical result that the social returns to R&D exceed the returns to other forms of investment, which are less susceptible to the erosion of profits caused by future competition.

Edwin Mansfield and Frederic Scherer have illuminated this matter by estimating the effect of one industry's R&D on the rate of growth in productivity in other industries that use its products.[2] Mansfield, using the ratio of supplier R&D to value added, finds that upstream R&D is about one-third to one-half as important as downstream R&D. Scherer uses a more complicated procedure that disaggregates R&D by line of business so that only the R&D pertinent to a specific product counts in affecting downstream productivity. His findings are more dramatic: upstream R&D is about three times as important as the downstream firm's R&D in influencing productivity growth. Scherer's work is still preliminary, and

2. Edwin Mansfield, "Basic Research and Productivity Increase in Manufacturing," *American Economic Review*, vol. 70 (December 1980), pp. 863–73; and Frederic M. Scherer, "Inter-Industry Technology Flows and Productivity Growth," *Review of Economics and Statistics*, vol. 64 (November 1982), pp. 627–34.

he cautions that his findings depend on several assumptions that may prove incorrect. Nevertheless, Mansfield's and Scherer's studies indicate that much of the benefit of technical advance is passed on to customers, and hence would not enter into the profitability calculations of a firm that was deciding whether to undertake a research project.

Another factor truncating the private return to investment in R&D is the organization and technical environment of the firm in which research is undertaken. Richard Nelson and Sidney Winter argue that the research a firm undertakes is influenced by its technical history: where it has found success in the past and its perceptions about likely areas for further advance.[3] Its ability to use new knowledge depends on its overall strategies, plans, and resources. Because the knowledge R&D will discover is unpredictable, a project may produce new technical information that is useful in other organizations and contexts but not particularly valuable to the discoverer. Indeed, the new knowledge may be so unrelated to the experiences of the discoverer that its value is hidden. And selling new knowledge that a firm regards as inappropriate or fails to recognize as commercially interesting is difficult if not impossible. Besides, to sell it a firm must reveal it, which may enable the prospective buyer to learn enough to duplicate it at a lower cost than the cost of the original R&D. Thus a firm may not be able to benefit from some of the knowledge it produces, even though others may benefit by an amount that more than justifies its expenditures. This effect has yet to be quantified, but some anecdotal evidence attests to its importance. In the early history of the computer industry, for example, many of the innovations that turned out to be of greatest commercial importance were not successfully applied in the firm that developed them or were rejected as not sufficiently promising to pursue.[4]

In addition to being unable to profit from some of their investment in R&D, private firms may underinvest because of the response of capital markets to the risks associated with research. All business investments face some risk because their profitability cannot be completely known in advance. Investors will thus demand a premium to compensate them for taking a chance. To calculate this premium, they must have some means

3. Richard R. Nelson and Sidney G. Winter, *An Evolutionary Theory of Economic Change* (Harvard University Press, 1982).

4. See Barbara Goody Katz and Almarin Phillips, "The Computer Industry," in Richard R. Nelson, ed., *Government and Technical Progress: A Cross-Industry Analysis* (Pergamon, 1982), pp. 163–232.

for estimating the magnitude of the risk. Usually this involves examining the market in which the investing firm intends to operate and the track record of the firm in similar situations.

R&D risks are especially difficult to evaluate because research projects are necessarily designed to attack problems that have not been solved and for which there exists no directly relevant track record. Indeed, the more revolutionary the project's objective, the more difficult risk assessment will be. The project's designers can provide technical information to show a potential investor that the research is worth pursuing, but this may entail revealing valuable private knowledge that could then be given or sold to another firm. In addition, if the project is in an arcane technical field, conveying the information could be expensive and susceptible to errors of interpretation.

These problems differ more in degree than kind from those associated with other forms of investment. All of the risks associated with an expansion of production capacity from a known technology are present in an R&D project. The additional risks in developing a new technology arise because private, specialized knowledge will yield a maximum return on investment to its discoverers only if they are the first to apply it. If the knowledge could be conveyed cheaply and potential investors could be depended on to retain its private status, the risk would then be much like that of other investments. But such low cost and secrecy are likely to be more difficult to achieve for investments in new technology and more of a problem the newer and potentially more valuable the discovery.

The problems of appropriability and underinvestment arising from the special risks of R&D exemplify some of the factors identified by Oliver Williamson as affecting the choice between markets and hierarchies: that is, whether to include related business activities within a single organization.[5] Financing research within an organization reduces the likelihood of opportunistic behavior (for example, that the person who could provide research funds might abscond with the information that was used to try to justify the project) and eases the problem of relying on reputation and past record to allocate funds and reward achievements. The implication is that the risks of R&D are more likely to be dealt with efficiently inside a firm that has many ongoing research efforts. But this

5. Oliver E. Williamson, *Markets and Hierarchies, Analysis and Antitrust Implications: A Study in the Economics of Internal Organization* (Free Press, 1975).

only partly solves the problem of inadequate R&D if people outside large organizations also have good ideas but cannot acquire funds to pursue them. In dealing with organizations large enough to be efficient in evaluating R&D risks, outside innovators face the same problems they would face in dealing with any other external investors.

Studies indicate that many important inventions come from individuals or very small companies.[6] For small innovators, intraorganizational methods to overcome the problems associated with R&D risk are not available. A common path of innovation is for an original idea and perhaps a small production effort to come from a very small enterprise; full-scale implementation is then undertaken by a much larger enterprise, perhaps after it acquires the smaller one, but often after displacing it. For example, in the automatic bank teller industry, all but one of the initial entrants were displaced by large, integrated manufacturers within a decade after the product was introduced.[7] Obviously the original ATM firms did not capture all the benefits of their innovation. Thus, although a common pattern of industry evolution is one in which small firms innovate and large firms follow, this path nevertheless is probably less common than efficiency would demand, because small firms receive too little of the benefit from their innovations.

Even in large organizations, avoiding the risks of opportunism may come at the cost of bureaucratic inertia. Relatively expensive, long-term projects normally must be approved through a chain of command with several veto points. One reason many innovations come from small firms, despite their problems of financing risky projects and maintaining control of their inventions, is the greater difficulty of getting innovative ideas approved in a large bureaucracy. Many major innovations have been conceived in large organizations but rejected by them and have then been brought to fruition when the people who developed the idea formed their own company.[8] (This evidence is not completely convincing because it focuses on the successes and not the numerous firms that were started for similar reasons but failed.) Bureaucracy may also be the reason that large firms in most

6. John Jewkes, David Sawers, and Richard Stillerman, *The Sources of Invention* (Norton, 1969).

7. Sarah J. Lane, *Entry and Industry Evolution in the ATM Manufacturers' Market*, Ph.D. dissertation, Stanford University, 1989.

8. See Jewkes, Sawers, and Stillerman, *Sources of Invention*; and Nelson, ed., *Government and Technical Progress*.

industries use R&D to create commercial successes of ideas produced elsewhere.[9] Apparently one aspect of risk-related problems not completely overcome in large corporations is that the employee with a new idea has the same difficulty in communicating with a bureaucracy as an independent innovator has in communicating with suppliers of financial resources. Innovation by committee can proceed only at the pace of the most skeptical person who can veto an idea, which is antithetical to radically creative projects.

The preceding analysis provides a strong theoretical and empirical argument that a market economy provides insufficient incentives for an efficient rate of expenditure on R&D, especially of the most innovative kind. And it is precisely this kind of research that is the most important single source of growth in output per worker in an advanced, high-wage society. The economics literature makes a convincing case that increasing R&D beyond the amount that the private sector is willing to support has large potential benefits.

The Case for Targeted Programs

To complete the case for policies designed to promote R&D requires one last step: to determine which policies are likely to produce sufficient increments in useful R&D to be worthwhile. Our aim in this book is to do just this for one class of programs: R&D commercialization projects that are financed and managed by the government. The justification for these programs must be not only that the private sector underinvests in research, but that certain industries and technologies are especially prone to be undervalued by private investors.

Thus far the discussion of rationales for government R&D policy has been very general: society depends on technical progress for economic growth; R&D produces this progress; and left to its own devices the private sector performs too little research. The case for targeted programs rests on arguments that underinvestment in R&D varies in a systematic, observable way across industries and varies according to the kind of project. If this is true, a general policy to support research would be inefficient. Industries

9. See Daniel Hamberg, "Invention in the Industrial Research Laboratory," *Journal of Political Economy*, vol. 71 (April 1963), pp. 95–115; R. G. Cooper, "The Dimensions of Industrial New Product Success and Failure," *Journal of Marketing*, vol. 43 (Summer 1979), pp. 93–103; and Edwin Mansfield and others, *Research and Innovation in the Modern Corporation* (Norton, 1972).

or projects with substantial underinvestment would still be relatively starved, while those in which the rationales for federal support are the least applicable might have overinvestment.

Targeted public support might cure three distinct types of problems. One is the classic problem of third-party benefits or externalities, which includes the quasieconomic issues discussed in chapter 1. Obviously, private investors have little incentive to take account of such factors as the contribution of innovation to political stability or environmental quality. To the extent these matters vary in their importance across industries and technologies, they cause differences in the extent of underinvestment in R&D.

The second problem is insufficient appropriability: the inability of an innovator to capture the benefits of innovation. Scholars and industrial innovators agree that the appropriability of inventions varies considerably among industries and technologies.[10] In some cases the protection that patents, copyrights, and so forth accord innovations may be so great that excessive R&D is induced as competitors attempt to invent around a lucrative innovation. At the same time, more fundamental research, with its high propensity to produce unanticipated results and to cause major changes in the underlying technology of an industry, is often especially difficult to protect. Usually its product opens a whole new area of potential innovation. The original innovator may appropriate the gains from the first application of the new knowledge, but others will be attracted to produce further advances that are not appropriable to the original innovator. Perhaps this explains why so much basic research is undertaken in universities, where career rewards depend primarily on success in disseminating new knowledge rather than on benefiting directly from its ultimate commercial use.

The third problem arises from R&D risk and, like appropriability, its importance depends on the novelty of an idea. Risk also depends on the size of the R&D effort required to develop an idea in relation to the size of the sales and research activities of firms in the industry. If the innovation requires a relatively large investment compared with the size of the industry, the innovating firm is more likely to face other significant entry barriers, such as its possible lack of sophisticated marketing capabilities.

10. Richard C. Levin, Wesley M. Cohen, and David C. Mowery, "R&D Appropriability, Opportunity, and Market Structure: New Evidence on Some Schumpeterian Hypotheses," *American Economic Review*, vol. 75 (May 1985, *Papers and Proceedings, 1984*), pp. 20–24.

A specific technology may warrant targeted support for many reasons.[11] First, the market price of a product may not reflect its social value or its production costs, and immediate corrective actions through price controls and other regulations may not be effective or may cause a politically unacceptable change in income distribution. Second, the new technology may be likely to present the innovating firm with substantial learning curve effects that spill over to noninnovating firms. Third, some start-up costs of adopting the technology, such as educating users about the existence of a new product or overcoming regulatory barriers, benefit late entrants, not just the innovating firm. Fourth, part of the private reluctance to undertake a project may be due to the potential government imposition of unusually stringent environmental regulations, the possibility of price controls to prevent windfall gains, or the strategic response of other governments to a new technology that threatens to disrupt their markets. Because government cannot guarantee that any of these events will not occur, it may have to use subsidies to overcome the fear of them.

Two other questions about the adequacy of private R&D are whether the competitiveness of an industry encourages its propensity to innovate, and whether certain so-called strategic industries are so essential to long-term economic growth that they should be supported to ensure that the nation is a leader in them.

Market Structure

Judging from the attention received from economists, the most important source of differences among industries in their support for R&D is competitiveness within an industry. Many theoretical and empirical studies have examined how R&D effort and the rate of technical progress are affected by the absolute size and the market share of a firm. Single-effect, single-cause arguments have been developed that support logically almost any conjecture about this relationship.[12]

11. Richard Schmalensee, "Appropriate Government Policy toward Commercialization of New Energy Supply Technologies," *Energy Journal*, vol. 1 (April 1980), pp. 1–40, provides a detailed list of situations in which private underinvestment is especially likely.

12. For interesting and comprehensive discussions of this issue, see Richard R. Nelson, Merton J. Peck, and Edward D. Kalachek, *Technology, Economic Growth, and Public Policy* (Brookings, 1967); Morton I. Kamien and Nancy L. Schwartz, "Market Structure and Innovation: A Survey," *Journal of Economic Literature*, vol. 13 (March 1975), pp. 1–37; and Jennifer F. Reinganum, "Practical Implications of Game-Theoretic Models of Innovation," *American Economic Review*, vol. 74 (May 1984, *Papers and Proceedings, 1983*), pp. 61–67.

Several factors support the proposition that monopolists have a greater incentive to innovate. First, innovations by a monopolist may be more appropriable because others would have to overcome industry entry barriers to copy an invention. Second, firms with larger market shares can capture more gains from an innovation before others imitate it without having to increase market share. Hence, because the costs of discovering new knowledge are independent of the sales of the firm, larger firms are more likely to recover fixed R&D costs. Third, if R&D exhibits economies of scale and scope, the largest firms will tend to specialize in innovation and the smaller firms in imitation. Fourth, monopolists may have an additional incentive to innovate so as to preempt entry; the larger the market share of a firm, the more appropriable the gains from keeping out an entrant.

Other factors work against the primacy of monopoly. Firms with small market shares can increase sales through innovation to a greater extent than can firms with larger market shares (the Arrow effect).[13] In addition, larger, more bureaucratic firms may be less willing to undertake the most innovative R&D projects. Finally, in more competitive industries the prospect of shrinking sales and profits generates a greater chance of loss from shirking on R&D effort (the "hidden foot" described by Burton Klein).[14]

The empirical research on the effect of market structure on R&D is equivocal. Some industries seem to have a minimum efficient scale of R&D effort that causes the very smallest firms to undertake little or no research. In other industries a disproportionate share of major innovations comes from small firms. The data show that when all else is equal, the very largest firms may be somewhat less research intensive than medium-sized firms, and that the productivity of their R&D effort may be a little lower; however, AT&T, IBM, and some other corporate giants have highly effective research programs. The most plausible conclusion is that competition among a handful of large firms and a fringe of smaller ones is probably the market structure most conducive to rapid technological advance, but even this is not without exception, because there is also evidence that the observed relationships are actually caused by interindustry differences in the technical foundations of production processes and the degree of appropriability

13. Kenneth J. Arrow, "Economic Welfare and the Allocation of Resources for Invention," in National Bureau of Economic Research, *The Rate and Direction of Inventive Activity: Economic and Social Factors* (Princeton University Press, 1962), pp. 609–26.

14. Burton H. Klein, *Dynamic Economics* (Harvard University Press, 1977).

of innovations.[15] The primary lesson to be learned from this literature is that other empirical factors are likely to be more important than market structure in determining an industry's investment in R&D and technical progress, unless the industry is at the extreme of either secure monopoly or highly fragmented competition. In any case, studies of the relationship between market structure and innovation do not provide much in the way of robust policy implications.

Strategic Industries

In the practical world of the policymaker, a targeted R&D program is most commonly justified when there are exceptionally large economic spillovers from new technology that an innovating firm cannot capture. In the 1980s the concept of "strategic industry" described this situation. An industry is strategic, and therefore a candidate for special attention, if it has two characteristics.[16] First, progress in the underlying science and technology of the industry makes it ripe for rapid technological advance. That is, it is one of the high-technology industries that must form the foundation for productivity growth in a high-wage society. Second, the industry is closely linked to other industries in a way that causes technological progress in one to affect the rate of progress in others. An example is the close linkage among the semiconductor, computer, and telecommunications equipment industries. Technological progress in a linked industry is likely to be more important to the overall economy than similar rates of progress would be elsewhere.

The special importance of strategic industries has not been convincingly established, although it is plausible. Indeed, the idea has a noble heritage in Joseph Schumpeter's concept of leading industries based on waves of technological revolution.[17] The key issue is whether these industries are unusually likely to be susceptible to the problems of insufficient appropriability and imperfections in the market for financing R&D. Broad technological revolutions based on rapid advance in basic scientific understanding could exhibit these characteristics.

15. See Almarin Phillips, *Technology and Market Structure: A Study of the Aircraft Industry* (Lexington, Mass.: Heath Lexington Books, 1971); and Levin, Cohen, and Mowery, "R&D Appropriability, Opportunity, and Market Structure."

16. Congressional Budget Office, *Federal Financial Support for High Technology Industries* (Washington, June 1985).

17. Joseph A. Schumpeter, *Business Cycles: A Theoretical, Historical and Statistical Analysis of the Capitalist Process* (McGraw-Hill, 1939).

Table 2-1. *Shares of Major Industrialized Countries in Total International Trade, Selected Years, 1965–84*

Year	*France*	*West Germany*	*Japan*	*United Kingdom*	*United States*
			High-technology products		
1965	7.5	16.5	7.2	12.2	28.4
1970	6.9	16.2	11.8	9.9	28.8
1975	8.2	16.5	12.1	10.0	26.3
1980	8.2	15.8	15.2	11.0	25.5
1982	8.2	10.6	18.3	10.1	28.7
1984	7.3	13.6	22.2	8.4	26.4
			All products		
1965	5.9	10.3	4.9	8.0	15.9
1970	6.2	11.8	6.7	6.7	14.9
1975	6.5	11.0	6.8	5.3	13.2
1980	6.1	10.2	6.9	5.8	11.6
1982	5.6	10.2	8.0	5.6	12.3
1984	5.5	9.6	9.5	5.3	12.2

Sources: National Science Foundation, *Science and Engineering Indicators, 1987* (Washington, 1988), p. 325; and International Monetary Fund, *International Financial Statistics Yearbook* (Washington, 1987), pp. 116–17.

Nevertheless, there are good reasons to believe that an R&D policy focusing on such industries is unwarranted.[18] The importance of high-technology industries in high-wage economies does not mean that each advanced society must be at the forefront of all or most strategic industries to experience rapid growth. For one thing, no nation can monopolize progress: the modern world economy is fairly closely integrated, so that the spillover effects of technological advance are not confined within national boundaries. Moreover, in recent history advanced societies have not needed to succeed in all strategic industries. West Germany and Japan, with per capita income growing more rapidly than it is in the United States, do not rely more extensively than the United States on exporting products from these industries. Table 2-1 shows the share of world trade in high-technology products for the five leading Western nations. All five have a higher share of trade in high-technology industries than their share of trade in all products combined, yet in all five both shares tend to rise and fall together. The United States, West Germany, and the United Kingdom have had a declining share of world trade in all products, including high-technology ones. Japan substantially increased its share, indicating that it alone seems to have experienced parallels between rising income and

18. For a more thorough analysis, see Richard R. Nelson, *High-Technology Policies: A Five-Nation Comparison* (Washington: American Enterprise Institute, 1985), chap. 3.

Table 2-2. *Research and Development as Percent of GNP, Five Industrialized Countries, Selected Years, 1964–85*[a]

Year	*France*[b]	*West Germany*	*Japan*	*United Kingdom*	*United States*
			Total R&D		
1964	1.8	1.6	1.5	2.3	2.9
1968	2.1	2.0	1.6	2.3	2.8
1972	1.90	2.20	1.86	2.11	2.35
1975	1.80	2.22	1.96	2.19	2.20
1978	1.76	2.24	2.00	2.24	2.14
1981	2.01	2.44	2.38	2.41	2.35
1983	2.15	2.54	2.61	2.25	2.56
1985	2.31	2.67	2.77	2.42	2.69
			Nondefense R&D		
1972	1.58	2.08	1.84	1.56	1.60
1975	1.46	2.08	1.95	1.55	1.63
1978	1.41	2.10	1.98	1.61	1.63
1981	1.50	2.34	2.37	1.72	1.81
1983	1.69	2.43	2.60	1.60	1.87
1985	1.85	2.53	2.75	1.71	1.86

Sources: National Science Foundation, *Science and Engineering Indicators, 1987*, p. 234; and NSF, *International Science and Technology Data Update, 1987*.

a. Years selected correspond to years in which U.K. data are available.

b. Percent of gross domestic product.

growth in its share of high-technology trade. And even in Japan, much of the improvement in trade was in industries such as automobiles that are not regarded as leaders in technology.

Another element of the strategic industries rationale is that other advanced nations are targeting R&D in these industries and that the United States is likely to fall behind if it does not follow suit. The data show that others have expanded their total R&D effort but that it is not especially targeted on high-technology industries.

Table 2-2 shows that in recent years the five leading nations have spent roughly the same fraction of gross national product on R&D, but because nearly all R&D in West Germany and Japan is for nondefense purposes, their share of GNP devoted to nondefense R&D has risen from rough equality with the United States in 1970 to a much larger portion by the 1980s. Even though defense R&D has commercial spinoffs, its commercial significance is certainly less than that of research that is explicitly commercial. Consequently, since the early 1970s the United States seems to have fallen significantly behind these two major competitors in conducting commercially useful R&D. The number of innovations originating in U.S.

Table 2-3. *R&D Distribution, by Industry, Selected Years, 1969–85*[a]

Industry	United States 1970	United States 1985	Japan 1970	Japan 1985	West Germany 1969	West Germany 1984	United Kingdom 1969	United Kingdom 1985	France 1984
Aerospace	28.9	22	0	0	7.0	5	23.1	17	17
Electrical and electronics	23.4	22	24.8	27	28.1	24	20.7	31	25
Instruments	4.1	7	2.3	3	1.5	2	2.5	2	1
Machinery and computers	9.6	14	11.6	12	7.6	14	10.5	15	8
Chemicals and allied products	9.8	11	21.3	16	28.2	20	13.4	21	16
Motor vehicles	8.8	9	9.5	14	14.7	15	7.2	6	11

Sources: National Science Foundation, *International Science and Technology Data Update, 1987*, p. 14; and NSF, *International Science and Technology Update, 1985*, pp. 11–12.

a. In all countries, these industries account for 70 percent or more of industrial R&D; in all but Japan, they account for more than 80 percent.

firms has not fallen because U.S. R&D constitutes a share of a much larger total gross national product, but the nation's relative position among the most advanced nations will probably continue to decline.

Although commercial R&D has become more important in some advanced economies, the data do not indicate the kind of targeting implied by the strategic industries argument. Table 2-3 shows the distribution of R&D among six industries in the same five countries. Japan and West Germany are active in R&D in motor vehicles and chemicals, which were Schumpeterian leading industries during the first half of the century but are not now. All countries are extensively involved in computers and electronics, but only the United States and United Kingdom are also heavily investing in aerospace R&D (reflecting, no doubt, their relatively large amounts of defense R&D). In another high-technology field, instruments, the United States allocates substantially more R&D than anyone else, as it does in the medical care sector (not shown in the table). Thus the larger share of GNP going to industrial R&D in Japan and West Germany apparently is not particularly focused on strategic industries but is more general and sensitive to the overall industrial structure of these countries. If any single fact stands out, it is that the United States seems to be spending relatively less than the others and perhaps underinvesting in two traditional strengths, automobiles and chemicals, and in one high-technology strength, electronics.

Do these differences in the patterns of R&D spending reflect differences in the strategic orientation of public expenditures to support research?

Table 2-4. *Distribution of Federal Government R&D, by Function, 1983*
Percent

Country	Government share of R&D	Function					
		Defense	Space	Energy	Health	Industry	Other
United States	50	64.3	5.5	6.6	11.5	0.3	11.8
Japan	36	2.4[a]	5.8[a]	12.5[a]	2.6[a]	6.0[a]	70.7[a]
West Germany	42	9.4	4.3	16.9	4.1	11.0	54.3
United Kingdom	59[a]	50.0	1.9	5.4	1.3	6.5	34.9

Source: National Science Foundation, *International Science and Technology Data Update, January 1985*, pp. 8, 10.
a. 1981 data.

Table 2-4 shows the fraction of total research accounted for by the central government and its distribution across categories. The two most important categories in the United States, defense and health, are relatively unimportant in Japan and West Germany; space research is about equally significant everywhere. Energy receives considerably more attention in Japan and West Germany, each of which is poorer in energy resources than the United States. Industrial R&D for purposes other than defense, space, energy, and health is more important everywhere than in the United States, but no country emphasizes this category. Nearly all the "other" R&D is for advancement of knowledge; the remainder is oriented toward nonmanufacturing sectors (agriculture, mining, the public sector, and infrastructural industries such as utilities and transportation). This remainder is far more important in other nations than it is in the United States: only about 5 percent of U.S. government R&D falls into this category, while other countries allocate between 20 and 30 percent to it. Most of this difference is accounted for by the fact that in most other nations universities are infrastructural industries that are almost entirely government operated.

Thus none of the countries are very heavily involved in direct expenditures to improve technology in nondefense industries, although each country has apparently singled out a few industries or technologies for special support. Other than the effects of defense R&D on aerospace and perhaps electronics, government expenditures do not appear to be driving the distribution of R&D across industries in any of these countries and do not appear to reflect a policy to emphasize strategic industries.

Targeting in the United States

In the United States, targeted R&D support is especially important in the space and energy sectors (table 2-4). In 1989 these programs accounted for about 12 percent of federal R&D, a third of federal expenditures that were not directed at defense, and 5 percent of the total national nondefense research effort. Expenditures for space and energy were lower than they had been only a decade earlier. Table 2-5 shows the distribution of federal R&D by sector from 1961 to 1989. The heyday of federal support for commercial R&D in energy and space was from 1975 to 1980, when the nation was developing the space shuttle and several new energy technologies.[19] During these years research in energy and space averaged about 23 percent of federal expenditures on R&D and 14 to 18 percent of total R&D for nondefense purposes. The budget for space and energy fell in the first few years of the Reagan administration. In space, the cut was more apparent than real because the space shuttle was declared commercially operational in 1983 and was removed from the R&D budget. But the transatmospheric aircraft, improvements in the space shuttle after the 1986 *Challenger* disaster, the space station, and other projects caused the budget for space R&D to climb steadily after 1984. In energy, most commercial development projects were killed in the early 1980s. In 1989, however, nine "clean coal" demonstration projects were added to the budget.

Although data have not been collected in a manner that allows a precise estimate, a large part of the budget for energy and space R&D was accounted for by commercial projects such as the space shuttle, the Clinch River breeder reactor, the development of synthetic fuels, and the photovoltaics commercialization program. Thus these programs constituted something of a large-scale test of targeted R&D programs comparable in magnitude to government programs in other advanced nations to promote new commercial technology. Indeed, if anything, table 2-5 understates the case; in constant dollars, total R&D in the United States was virtually unchanged from 1968 ($29.8 billion in 1972 dollars) to 1977 ($30.5 billion). The real value of federal R&D fell in the early 1970s and was about equal in 1971 and 1978. Thus the space shuttle and the various energy demonstration programs were squeezed into a constant national effort; they were not incremental.

Another important aspect of the pattern is that until the early 1970s almost all federal R&D was allocated to defense and noncommercial aspects of space. Before the space program defense typically accounted for about

19. The previous boom in space R&D, during the 1960s when the sector accounted for 20 to 25 percent of total R&D, was due to the Apollo program, which had very little commercial importance.

Table 2-5. *Federal R&D, by Function, 1961–89*
Billions of current dollars

Year	Total R&D[a]	Federal R&D									
		Total	*Defense*	*Health*	*Space*	*General science*	*Energy*	*Resources*[b]	*Transport*	*Agriculture*	*Other*[c]
1961	14.3	9.1	7.0	0.4	0.8	0.1	0.4	0.1	0.1	0.1	0.1
1962	15.4	10.3	7.2	0.6	1.4	0.2	0.4	0.1	0.1	0.1	0.1
1963	17.1	12.5	7.8	0.6	2.8	0.2	0.5	0.1	0.1	0.1	0.1
1964	18.9	14.2	7.8	0.7	4.2	0.3	0.6	0.1	0.1	0.2	0.2
1965	20.0	14.6	7.3	0.8	4.9	0.3	0.6	0.2	0.1	0.2	0.2
1966	21.8	15.3	7.5	0.9	5.0	0.4	0.6	0.2	0.3	0.2	0.3
1967	23.1	16.5	8.6	0.9	4.8	0.4	0.6	0.3	0.4	0.2	0.3
1968	24.6	15.9	8.3	1.0	4.3	0.4	0.7	0.3	0.3	0.2	0.4
1969	25.6	15.6	8.4	1.1	3.8	0.4	0.6	0.3	0.4	0.2	0.4
1970	26.1	15.4	8.0	1.1	3.6	0.5	0.6	0.3	0.5	0.2	0.5
1971	26.7	15.5	8.1	1.3	3.0	0.5	0.6	0.4	0.7	0.3	0.6
1972	28.5	16.5	8.9	1.5	2.9	0.6	0.6	0.5	0.6	0.3	0.6
1973	30.7	16.8	9.0	1.6	2.8	0.7	0.6	0.6	0.6	0.3	0.7
1974	32.9	17.4	9.0	2.1	2.7	0.7	0.8	0.5	0.7	0.3	0.6
1975	35.2	19.0	9.7	2.2	2.8	0.8	1.4	0.6	0.6	0.3	0.6

1976	39.0	20.8	10.4	2.4	3.1	0.9	1.6	0.7	0.6	0.4	0.7
1977	42.8	23.5	11.9	2.6	2.8	1.0	2.6	0.8	0.7	0.5	0.7
1978	48.1	26.0	12.9	3.0	3.0	1.1	3.1	0.9	0.8	0.5	0.8
1979	55.0	28.2	13.8	3.4	3.1	1.1	3.5	1.0	0.8	0.6	0.9
1980	62.6	30.0	14.9	3.7	2.7	1.2	3.6	1.0	0.9	0.6	1.1
1981	71.8	33.7	18.4	3.9	3.1	1.3	3.5	1.1	0.9	0.7	0.9
1982	79.3	36.1	22.1	3.9	2.6	1.4	3.0	0.1	0.8	0.7	0.8
1983	87.2	38.8	24.9	4.3	2.1	1.5	2.6	1.0	0.9	0.7	0.7
1984	97.6	44.2	29.3	4.8	2.3	1.7	2.6	1.0	1.0	0.8	0.8
1985	107.5	49.9	33.7	5.4	2.7	1.9	2.4	1.1	1.0	0.8	0.9
1986	116.8	53.2	36.9	5.6	2.9	1.9	2.3	1.1	0.9	0.8	0.9
1987	124.3	57.1	39.2	6.6	3.4	2.0	2.1	1.1	0.9	0.8	1.0
1988[d]	n.a.	59.2	40.3	7.1	3.6	2.2	2.1	1.2	0.9	0.9	1.0
1989[d]	n.a.	62.9	41.3	7.8	4.7	2.5	2.5	1.1	1.1	0.9	1.0

Sources: National Science Foundation, *Federal R&D Funding by Budget Function, Fiscal Years 1987–1989* (Washington, April 1988), pp. 108–110; and NSF, *Science and Engineering Indicators, 1987*, p. 238.

n.a. Not available.

a. Total R&D for calendar year.

b. Includes environment.

c. Includes all research related to social services and international affairs.

d. Estimated.

three-fourths of the federal R&D budget, and defense and space together sometimes accounted for more than 80 percent of federal R&D during the 1960s. In the early 1970s, however, as other programs grew substantially, defense and space expenditures were held approximately constant in nominal terms (and declined in real terms). By 1979 and 1980, expenditures for the two sectors had fallen to about 60 percent of federal R&D. The last Carter budget (fiscal 1981) and the early Reagan budgets reversed this trend, and by the mid-1980s the federal government was once again spending more than 70 percent of its R&D budget on space and defense. Nevertheless, the recovery was distinctly not at the expense of most other types of federal R&D. Except for energy, the other sectors continued to receive increased budgets during the 1980s; defense and space simply grew faster. In the late 1980s, defense R&D declined in real terms (growing less than 3 percent a year in nominal dollars). Meanwhile the budgets grew faster than the rate of inflation for research in health, space, transportation, agriculture, and general science. Comparing expenditures at the end of the Reagan era with those at the end of the Johnson administration twenty years earlier shows the total federal R&D budget had increased fourfold. The increase in space and defense had been slightly less than this much. Meanwhile, greater than fourfold increases had been allocated to health, general science, agriculture, and even energy. R&D for resources and environment has grown roughly as fast as space and defense. Only transportation R&D has grown much less rapidly than the other major categories, and here the figures are somewhat misleading. In the base year the American supersonic transport had entered the prototype production phase. The contemporary counterparts to the SST are the advanced high-speed aircraft (SST-II) and the transatmospheric aircraft (National Aerospace Plane). Whereas the SST was undertaken by the Federal Aviation Administration, the transatmospheric aircraft, which received more than $295 million in fiscal 1989, is primarily in the defense budget, with the remainder part of NASA.[20] Had it been allocated to transportation, this category, too, would have grown as rapidly as defense and space.

The transatmospheric aircraft is not the only element of defense R&D that has potential commercial significance. The Department of Defense is an important source of support for research in microelectronics, computers, and telecommunications equipment. Some of these projects are oriented

20. National Science Foundation, *Federal R&D Funding by Budget Function, Fiscal Years 1987–1989* (Washington, April 1988), pp. 14, 79.

toward specific military systems, but many are not. For example, the $100 million annual subsidy to Sematech, the industry joint research effort for semiconductor production technology, is financed by the defense budget.

The story implied by the data in tables 2-1 through 2-5 can be summarized as follows. Throughout the 1970s and into the 1980s, several other nations were increasing their R&D efforts, while the share of U.S. GNP devoted to research, after a brief dip, remained constant. By the end of the 1980s the United States had fallen behind its two biggest challengers, Japan and West Germany, in the share of GNP allocated to nondefense R&D. In all countries, most R&D was privately financed; the share of government programs in the total research effort declined everywhere, but the decline was by far the greatest in the United States. The distribution of R&D also differed among countries, but no country concentrated resources on high-technology industries. Meanwhile, the United States carried out some significant experiments in federally financed R&D commercialization programs that accounted for a significant part of total R&D expenditures. And although the Reagan administration initially cut the federal budget for commercial R&D, it later adopted such commercial projects as the transatmospheric aircraft, clean coal electric generation facilities, and Sematech. The result is that federal support for basic research and for the development of commercially relevant nondefense technologies was more important in 1990 than it was in the 1960s.

Conclusions

There is no simple rule for identifying industries and technologies that may be especially prone to underinvest in R&D relevant to society's needs. Although market structure and the concept of strategicness might affect the adequacy of R&D, neither is an unambiguous indicator. Both competitive and monopolistic industries may underinvest or overinvest, depending on the details of the technology and institutional factors affecting them. A high-income society probably needs to be progressive in some high-technology industries with strong linkages to the rest of the economy, but it does not have to be a leader in all of them, and in any case it will not necessarily underinvest in R&D if left alone. Other nations do not seem to be targeting strategic industries; they seem instead to be pursuing more balanced programs of investing government R&D funds. Thus for the United States the necessity and appropriateness of a targeted R&D effort aimed at a specific industry or technology is not obvious. Such an effort depends on case-by-case analysis that focuses on whether the combination of competition, appropriability, risk, and national spillovers is important enough to warrant assistance.

As a practical matter the United States has supported a variety of commercial R&D activities in the postwar era, and despite the recent rhetoric about scaling back these programs, it has embarked on several new projects. Of the six R&D commercialization projects examined in detail in this book, five are oriented toward a leading high-technology industry: the SST (aerospace), communications satellites (aerospace and electronics), the space shuttle (aerospace), the Clinch River breeder reactor (nuclear energy), and the photovoltaics commercialization program (electronics). The sixth, synthetic fuels research, is oriented toward a more traditional industry (hydrocarbon fuels) and generally is aimed at pushing ahead relatively old technology. All six are closely related to projects pursued in the late 1980s. The advanced high-speed aircraft and the transatmospheric aircraft have replaced the SST and the space shuttle; the communications satellite program, after being killed in 1974, was reinstituted in 1985. R&D continues on nuclear technologies, with a major commitment having been made to building fusion reactors. Research on photovoltaics manufacturing was replaced by research on semiconductor production, with important similarities in materials processing; however, the 1990 budget initiated a new R&D program to support advances in photovoltaic manufacturing. And the old synthetic fuels program was the source of the technologies being used in the nine new clean coal demonstration projects. Thus the cases examined in this book have contemporary relevance. They provide an opportunity to examine the likely effectiveness of a limited strategic industries program in the American political context and to assess the likely performance of current federal commercial R&D projects.

Chapter 3

Efficient Management of R&D Programs

LINDA R. COHEN & ROGER G. NOLL

THE NORMATIVE rationales for public support of private R&D can be construed as constituting a positive theory of research policy. An R&D program can be evaluated on the basis of the contribution it makes to correcting the misallocation of private resources to research and, hence, by the societal benefits that flow from it. The positive theory of government action that is suggested by this normative rationale is the economic "public interest" theory applied to research: government decisionmakers seek to maximize the net societal benefits of R&D programs. This implies that public decisionmakers can meaningfully compare, and make trade-offs among, the various social objectives that can be served by promoting the development of new technology.

The relevance of the economic public interest theory to R&D programs is as follows. Whereas few analysts believe that a public interest theory of government is scientifically accurate, most believe that efficiency is and ought to be an important element of policy choice.[1] More important, whatever the objectives of government decisionmakers are, enhanced

1. See, for example, Stephen Breyer, *Regulation and Its Reform* (Harvard University Press, 1982), chap. 1. A somewhat stronger view is that, in the long run, government policy tends to converge toward efficiency because all sides will recognize that, in principle, a more efficient policy can be orchestrated to benefit everyone. See Michael E. Levine, "Revisionism Revisited? Airline Deregulation and the Public Interest," *Law and Contemporary Problems*, vol. 44 (Winter 1981), pp. 179–95; and Gary S. Becker, "A Theory of Competition among Pressure Groups for Political Influence," *Quarterly Journal of Economics*, vol. 98 (August 1983), pp. 371–400. For a more detailed review of this theory and its applicability, see Roger G. Noll, "Economic Perspectives on the Politics of Regulation," in Richard Schmalensee and Robert Willig, eds., *Handbook of Industrial Organization*, vol. 2 (New York: North-Holland, 1989), chap. 22.

efficiency is almost always going to be a useful instrument for achieving them. For example, even if government is nothing more than a self-interested pursuit of favorable income redistribution, those in control, all other things equal, will always select an efficient means of helping themselves over an inefficient one.

In chapter 4 we explore how some of the institutional features of the American political system may deflect government officials from the pursuit of efficiency. For our purposes, the import of the theory of efficiency maximization is twofold. First, it provides a set of benchmarks and alternative hypotheses against which to compare other theories of government action and the actual operation of government R&D programs. Second, a comprehensive exploration of the implications of efficiency maximization provides guidelines for achieving greater efficiency by changing the structure of programs and other institutional arrangements.

The purpose of this chapter is to examine in some detail how a conscious attempt to maximize net social benefits would affect the selection and management of R&D projects. Logically, the problem of selecting an optimal scale and mix of programs has two components: first, for each program, the most efficient way to execute it is identified; second, programs are selected that together maximize the net social benefits from the overall R&D policy.

Optimal Project Management

The focus of this book is on comprehensive programs that are intended to deliver commercially useful technology to private users. These are sometimes called demonstration projects, in reference to the final phase of the project in which a full-scale, operational example of the technology is constructed as final proof that it is technically and economically worth adopting for commercial use.

Two aspects of commercial R&D programs need to be understood in order to comprehend fully their role in a portfolio of R&D policies. The first is that they are appropriate only for a limited range of activities that lead to technological progress. The second is that the name "demonstration project" is somewhat misleading, for it masks the reality that these programs normally involve substantial research before work is begun on the demonstration phase.

Demonstration projects refer to programs that attempt to develop such a radical change in technology that an entire production process must be set up de novo to test the commercial value of the technology. Although such wholesale transformations of technology are important, they probably do not account for most of the advances in productivity. Much technological progress takes the form of relatively minor adjustments to production methods as experience is gained in their operations, and is implemented by such actions as small design changes in equipment, reorganization of tasks in a complex assembly process, or redefinitions of the tasks assigned to employees.[2] Achieving these productivity advances does not require redesigning a production system, let alone a demonstration of commercial feasibility. Adam Smith's famous example of technical progress in the pin factory is an illustration of this rather mundane but important source of productivity advancement. It also accounts for the commonly observed phenomenon of a "learning curve."[3]

Demonstration projects are usually intended to accomplish one of two ends: to introduce a production process that produces an entirely new product or to initiate a new method of producing an established product that, for a given quantity of inputs, yields more output. But at least two other kinds of changes may require demonstration as well. Both reflect real-world complications that are usually ignored in the economic theory of production. Specifically, changes in either the scale or the relative proportions of inputs using an established technology commonly require some R&D and entail some uncertainty about the efficiency of the redesigned production process.[4] But regardless of the nature and source of technical uncertainty, the key point is that a demonstration project is the last phase of a broader program that begins with attacking a new technical problem, the solution to which must be tested by embedding it in a new production system before its commercial utility can be established.

Underpinning a commercial R&D project in either the public sector or private industry is a prior belief that a radical change in production methods

2. See Nathan Rosenberg, *Perspectives on Technology* (Cambridge University Press, 1976), chap. 5.

3. See Harold Asher, *Cost-Quantity Relationships in the Airframe Industry*, Rand Report R-291 (Santa Monica, Calif.: Rand Corporation, 1956); Leonard Dudley, "Learning and Productivity Change in Metal Products," *American Economic Review*, vol. 72 (September 1972), pp. 662–69; Leonard Rapping, "Learning and World War II Production Functions," *Review of Economics and Statistics*, vol. 47 (February 1965), pp. 81–86; and Eytan Shishinski, "Tests of the Learning by Doing Hypothesis," *Review of Economics and Statistics*, vol. 49 (November 1967), pp. 568–78.

4. See Rosenberg, *Perspectives on Technology*, chap. 4; and William R. Hughes, "Scale Frontiers in Electric Power," in William M. Capron, ed., *Technological Change in Regulated Industries* (Brookings, 1971), pp. 44–85.

or a radically new product is technically feasible and economically attractive. The source of this belief can be either new scientific and engineering knowledge developed in another context, perhaps without regard to its ultimate commercial value,[5] or practical experience that provides hints about promising avenues for further development.[6] These hints may involve new applications of existing scientific knowledge, or may even suggest areas of fundamental knowledge that are ripe for advance in commercially important ways. In all cases, however, a prior belief about a technology's promise is subject to uncertainties over exactly how the new technical idea should be implemented and whether it will be commercially successful.

Stages of Commercial R&D Projects

As implied by referring to demonstration as the last phase of a project, commercial R&D programs, whether public or private, are undertaken in stages. These activities can be usefully separated into four categories.

RESEARCH. The first stage consists of R&D that is necessary to determine whether the basic idea is technically sound; namely, exploratory research that either expands the base of fundamental knowledge or applies the existing base to some new set of problems. These activities are subject to considerable uncertainties about their outcome, especially if they explore new areas of fundamental knowledge and expand the science base of a technology. They are also fairly inexpensive to perform, involving a relatively small number of research personnel undertaking either paper studies or small experiments in laboratories and testing facilities. Finally, these activities are fairly difficult to speed up or slow down by changes in the amount of resources devoted to them. Whereas the overall pace at which new technical knowledge is produced is affected by the total effort in exploratory research, a specific study or experiment has a more predetermined cost and duration. Researchers must design and carry out a sequence of procedures to validate their technical hypotheses. The extent to which experimental procedures are replicated, and hence the extent of

5. See Almarin Phillips, *Technology and Market Structure: A Study of the Aircraft Industry* (Lexington, Mass.: Heath Lexington Books, 1971), for a general theory of how technological progress is driven by seemingly independent progress in basic scientific knowledge and an application of this theory to the history of the aircraft industry.

6. For two quite different views about how recent practical experience affects the search for new technology, see Burton H. Klein, *Dynamic Economics* (Harvard University Press, 1977); and Richard R. Nelson and Sidney G. Winter, *An Evolutionary Theory of Economic Change* (Belknap Press of Harvard University Press, 1982).

confidence in the results, can be varied by allocating more or less time and resources to verification; however, the task of dcvcloping and testing hypotheses is pretty much fixed in cost and time requirements.

DEVELOPMENT. The second category of activities consists of designing, building, and testing components and even small-scale versions of the new technology. Although developmental activities are also subject to uncertainty because there are normally surprises in trying to put new knowledge to work, at least this type of activity is built on a firmer scientific base than the first type, and so the uncertainty is usually less. Developmental work, while using the skills of research personnel, also typically entails a greater proportion of expenditures on more prosaic endeavors such as the procurement of equipment and materials, the manufacture of system components, and perhaps even some construction of facilities for operating subsystems and prototypes. Because developmental research includes these latter kinds of activities, it is somewhat more susceptible than the initial research to being speeded up or slowed down by variations in the resources made available to it.

DEMONSTRATION. The third category of activities is the construction of an operating example of the new technology to prove its technical and commercial feasibility. This phase is normally the most expensive. It requires the procurement of system components and their assembly into the full-scale demonstration model. Whereas much can be learned about the technology at this stage, further improvements run more along the lines of learning by doing: tinkering with elements of the new technology as experience is gained using it. Because of the expense of the demonstration phase, it is unlikely even to be attempted unless the uncertainties surrounding its performance are considerably less than those associated with the previous two stages.

ADOPTION. The fourth set of activities in a commercial R&D project is the adoption of the new technology for commercial use. In the private sector, the firm that develops a new technology is often its first user. But when government develops a new commercial technology, and even sometimes when the developer is in private industry, the project can be considered a success only if other organizations adopt it. That requires educating these organizations about the characteristics of the technology and its possible applications. Thus an important component of a government demonstration program is the process of disseminating knowledge about the new technology to those who will use it. Concerted efforts to transfer new technology can take place throughout an R&D project, but they are especially likely to be important in conjunction with the completion of the demonstration stage.

Eliminating Options and Reducing Uncertainty

An important characteristic of demonstration projects is that initially several alternative paths of technical development could be chosen to achieve the ultimate commercial objective. One facet of the uncertainty surrounding a program is that the initial technical hints may suggest several ways in which the new scientific base may be built, or they may be unspecific about precisely how the new knowledge, once developed, can best be applied. For example, the technology for using nuclear power to generate electricity was initially, and to some degree still is, unspecific about the composition of the nuclear fuel, the kinds of nuclear reactions that would be used to generate heat, and the medium that would be used to transfer the heat from the nuclear reactor to turbine generators. One of the uncertainties that is resolved in the course of a successful commercial R&D program is which of the alternative ways of developing the technology are the most promising for commercial application.

Initially, expectations about the overall promise of the new technology and the relative virtues of alternative approaches are extrapolations of incomplete and only partly related knowledge, and are likely to be based on the specialized training and experience of the people advocating a program of research. These expectations may be quite subjective and can be the source of considerable disagreement among otherwise equally expert individuals with different knowledge and experience. Each of the programs examined in this book had important scientific controversies; for example, whether geosynchronous satellite orbits were possible (they were) and whether certain lightweight, heat-resistant alloys for supersonic planes were technically feasible (they were not). In each case, the subjective beliefs that led to scientific controversy produced objectively testable hypotheses. Hence, as research proceeded, differences of opinion narrowed, uncertainties were reduced, and expectations became more objectively based.

The problem of optimizing the management of a project includes picking the technical alternatives that will be investigated, arranging the various activities that are to be undertaken in the proper sequence, and, as information is developed from activities as they are completed, changing and even eliminating subsequent activities in light of the new knowledge.

Initially, the decisionmaker is faced with a list of ways to develop a new technology, each of which could ultimately end in an expensive demonstration and each of which presents uncertain prospects of success. Fortunately, each technical hypothesis has objectively testable implications that can be examined in the relatively inexpensive research stage. The best strategy for the decisionmaker, assuming that the development appears to be sufficiently promising to pursue, is usually to undertake parallel research on several alternatives, while leaving others unexplored. The results of the first wave of research are used either to narrow the list of candidates before the more expensive developmental and demonstration phases are entered, or, if the early research is not promising, to redirect effort toward options that were not studied initially.[7] Early research can be viewed much like the purchase of an option: at a small cost, the project manager buys information that will permit a more informed decision later on whether to commit to the cost of development. Because this option price is relatively small, exploratory research can be justified even if the new technology is extremely risky—a quite low probability of a very significant advance. Hence an optimally managed R&D program ought to have many projects that are regarded as failures at the end of the exploratory research phase and are then canceled before the commitment to develop them further.

An important element of the optimal program is the decision to move from stage to stage. Research is relatively cheap, but time-consuming. Hence if one does more research to narrow further the uncertainties associated with a technical option, one not only spends more now on research but postpones the ultimate completion of the demonstration.[8] Both effects lower the net expected benefits of a project. At the same time, the expected commercial value of the new technology will be higher as its technical foundation is more firmly grounded in research. At some point, the extra value of additional information is insufficient to justify the costs of additional research and further delay. One proceeds with more research than was initially anticipated only if the early research results are somewhat less promising than expected—and then only if they are not so unpromising as to warrant cancellation of the program.

7. For some interesting examples, see F. M. Scherer, *Innovation and Growth: Schumpeterian Perspectives* (MIT Press, 1984), chap. 4.

8. The formal theory of switching from stage to stage, and of the ultimate decision whether to adopt the technology, to do more research, or to stop, is developed in Tom K. Lee, "On the Reswitching Property of R&D," *Management Science*, vol. 28 (August 1982), pp. 887–99; and Tom K. Lee, "On the Joint Decision of R&D and Technology Adoption," *Management Science*, vol. 31 (August 1985), pp. 959–69.

Similar considerations enter into the decision to proceed from applied and developmental work to a full-scale demonstration. The added cost of demonstration compared with developmental work means that the latter ought to be used to narrow the range of options that are pursued to the demonstration stage, as well as to reduce uncertainties associated with demonstrating a specific option. Once again, the developmental work proceeds until the costs of continuation—both direct expenditures and the concomitant delay of the adoption date—exceed the benefits in expected performance to be gained from additional development. Alternatively, a project may be canceled during developmental work if the results in this stage are sufficiently discouraging. Or discouraging results can lead to a decision either to switch back to basic research to overcome an unanticipated problem or to reopen investigation of a previously abandoned alternative.

At each decision point during the program, the decision whether to continue, redirect, or cancel the program is based solely on prospective costs and benefits. Only at the beginning are all of the costs through all stages considered. Once an activity is completed, the decision whether to continue further work is based exclusively on the prospective benefits and on the additional costs likely to be needed to complete the program, both of which are updated by the information gained from the completed activities. A program that turns out poorly may not have been worthwhile according to a retrospective analysis of all its costs; however, at each stage it can still have been worth continuing when only its prospective costs and benefits are considered.

Some implications of this analysis for a successful program are illustrated in figure 3-1, which is taken from a study of the flow of research and development costs associated with the commercialization of nylon.[9] DuPont's early research on synthetic fiber was relatively inexpensive and proceeded for several years. Eventually, the results were sufficiently promising for the company to begin investigating whether mass production of a sufficiently high-quality product was technically feasible and economically attractive. This required substantially increased expenditures, some for developmental research and others for experimental facilities. Figure 3-1 shows a steeply rising cost both for R&D and other costs after the first successful synthesis of the nylon superpolymer in 1934.

The distinctive features of the curve are apparent: a huge bulge in annual costs as DuPont moved toward the creation of commercial production

9. Scherer, *Innovation and Growth*, chap. 1.

Figure 3-1. *DuPont's Annual Expenditures to Develop Nylon, 1928–48*

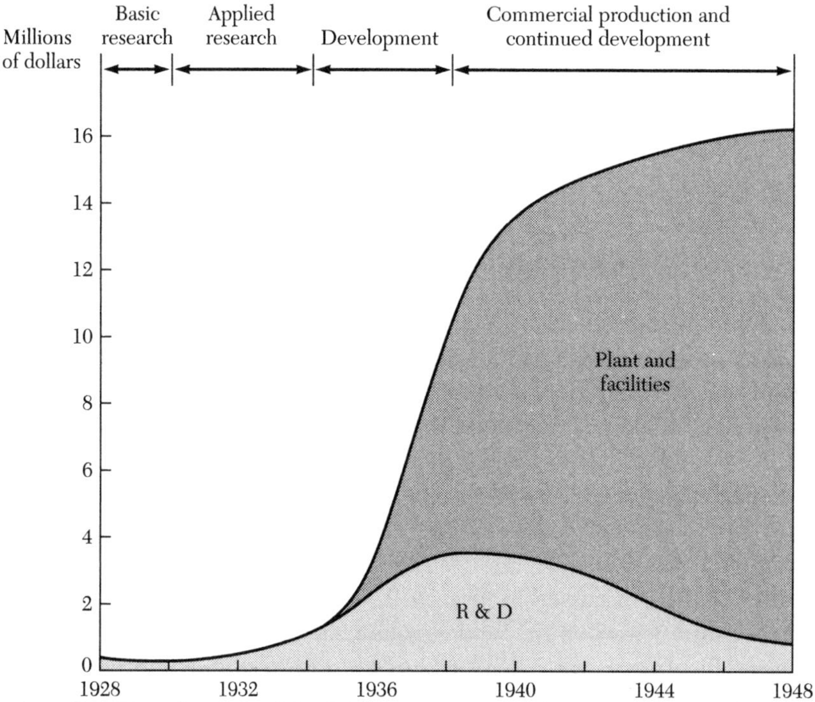

Source: F. M. Scherer, *Innovation and Growth: Schumpeterian Perspectives* (MIT Press, 1984), p. 4.

capacity and a peak in annual R&D expenditures a decade before the peak in annual total costs. As depicted, the phases of the program begin and end at specific dates, but in reality, of course, the transitions are not so distinct. In particular, the R&D component, though initially almost all basic exploratory research, will exhibit a gradual shift in composition toward applied and developmental work as the program proceeds, especially so after the original concept had been proven to be technically feasible. It is also worth noting that no one at DuPont was likely to have been able to draw this curve in the late 1920s when the project began, nor even to have fully anticipated that 1934 would be the magic year when the project moved into the phase of developing a commercial production method. To the contrary, the inflection point in the cost function in 1934–35 was the result of a contemporaneous internal decision that exploratory research results were sufficiently favorable to warrant moving on to the next stage. Nevertheless, decisions to proceed first with exploratory research and eventually with the development of a full-scale production process incorporate expectations about the range of possible shapes of cost functions in the future, as well as expectations about the ultimate commercial value of the process once it is adopted. These expectations are a form of technological optimism, for they are based in part on subjective expectations that a new line of R&D is likely to be productive and in part on the belief that a new project will have a cost profile that parallels the cost history of other R&D programs as depicted in figure 3-1.

Managing Public R&D

Optimal program management is not conceptually different in the public and private sectors. Nevertheless, some differences in details are potentially significant. These pertain to the problem of evaluating projects. They arise because of differences between the public and private sectors over the appropriate method for discounting the future and in the availability of information about the prospects of proposed programs.

Detecting Attractive Technologies

For reasons discussed in chapter 2, undertaking efficient research in one organization when the fundamental ideas are generated in another is more difficult than when both take place within the same organization. Government support of new technologies for the private sector generally requires interorganizational transfers of information not just in the adoption phase but in earlier stages of project design. At the beginning of a project, this information is the more subjective kind that pertains to the technical hints that guide choices regarding basic research. Because government officials typically lack relevant experience in the industry that will adopt the new technology, one would expect that government decisions to support private commercialization projects are subject to more uncertainty than similar decisions in the private sector. Even for projects to improve the technology of government goods, such as weapons, the private sector is likely to be better informed about production technologies and hence the costs of new products.

The uncertainty in public R&D programs should be greatest at the early phases of projects, especially if the underlying technical idea is radical. As research proceeds, objective technical results will narrow the uncertainties in the more subjective early expectations. For these results to narrow the uncertainties in government decisions, however, accurate interorganizational transfer of information must take place. If accurate information is transferred, government decisions can be expected to be better in the developmental and demonstration stages than in the early research phases. But some new technical knowledge is likely to be appropriable by the firm undertaking the research. Moreover, a firm may want to conceal "bad news" in order to keep the project alive, realizing that subsequent "good news" may be forthcoming that will cause adoption of the technology. In either case, firms may be reluctant to reveal complete, accurate information, so that government decisions will always be subject to somewhat more uncertainty than comparable decisions in the private sector.

Historical experience seems to confirm the hypothesis that government decisions are somewhat better at the developmental stage than at the early research stage. For example, when the Department of Defense decided to accelerate the development of computers, communications satellites, and semiconductors, it did not correctly identify and support the most important early basic research activities; however, once the most promising lines of development had been pursued long enough to produce interesting, objective results, the department quickly adopted them, and in some cases even reimbursed the researchers for work done before it took over the project.[10] Although all three examples are drawn from defense, they are similar to private commercialization projects in that what the government sought was not a weapons system or other product that is exclusively used and mostly developed by government. Instead, the Defense Department sought advanced capabilities in goods that had potential uses in defense, but with which government officials had little previous experience and which would be built on a scientific base that had been developed in the private sector. Hence technical capabilities developed at institutions such as Bell Labs, Hughes Aircraft, the University of Pennsylvania, Harvard University, and the Massachusetts Institute of Technology had to be assessed for support by imperfectly informed officials in the government. The result was a spotty record in identifying in advance what turned out to be the best initial research efforts.

Discounting the Future

The choice of a discount rate is also a crucial matter in designing a commercialization program. As apparent in figure 3-1, R&D programs

10. See Barbara Goody Katz and Almarin Phillips, "The Computer Industry," in Richard R. Nelson, ed., *Government and Technological Progress: A Cross-Industry Analysis* (Pergamon, 1982), pp. 162–232; Delbert D. Smith, *Communications via Satellite, A Vision in Retrospect* (A. W. Sijthoff-Keyden, 1976); and Richard C. Levin, "The Semiconductor Industry," in Nelson, ed., *Government and Technological Progress*, pp. 9–100.

often have a long gestation, with the commercial adoption of a new technology coming more than a decade after the initial research is begun.[11] This requires comparing streams of costs and benefits over a lengthy period.

One of the normative rationales for government support of private R&D is that the private sector uses too high a discount rate when evaluating R&D projects, as reflected in the research findings that private returns to R&D investment are substantially higher than the returns to other investment. In part, this means that government should support some "losers"—that is, projects which have rates of return that are too low to be pursued in the private sector.

One characteristic that can cause a project to be regarded as a loser in the private sector is a long gestation period. The benefits of a time-consuming project can have a low discounted present value despite a high nondiscounted value. If the private sector uses a discount rate that is too high, it might reject the project when the government, using a lower discount rate, might adopt it. Moreover, the character of the project could be changed by using a different rate. Low discount rates imply more patience with the research phase, and hence the construction of a firmer scientific base for the technology. That would lead to a later date of adoption, but a greater magnitude of technical advance when commercialization is achieved. Indeed, lower discount rates in the public sector cause a tension even with projects the private sector would adopt, for the socially optimal program would go slower, undertake more research, and produce a more revolutionary final product. The same line of reasoning implies that, in some cases, the optimal decision for the private sector is to drop a project after early bad news from research points toward a longer period of development than was initially planned; however, if the stretchout is not too long, social optimization calculation might conclude that the project is still worth undertaking.

An important caveat to the preceding argument is that the appropriate social discount rate is not necessarily lower than the private rate.[12] The preceding argument rests on the assumption that government programs are net additions to the total research effort. In practice, if the government decides to pursue a new technical development, the resulting program might simply substitute for private research. The extent to which federal R&D substitutes for, or actually encourages, privately sponsored R&D is

11. See Rosenberg, *Perspectives on Technology*, pp. 68ff.

12. James P. Quirk and Katsuaki Terasawa, "Sample Selection and Cost Underestimation Bias in Pioneer Projects," *Land Economics*, vol. 62 (May 1986), pp. 192–99.

a matter of considerable dispute, but the most recent work indicates a reasonably high level of substitution.[13] If so, part of the opportunity cost of federal R&D projects is the loss of some private projects that have a high return. All else equal, it does not make sense to replace high-return projects with projects producing low returns. To avoid this error requires accounting for the opportunity cost of the investment capital that is switched from private to public projects in the calculation of the discounted net present value of the public program. Hence a necessary ingredient in the development of the optimal portfolio of activities in a public R&D program is to assess and to take into account the extent to which projects drive out private R&D.

Benefits Assessment

An important difference between assessments of R&D programs in the public and private sectors arises from the broader conception of the benefits of technological change included in the normative rationales for public support of R&D. The private calculus considers the effects of R&D on the profits of the innovative firm or on some correlate of profits such as sales or market share; the public calculus is supposed to consider the economic value of R&D that is not appropriable by the innovator and the noneconomic spillover effects. Indeed, the rationale for targeted demonstration programs is that support should be concentrated on some important cases for which the latter consequences of innovation are especially large compared with the former, making R&D publicly beneficial while privately unprofitable. In addition, government support can serve to enhance the social value of innovation simply by making it widely available to competitors. Eliminating the possibility that most of the benefits of the innovation will be retained by its adopter ensures that prices for the product of the new technology will be lower. Consequently, more of its economic benefits will be passed on to its users, enhancing the utilization efficiency of the new technology.

If government does pursue targeted policies according to these objectives, a rather serious measurement problem arises in assessing the net benefits of a program. Recall the observation that a large fraction of all innovation consists of product improvements, so that the form of the private profit inducement to innovation is not cost reduction but a (perhaps short-run) monopoly in a superior product. If government R&D commercializa-

13. Frank R. Lichtenberg, "The Relationship between Federal Contract R&D and Company R&D," *American Economic Review*, vol. 74 (May 1984, *Papers and Proceedings, 1983*), pp. 73–78.

tion programs eliminate this monopoly, the program will have no observed effect on profits or productivity in the innovating industry. All the benefits will accrue to users.

Empirical work on the effects of government R&D support on productivity in the contracting firm is consistent with this view.[14] Many studies have found that the effect of government-financed, private R&D on productivity in the contracting firm is at or near zero.[15] Even though this is consistent with the view that federally supported R&D is a total waste, the opposite conclusion is equally well supported: that federal R&D is accurately targeted at the kinds of revolutionary product innovations that have large spillover effects on downstream firms and society generally. Certainly part of this argument is correct, in that a very large portion of government research is directed at product innovations, not cost reductions. The large expenditures on defense R&D are almost exclusively oriented to the development of weapons systems and electronic components of military hardware, and here the overriding concern is virtually always to enhance performance by inventing a superior product.[16]

Whatever the underlying truth about the social productivity of government R&D, the important lesson to be gleaned from this discussion concerns the problems of measuring the benefits of the programs. Specifically, by their nature, government programs are intended to produce secondary effects. To assess the benefits of a program requires examining its effects on customers of the target industry and perhaps many other firms and individuals further downstream in the product chain. Thus the place to look for the principal beneficiaries of, for example, the power reactor development program is neither the industry that supplies nuclear steam systems for electric generators nor, given the regulated status of electric utilities, the firms that procure nuclear power plants. Instead, it is the users of electricity and the breathers of air that would be less polluted if nuclear energy replaced generators run on hydrocarbon fuels. The problem this assessment entails, of course, is that in a complex, interconnected economy, working out the flows of secondary effects as they cascade through the

14. Paul G. Kochanowski and Henry Hertzfeld, "Often Overlooked Factors in Measuring the Rate of Return to Government R&D Expenditures," *Policy Analysis*, vol. 7 (Spring 1981), pp. 153–67.

15. In addition to Lichtenberg, "Relationship between Federal Contract R&D," see also Nestor E. Terleckyj, *Effects of R&D on the Productivity Growth of Industries: An Exploratory Study* (Washington: National Planning Association, 1974); and Albert N. Link, *Research and Development Activity in U.S. Manufacturing* (Praeger, 1981).

16. See most of the studies in Nelson, ed., *Government and Technological Progress.*

economy involves an ever-increasing likelihood of error as one moves further away from the point of innovation. Hence whether the issue is ex ante decisions to proceed or ex post evaluations of a program, the kinds of targeted projects that exemplify the standard normative rationales are usually subject to higher uncertainty about their benefits than programs that are likely to be attractive to private organizations.

Implications of Efficiency Maximization

The preceding discussion highlights several features of a well-managed government R&D commercialization program that deserve emphasis. As we examine specific projects and programs, we will refer to these characteristics of an efficient program as a benchmark in trying to assess whether and, if so, how a public program jumped off track.

The key features of government R&D commercialization programs should be these.

—The projects generally should be more research intensive, more radical in concept, and take longer than similar types of projects in the private sector.

—The principal direct economic benefits of the program should accrue to the users of the product of the program, not to the firms contracting with the government and producing the innovation.

—Noneconomic effects and external economies should play a significant role in the evaluation of government programs.

—Government R&D projects should be subject to more uncertainty, in part because private R&D decisions are likely to be excessively risk-averse from a societal point of view, in part because of the greater measurement problems associated with estimating the benefits of government programs, and in part because of problems associated with transferring information from private firms.

—The optimal management strategy for a government project is to be more flexible and unpredictable as the program proceeds from stage to stage: greater uncertainty leads to greater likelihood of unanticipated changes in the expected performance of the program, and hence a greater likelihood of redirecting, canceling, or even speeding up the program in relation to original expectations.

For a commercialization project to be efficient, the rationales for government action must apply at all phases of the program. The problems of interorganizational transfer of information will systematically be a source of inefficiency at all stages: at the research end because of the difficulty of taking into account biases arising from specialized personal experience and knowledge, and at all stages because of the reluctance of holders of information to risk losing their benefits from the program by revealing accurate assessments of the technology as it develops. For government to undertake a phase in a commercialization program, a positive efficiency rationale must be forthcoming that plausibly offsets this inherent problem of government projects. Specifically, at each point in the execution of a program, the remaining magnitude of the problems associated with risk and spillovers must continue to inhibit private pursuit of the project. Or, if they do not, the gains from displacing private management must outweigh the attendant efficiency loss that is sure to come from imposing a decision process that depends on obtaining accurate and complete information from reluctant private parties.

Chapter 4

The Politics of Commercial R&D Programs

JEFFREY S. BANKS, LINDA R. COHEN & ROGER G. NOLL

THE SELECTION and management of commercial research and development projects differ between the public and private sectors in two important respects. One is the inclusiveness of the economic effects considered in evaluating a project. Private decisions are largely motivated by prospective profitability, whereas a wider array of social benefits and costs are relevant in the public sector. Specifically, public decisions would normally take into account economic benefits accruing to parties other than the innovating firm (such as competitive copiers of the technology and the customers of the industry), as well as any external costs from adopting the innovation (such as increased environmental pollution).

The second difference between public and private decisionmaking is the institutional structure in which decisionmakers are evaluated. Although retrospective evaluation of R&D is difficult and imperfect in the private sector, it is facilitated by the shared recognition that R&D is intended to provide financial returns to the company and by the presence of quantitative, quite easily observed, indexes of success, such as sales, unit costs, accounting profits, and evaluations of the firm in capital markets. In the public sector, the ultimate external test of an R&D program is its ability to generate more political support than opposition.

This chapter explores the implications of political evaluation on the selection and management of R&D projects. The principal conclusion is that American political institutions introduce predictable, systematic biases into R&D programs so that, on balance, government projects will be susceptible to performance underruns and cost overruns. Moreover, the importance of these problems is related to characteristics of the industry, technology, and implementation strategy of the program. In some cases, the causes of the "market failure" that is used to justify a federal R&D project also produce a corresponding "political failure"—that is, the same factors that cause inefficient investment in private R&D also undermine the efficiency of a government program.

Electoral Influences on R&D Programs

The design, adoption, and execution of a government R&D project are the result of a complex interaction among a large number of government officials, each of whom has the power to delay or even to derail a proposal. With rare exceptions, programs are conceived and designed in executive branch agencies by professional civil servants who are dedicated to the mission of their organization. The first hurdle they must clear is approval from the presidency—the president and the relevant officials in the Executive Office of the President (including the Office of Management and Budget). Then, to obtain authorization for the project and appropriations to pay for it from Congress, the proposal must clear a series of congressional hurdles involving subcommittees, committees, the floor of each branch, and a conference committee.

Elections shape this process to the extent that a proposed project is electorally significant to the president, to Congress in general, or to individuals in Congress who are in control of the veto points in the process, such as committee chairs and the leadership of the majority party in each branch. The purpose of this section is to explain why and how electoral politics can affect the design and execution of an R&D project.

The Electoral Saliency of Policy Decisions

Much of the research on decisions by elected politicians, whether in making policy decisions or adopting campaign strategies, begins with the hypothesis that politicians try to maximize the number of votes they receive or the probability that they will continue to be reelected. But this assumption is more extreme than is needed to justify the conclusion that politicians will take into account the response of the electorate in the adoption and design of policies.

As long as politicians care at all about job security, their optimal strategy will be to find the best trade-off between single-minded dedication to their personal policy objectives and success with the electorate. Indeed, the ability to retain office is a necessary part of continuing to be able to influence policy in directions the politician perceives to be desirable. Even a totally altruistic politician, on perceiving a policy disagreement with a challenger, will prefer to shade his or her policies to increase reelection probabilities and the chance that a policy closer to the politician's ideal will be pursued. Moreover, a politician who adopts an unpopular position and refuses to make any compromises will be most vulnerable to successful challenge.

One of the interesting properties of majority-rule elections is "monotonicity," that is, if the preferences of voters shift in a particular direction and if new alternatives can be freely introduced, the outcome of the electoral process will shift toward the change in voter preferences.[1] The practical interpretation here is that politicians who "lose touch" with the electorate are susceptible to successful challenge in the next election. Hence, on balance, the higher survival rate of politicians who strike a compromise between their own policy preferences and those of their constituents should produce a group of elected officials who are responsive to the electorate or, alternatively, who have policy ideals that are near the position that would achieve maximum electoral support, thereby eliminating the need for compromise.

Whereas establishing the plausibility of electoral responsiveness by elected politicians is easy, the argument that the electorate will be responsive to R&D projects is much more difficult to establish. Votes carry little information about the preferred policies of voters. Votes are a simple trichotomous signal of approval, indifference, or disapproval in response to a long, complex list of attributes and positions of candidates. Moreover, a single vote has almost no chance of deciding an election and so has little functional value in determining policies. Consequently, a voter has little incentive to undertake sophisticated calculations about the relative merits of candidates or to devote much time and money to learning about their likely performance in office.[2]

Research on voting behavior bears out the theoretical expectation that voters use a few simple criteria to evaluate the performance of candidates as national policymakers. Attitudes about incumbents are shaped by the general state of the nation, by personal assessments of the personality and character of the candidates derived primarily from the media, and perhaps

1. William H. Riker, *Liberalism against Populism: A Confrontation between the Theory of Democracy and the Theory of Social Choice* (San Francisco: Freeman, 1982), pp. 45–51.

2. The theory of "rational ignorance" by voters was first developed extensively in Anthony Downs, *An Economic Theory of Democracy* (Harper, 1957).

by a few policy positions associated with the candidates because of either their own well-publicized actions or the strong positions taken by their party. Only rarely will an R&D project bear any important relationship to these phenomena.

Under only two circumstances might an R&D project significantly affect voter assessments of the policy role of candidates. The first occurs when an issue relating to the state of the nation has a strong R&D component. One example is the space race with the Soviet Union that was initiated by Sputnik in 1957 and that came to an end with the successful landing of Americans on the moon in 1969, at the end of the Apollo program. Another is the energy crisis of the 1970s, and the accompanying attempt to wean the American economy from dependence on erratic supplies of foreign oil. A third example is the strategic defense initiative (Star Wars) by which the Reagan administration sought to use defensive weapons systems to reduce the threat of nuclear war.

The second circumstance under which R&D might affect voter assessments is when a program becomes a scandal because of either gross mismanagement or abject corruption. Here the issue is not the validity of the R&D project itself but the effect of scandal on assessments of the competency and integrity of the political officials who are regarded as responsible for it. In practice, there are no clear examples of a politically relevant scandal in an R&D program. Although claims of gross mismanagement have been leveled against R&D programs, notably the Clinch River Breeder Reactor and the space shuttle after the disastrous loss of the *Challenger*, they apparently have not inflicted political costs on either the president or the members of Congress who oversee the programs. R&D programs have also been free of major instances of overt political corruption.

In sum, R&D programs are unlikely to play an important role in national debates over policies that are electorally significant. Only rarely will their existence have widespread political relevance. But that does not mean they are unimportant to the reelection fortunes of at least some political leaders. R&D projects can also have electoral significance because of their distributive effects.

The Distributive Politics of R&D

The distributive aspects of a policy are normally regarded as the geographic pattern of expenditures on a program, but the principle is much broader and refers to the concentration of the benefits and costs of a program among an identifiable group within the electorate. Although only a small fraction of the population collects the expenditures required to implement any program, this group's economic status can depend heavily on the continuation of appropriations for the program, especially in the short run. If voters hold elected political officials responsible for the fate of a program that is providing local jobs, politicians can be expected to take into account the likely recipients of a program's expenditures when evaluating it.[3] For members of Congress, this means considering whether a program "creates jobs" in the home state or district. As one observer of Congress has noted: "Congress is predisposed by its electoral and geographic bases to distribute benefits to constituents. Some members garner electoral support by practicing public parsimony; many more do so by spending public funds. A few members are indifferent to the flow of federal dollars to their states and districts; most actively seek to increase those funds."[4]

Of course, the geographic representation of Congress is not the only source of distributive concern. Successful campaigns require funds and volunteers, and these need not be obtained from within politically relevant boundaries. Even the president is not immune to distributional concerns. The president is elected by a constituency that is not likely to be representative of the entire nation in terms of geography, economic interest, or political philosophy, and in addition a presidential candidate requires contributions to wage a successful campaign. Moreover, the president can be held responsible for the loss of jobs resulting from a canceled program. Finally, presidential success depends in part on the composition of Congress and the reelection success of the members who are most likely to support the administration's program.

The benefits of the program, as well as the expenditures, can be susceptible to distributive considerations. In some programs, the benefits fall unevenly among regions and communities. In others, the value of the benefits may be controversial. Two examples illustrate the point. In the 1970s research programs to develop renewable energy sources would probably have been evaluated differently in different parts of the country because of regional variations in the availability of renewable energy resources. For example, geothermal energy is limited to a few areas having readily accessible underground reserves of hot water, and solar technologies

3. Barry R. Weingast, Kenneth A. Shepsle, and Christopher Johnsen, "The Political Economy of Benefits and Costs: A Neoclassical Approach to Distributive Politics," *Journal of Political Economy*, vol. 89 (1983), pp. 642–64.

4. Allen Schick, "The Distributive Congress," in Allen Schick, ed., *Making Economic Policy in Congress* (Washington: American Enterprise Institute, 1983), p. 260.

are most valuable in regions that have little cloud cover. Hence R&D in renewable energy has important distributive aspects in both benefits and costs. By contrast, the U.S. space program provides an almost perfect example of a pure public good. All regions shared in the experience of Neil Armstrong setting foot on the moon, and in the reduction in the costs of intercontinental telecommunications services owing to communications satellites. Nevertheless, the values citizens place on these benefits differ greatly. Not everyone cares about making cheap telephone calls to Europe or viewing live coverage of international news and sports events, and not everyone agrees that the pictures of Americans on the moon or from spacecraft visiting other planets are worth the cost.

The electoral significance of the distributive effects of policies depends on more than simply the number of dollars spent on the program and the number of people who enjoy its benefits. It also depends on whether the beneficiaries are likely to take political action—voting, volunteering, contributing—because of their personal stakes in the issue. Like more general national policy issues, to be electorally significant the distributive aspects of a policy must be at the top of the list of factors that affect participants in the political process. Otherwise they are unlikely to motivate politically relevant behavior.

Two factors facilitate the importance of a program to a specific, targeted constituency.[5] One is the absolute size of each constituent's personal stake in the issue, and the other is the extent to which the relevant constituency is already organized for effective political participation. All else equal, the greater a person's stake in an issue, the more likely it is that the issue will become one of the handful by which the person evaluates candidates. And if a group is already organized—say, as part of a firm, a trade union, or a social or political organization—the leaders of the group can inform and mobilize its members without having to incur the costs of organizing them.

These observations suggest several hypotheses about the way in which distributive concerns affect decisions by elected politicians. First, the political significance of a program is not strictly proportional to its size. Consider the distributive aspects of a policy in a specific congressional district. If the district receives a very small contract, only a small part of the community will become economically dependent on the program. These people will obviously care a great deal whether the contract is continued. For the larger constituency, the overall economic impact—the

5. Mancur Olson, *The Logic of Collective Action* (Harvard University Press, 1965).

spillover effect on the local economy—is very small per person and so unlikely to be detected, let alone to motivate political action.

As the size of the contract grows, and the proportion of jobs and income in a community that derives from the program increases, the spillover effects become more important and eventually can motivate political behavior by people other than the recipients. Moreover, the effects can spill across congressional districts, being of concern to senators and even presidential candidates. In this range, a project can be regarded as exhibiting increasing political returns to scale. Eventually the contract can become so big that further growth has little or no electoral significance; that is, it begins to exhibit decreasing political returns. An entire state or region becomes so politically sensitized to the program that further increases in it have no additional effect on how the program motivates constituents. The implication is that politicians are not likely to be much influenced by the distributive aspects of small programs, but that once a certain threshold of visibility in their constituencies is attained, distributive concerns become important—but not necessarily of greatly increasing importance as their constituents' stake in the program grows well beyond the threshold. Moreover, because of differences in the size of constituencies, the threshold is higher for senators than for members of the House, and highest for the president.

The second hypothesis is that the growth and contraction of a program are politically asymmetric. When a program is proposed, the identities of the contractors and, especially, of the people who will be hired to undertake the project are usually uncertain, if not unknown. Citizens who are unaware that they are destined to be employed in the new program are unlikely to engage in political behavior motivated by its enactment. Once contracts are awarded and workers are hired, however, continuation of the flow of funds is an issue in which identifiable and organized groups (firms, unions, local governments) have a clear stake. Consequently, the distributive aspects of a program are likely to be more important after it is in operation than when it is initially authorized. Moreover, distributive aspects are apt to be more important in motivating decisions to cut a program than in influencing the outcome of proposals to increase them. The former impose direct costs on all the relevant organized groups, whereas the latter will deliver at least some benefits to people who do not yet realize they will be beneficiaries.

So far the discussion of the distributive aspects of a program has focused on its benefits: the public good it produces and the pecuniary gains to the people who provide it. But the cost side of the program can also have distributive elements. For example, the project may be a LULU—a "locally unwanted land use," meaning that the local economic benefits from construction and operation of a project are more than offset by other, negative effects.[6] A useful illustration is facilities related to nuclear energy: experimental, operational, waste disposal, and so forth.

Another potentially important distributive aspect of a program is its effects on the firms, industries, and areas that do not receive contracts or otherwise will not benefit from the program. In the case of an R&D program, firms receiving contracts to undertake research can be expected to obtain useful knowledge about the new technology if it proves commercially attractive and about the science and engineering base of their industry even if the project itself is unsuccessful. Competitors that do not receive contracts are likely to perceive the government program as giving contracting firms an unfair advantage. Hence firms that are disadvantaged by the program are likely to favor discontinuing it, scaling it back, or performing it only in open research facilities such as universities or national laboratories.

Impatience and Risk Aversion

Incumbent politicians, insofar as they are motivated by the desire to be reelected, will tend to be impatient and to avoid political risks. Such tendencies, if present, have significance for R&D programs, which are usually long-term projects that have considerable risk of failure and so are likely to be systematically undervalued by elected politicians.

A key feature of democratic systems of government is that political actors cannot make durable, long-term commitments to the continued implementation of programs. Candidates cannot sign contracts with voters or contributors that bind them to behave as promised during the campaign in return for votes and contributions. One session of Congress (or one administration) cannot bind subsequent sessions (or administrations) to continue to provide funds to complete a long-term project. Voters are therefore forced to evaluate candidates in part on their reputation and to punish fickle politicians by shifting their support in subsequent elections.

Militating against this, especially for members of Congress, is the difficulty a voter faces in detecting what role, if any, one representative had in deciding the fate of a program and perhaps leaving a promise unfulfilled. A favored program can be cut or killed for reasons having

6. Frank J. Popper, "LULUs," *Resources*, vol. 73 (June 1983), pp. 2–4.

nothing to do with either its success or the efforts of representatives from a benefiting area. A change in the presidency or in the ideological and party composition of Congress, for example, is likely to result in a reallocation of the federal budget in keeping with the policy preferences of the newly elected officials and their supporters. Because of the difficulties in measuring the quality and quantity of effort by elected representatives in carrying out their promises, and because of the presence of effects that are beyond the control of legislators from a specific state or region, citizens have reason to engage in "retrospective voting"—that is, to evaluate candidates, especially incumbents, on their recent accomplishments, rather than on the professed plans and future prospects that candidates say they will try to deliver.[7] Of course, the concept of retrospective evaluation applies to all politically relevant activities, including campaign contributions.

Retrospective evaluation of candidates has important implications for R&D programs. First, it reinforces the tendency for the distributive politics of a program to be based more on current spending than on expected future growth. Second, it creates a political disadvantage for long-term projects that will deliver most of their benefits in the distant future. In assessing the expected value of a new program, a citizen not only will use normal time discounting for evaluating future consequences but in each future period will multiply the expected net benefits by the probability that the program will survive that long.[8] This decreases the likelihood that the future effect of the program will become important enough to the voter to be among the handful of concerns that motivate current political participation.

From the perspective of the elected official, the implication of retrospective evaluation is that, all else being equal (the present value of the net economic benefits and the distribution of benefits and contracts), a project with earlier realization of politically relevant benefits will be preferred to longer-term projects. That is, to the extent that citizens heavily discount future plans of programs and engage in retrospective evaluation, they create an incentive for political officials to be too impatient in evaluating proposed programs. Because R&D projects are usually long term, they will normally face an uphill struggle in the battle for budgets with operating programs that provide current benefits. In general, political leaders will

7. Morris P. Fiorina, *Retrospective Voting in American National Elections* (Yale University Press, 1981).

8. Linda R. Cohen and Roger G. Noll, "The Political Discount Rate," Stanford University, Center for Economic Policy Research, 1990.

exhibit reluctance to adopt an R&D strategy for coping with a perceived national need. This means not that they never will pursue an R&D approach but that they must perceive a compelling reason to do so.

The source of political risk aversion among legislators arises from two features of the electoral process. One is the empirical regularity that members of Congress are extraordinarily effective in securing a very high probability of reelection. The other is the tendency for elections to be decided on a limited range of issues and, for legislators, on the ability of the incumbent to represent the home constituency effectively in the battle for federal contracts and other particularistic benefits.

If a politician ends up on the wrong side of a controversial issue, the challenger in the next election is likely to put considerable campaign effort into bringing this shortcoming to the attention of the voters. All else equal, a candidate would prefer to work on issues that have broad support. Establishing a record as an effective executive, for example, will be positively evaluated by all voters. For members of Congress, effective effort on behalf of the home constituency is likely to be a noncontroversial plus for an incumbent.[9] Of course, elected officials must take positions on controversial matters, but when given a realistic choice they would prefer not to do so. R&D projects are economic and technical risks, which in turn could translate into political risks. If a program fails and is canceled in midstream, the distributive consequences of cancellation can threaten not only the legislators who represent the areas receiving large contracts but also a president seeking reelection in a close contest. Moreover, in some cases technical and economic failure can mean that an important national objective is not attained, which in turn can undermine faith in the president's managerial skills.

Risks are undertaken, in politics and business, when the value of a good result is high enough to offset the losses from a bad result. If a politician is virtually certain to be reelected, a good result from a risky program cannot have much effect on the next election. But a bad outcome might give a challenger a much-needed issue that does reduce the chances of reelection.

9. The role of constituency service in the reelection strategy of a legislator was first fully elaborated in a series of books in the 1970s: David R. Mayhew, *Congress: The Electoral Connection* (Yale University Press, 1974); Morris P. Fiorina, *Congress: Keystone of the Washington Establishment* (Yale University Press, 1977); and Richard F. Fenno, *Home Style: House Members in Their Districts* (Little, Brown 1978). For the first systematic studies of the acquisition of contracts for home constituencies, see John A. Ferejohn, *Pork Barrel Politics* (Stanford University Press, 1974); and R. Douglas Arnold, *Congress and the Bureaucracy: A Theory of Influence* (Yale University Press, 1979).

Risk aversion related to the success or failure of an R&D project is not likely to be important for the president; rarely does a significant part of the president's program and reputation as an effective executive turn on the results of an R&D effort. For example, the failure of the TFX—a multipurpose aircraft intended to serve different needs for all the military services—did not undermine the popularity of the Kennedy-Johnson administration, and support for the Clinch River Breeder Reactor did not hurt the reelection of Ronald Reagan. One can only speculate whether the performance of *any* R&D project could have a notable effect on the president. Would a massive catastrophic failure of the Apollo project have hurt Richard Nixon in 1972, and would abject failure of Star Wars, the B-2 (Stealth), high-definition television, and the rocket plane combined prevent George Bush's reelection in 1992? Although these questions are probably unanswerable, each refers to an R&D project of massive proportions and high political salience. If the political significance to the president of even large, visible ventures is debatable, one must conclude that the vast majority of R&D projects do not involve policy-related political risks to the president.

The political risks associated with R&D projects are therefore primarily distributional, and primarily felt by members of Congress. They are tied to the technical risks of R&D only insofar as the uncertainties in the performance of an R&D project could potentially cause a project to be canceled or produce a result that damages an important constituency of some legislators.

A legislator can be harmed politically if an R&D project performs poorly and the home constituency thereby loses contracts, or if an R&D project succeeds in a way that causes firms in the home constituency to lose business because of the new technology. This suggests that, all else being equal, the legislature should favor R&D projects that have a narrow range of technical uncertainty and that, if successful, are unlikely to cause a significant redistribution of wealth within an industry. For example, legislators should prefer to pursue the next logical step in the advancement of technology in an industry rather than fundamental research that would dramatically alter its technological foundations but is less likely to succeed.

The Role of the President

The preceding argument suggests that important differences are likely to emerge between Congress and the president regarding R&D programs. Legislators, owing to impatience and risk aversion, have good reasons to be wary of long-term, risky programs, while the president need not be so cautious. A president must certainly be concerned that programs initiated in one term will not be carried to conclusion in the next; however, if a program is reasonably successful, and if expenditures are large enough to cross the threshold of political significance for a number of legislators, a president can be reasonably confident that a program will be difficult to kill in subsequent sessions of the legislature. Furthermore, because presidents face only one reelection and usually hold the office only at the end of a political career, their sensitivity to subsequent reelection campaigns ought to be far less than for members of Congress. Consequently, the president ought to be more favorably disposed to R&D projects than Congress.

In the American political system, the president can propose new programs, but only Congress can authorize them and appropriate the funds to carry them out. So how will the president's greater propensity to favor R&D be likely to influence the legislature?

Recent work by D. Roderick Kiewiet and Mathew D. McCubbins has shed light on the interaction between Congress and the president in determining the fate of a program.[10] Their analysis focuses on how the strategic positions of the president, with veto power, and Congress, with initiation power, affect the final budget allocation for a program. Their key insight is that the power of the president turns crucially on whether Congress is more or less anxious to spend money than the president, and on the relationship of the preferred budget in each branch to the expenditures in the previous year. If Congress wishes to spend more than the president and more than the previous appropriation, the president has more influence, because the presidential veto is more costly to Congress than to the administration. If the president wants a larger budget than the previous year and than Congress desires, the strategic advantage is with Congress.

The Kiewiet-McCubbins theory has strong implications for R&D programs. The president is more likely to want to initiate R&D than Congress is. Hence the probable result is that presidents will often fail to get Congress to initiate such programs, and when they succeed, the scale of the program is apt to be smaller than the president would desire. But once a program begins, the relative preferences of Congress and the president are likely to change. Once contracts are large enough to achieve political salience among

10. D. Roderick Kiewiet and Mathew D. McCubbins, "Presidential Influence on Congressional Appropriations Decisions," *American Journal of Political Science*, vol. 32 (August 1988), pp. 713–36.

constituents of some members of Congress, congressional support for the program is likely to increase. And the threshold of political salience is smaller for members of Congress than for the president. If the program becomes more popular with Congress than with the president, the strategic advantage switches to the latter, and programs will tend to resemble more closely the proposals of the administration.

If Congress likes a program much more than the president does, it has one remaining strategic possibility: it can bundle the budgets of the programs it favors with ones most favored by the president. For such a bundling to work, of course, Congress must give some support to the president's preferred programs, or else it cannot overcome the threat of a veto. Hence, when Congress and the president both seek their own pet projects while being less than enthusiastic about the pet projects of the other branch, a probable outcome is an omnibus program that increases spending across the board.

Disagreements between the branches over which commercial R&D projects have priority are likely to be based on their distributional and risk characteristics, not on perceptions of their intrinsic technical and economic merit. First, the president may represent a different constituency than the majority of Congress. Second, the president may want to emphasize longer-term projects, or to initiate new, risky projects, whereas Congress is more likely to want to keep old projects going after a change in administration, or after bad technical news has begun to arrive about a project that provides contracts for their constituents. In either case, the consequence is likely to be a larger portfolio of projects, and a larger budget, than either branch would prefer. The ability of the president to convince Congress to start new projects will then turn in part on the scope of existing projects that Congress wants to continue but the president wants to scale back or kill.

The Role of Research Agencies

Although political officials in the executive and legislative branches bear ultimate authority to allocate the budget among competing claimants, including proponents of R&D projects, they are at a great disadvantage in evaluating research proposals. Few if any elected officials and their political appointees are sufficiently sophisticated in science, technology, and the details of production in an industry to be able to make an independent assessment of the probable value of a proposed R&D project. Many students of bureaucracy have noted the informational advantage that agencies have and its implications for budgeting and program oversight.[11] R&D projects must be among the programs for which theories based on the agency's informational advantage are most applicable, for the technical expertise required to undertake them is both substantial and arcane. The central question is how this situation might affect the menu of alternatives considered by political authorities, and hence the performance of R&D programs.

Our analysis of the role of agencies focuses on the professional civil servants who design and implement proposals to develop new technology. Agencies capable of undertaking large-scale R&D programs consist of a thin veneer of political appointees at the top of a phalanx of professional scientists and engineers who develop detailed research proposals. Typically these agencies are either devoted to a well-defined "mission" or are subdivided into separate offices, each of which deals with a particular technology or industry, or both. The civil servants who populate these offices are self-selected by mission and technology not only in their employment but also typically in their technical education. On average, they are much better informed about technical opportunities in their domain than other government officials are. Moreover, they are likely to be more committed to the importance of technical advance in their field. And collectively they probably share a long-term vision of the most appropriate lines of development of new technology. Thus technologists in the Department of Energy offices that are responsible for research on nuclear energy are apt to be knowledgeable advocates of breeder or fusion technology, whereas the technologists in renewable resources programs are apt to be informed supporters of solar or geothermal technologies. Because of employment self-selection, these views usually reflect genuine beliefs rather than purely tactical political positions.

The job of government technologists is to generate politically acceptable plans for further development of technology within their domain, while recognizing their status as competitors with other proponents of research. This task has three main components. The first is to identify the "win set" of proposals, that is, the alternatives that have a reasonable chance of being approved by all the government officials who potentially have veto power

11. Aaron B. Wildavsky, *The Politics of the Budgetary Process* (Little, Brown, 1964); Charles L. Schultze, *The Politics and Economics of Public Spending* (Brookings, 1968); and William A. Niskanen, Jr., *Bureaucracy and Representative Government* (Chicago: Aldine, Atherton, 1971).

over a proposal.[12] The sequence of approvals runs from the Office of Management and Budget through the president and the White House staff, the relevant committees of Congress, and the two chambers of Congress.

Normally, if *any* R&D project is reasonably likely to be approved, several alternative approaches will be politically feasible. Hence the second part of the agency's job is to identify the alternatives in the win set that are preferred, on the basis of the knowledge, values, and mission embodied in the agency. The power of the agency to influence policy lies in its ability to select the alternatives to be considered from a larger set of feasible ones, with reasonable assurance, deriving from its superior information, that these other alternatives will not be seriously taken up by the decisionmakers outside the agency.

The third aspect of the agency's task is to develop an effective strategy for obtaining the approval of the relevant political actors for an alternative regarded as desirable by the agency. That amounts to providing the information and lining up the necessary external political support to convince those with veto power that the program is desirable from their points of view. Part of the agency's strategy for winning approval will clearly be to emphasize in detail the potential uses of the new technology it advocates, but that by itself is not likely to be sufficient. Although political actors will not be able to evaluate these claims directly, they are sure to be aware of the self-selection problem in staffing the agencies, and hence to understand that however genuine the agency's stated beliefs about its proposal, its views will probably be optimistic. Consequently, an agency's case for a new program will be enhanced if it can enlist the support of an outside group of potential users.

Establishing Agency-Industry Alliances

Many reasons have been advanced to explain the close alliances between agencies and external user organizations, all of which have some plausibility. First is the "revolving door" hypothesis, which maintains that after a period of government service civil servants and political appointees in an agency tend to move on to the industry with which they deal and so are motivated to maintain good working relationships with potential future employers. Second is the psychological argument that most people prefer to avoid

12. Kenneth A. Shepsle and Barry R. Weingast, "Structure Induced Equilibrium and Legislative Choice," *Public Choice*, vol. 37 (Fall 1981), pp. 503–20.

conflict, especially those who expect to be in frequent contact over long periods because of their organizational roles. Such people have reason to work out mutually accommodative arrangements that cause "debureaucratization"—the blurring of the functional boundaries between two putatively independent organizations, which encourages organizations to develop policies at the point of contact of two organizations and then transmit them upward to the organizational leaders, rather than downward through the top-down management structure that an organization might appear to have in place.[13]

A third reason for close alliances is that users must adopt the new technology if the technological vision of an agency is to be realized. Developing lines of communication with users early on is helpful in designing the technology to satisfy them and in setting the stage for eventual transfer of the technology. A fourth reason is that developing good relations with a user group provides an organized, politically relevant client that is useful in obtaining political approval. The client group provides additional confirmation of the reasonableness of the agency's technological vision as well as a means for generating a positive distributive element in the politics of the program.

Finally, the target industry is likely to have considerable expertise about the range of new technologies available to it. Hence it could upset the agency's plans by introducing alternatives to the agency's proposal that are also in the win set but that are much less attractive to the agency. Consequently, the agency has good reason to work out an accommodation with the industry so that both achieve an acceptable result.

To develop an accommodative relationship with a user group requires two ingredients. Most obviously, the technological alternative to be explored must be one that some organized user group will advocate publicly. For commercial R&D projects, this is usually either the industry that will use the new technology or the one that will be responsible for developing and eventually marketing it to the users. It is also essential that the implementation strategy of the agency not threaten the external developers and users. To undertake a successful R&D effort requires decisions about how a program will be organized and managed as well as which technical

13. James D. Thompson and William J. McEwan, "Organizational Goals and Environment: Goal-Setting as an Interaction Process," *American Sociological Review*, vol. 23 (February 1958), pp. 29–31; and Michael Aiken and Jerald Hage, "Organizational Interdependence and Intra-Organizational Structure," *American Sociological Review*, vol. 33 (December 1968), pp. 912–30.

options will be explored. The organization and management plan, in turn, requires decisions about the allocation of effort among in-house government research facilities, industrial contractors, and universities and other non-profit research centers. Most important among these are the industry people, for they are not controlled by the agency and are most likely to be organized for effective political opposition should the implementation plan threaten their interests.

An example is the decision about the degree to which the agency will use extensive contracting with the private sector to develop the new technology. Firms in the industry that fancy themselves to be potential developers of new technology will not want to be disadvantaged by a government R&D project. Each might prefer a sole-source contract, but that is not likely to be feasible because all the firms will oppose sole-sourcing for everyone else. Hence the industry is likely to prefer that the government implement the program either with extensive subcontracting or multiple-source contracting so that almost everyone can participate in the program, or through a central industrial research facility that allows all to participate and that will make its results openly available.

Multiple contracting, or a *fragmentation* strategy, is viable if the technology is easily decomposable into a number of independent components that roughly match the number of firms to be accommodated. It can also work if the most efficient R&D strategy is to pursue several technical alternatives simultaneously, such as might occur if there is much disagreement about the best alternative and if the cost of research, prototypes, and testing is small compared with the value of a technical success.

A government or industry laboratory, or a *centralization* strategy, is viable if the technology is readily transmitted across organizational boundaries and is not easily appropriated. Firms will then find it in their interests to participate in industrywide research and will not be able to engage in strategic behavior to keep useful, firm-specific information proprietary. This strategy will also be preferred if a single technical approach appears clearly dominant as the source of the next important commercial advance, or if the cost of research and development is large compared with the scale of use of the technology when it is adopted. The total expected fleet size of the space shuttle, for example, was never planned to exceed five, so that the first demonstration constituted 20 percent of the maximum long-run capacity to be provided by the technology.

Problems arise when the politically acceptable management structures are inappropriate for efficient development of the technology. For instance, if the act of developing the new technology produces technical knowledge that is easily appropriated and difficult to transmit across organizational lines except by using the people who developed it, firms in the industry are unlikely to be sufficiently cooperative to make an industry research facility effective. Or if commercial demonstration requires a very large facility that will account for significant industry capacity and give its builders appropriable technical knowledge, there may be no politically acceptable organizational method of constructing it in a competitive industry. Giving it to one firm will disadvantage the others, and undertaking the project in a government facility will put the government in direct competition with the industry. The final alternative, permitting an industry joint venture, may serve to create a cartel, which may then be opposed by the industry's customers as well as by other government agencies that advocate competition.

Effects on Performance

If the agency is confronted with a politically feasible set of implementation methods that does not include the most effective one, it must either scuttle its preferred path of research or adopt a plan that is less likely to be a technical success. Given the expected optimism and exuberance of technologists about their agency's mission, they will tend to conclude that even an imperfectly designed program is worth pursuing. Bending the design to acquire industry support will probably stifle any opposition even if the implementation plan is defective.

The essential point about the agency's information advantage is the control it gives the technologists over the agenda of alternatives that the political leadership will consider. Rational evaluation procedures—including benefit-cost analyses by staff economists in government and independent organizations—require that decisionmakers consider all points on the "efficiency frontier" (that is, the set of alternatives that plausibly could prove to be optimal). But because one must have considerable technical sophistication to define the relevant alternatives, government officials outside the agency usually cannot match the agency experts in their ability to identify them. Moreover, if the agency can work out an accommodation with its industrial collaborators, the principal source of alternatives to the agency's plans can be effectively co-opted.

The process just described is not necessarily invidious in that it somehow bamboozles political officials into taking actions that the agency knows to be against the interests of both the politicians and the national economy.

Instead, it is a process whereby agencies develop proposals for political leaders that satisfy two criteria: both the government technologists and the target industry believe the case for the project is plausible, and adoption of the project will not impose political costs on the politicians (and indeed may provide some benefits). Technical alternatives that are politically infeasible are of course uninteresting to political decisionmakers, so that from their point of view excluding such alternatives from the set to be considered makes sense. Moreover, if among the politically feasible set one alternative has the support of all the relevant parties, political leaders will not perceive much value in additional alternatives that have no external advocates and that they cannot really evaluate without help from the agency and its client industry.

Insofar as the process biases R&D programs, it does so in the following way: alternatives that satisfy the narrow mission of an agency and that are privately attractive to an industry may not produce the best national economic policy. Specifically, given the optimism of the agency and the incentive for the private sector to evaluate the program by its private benefits and costs (thereby ignoring the costs to the government and benefits to the public), the minimally acceptable program to the advocates will probably be less than minimally acceptable for national policy. Unfortunately, once the agency and the industry agree on a policy, political leaders have no equally knowledgeable source of information to assess whether the proposed program is of this minimally acceptable sort or whether a more dispassionate evaluation would still find it desirable. Indeed, for most programs no politically relevant opposition may be voiced at all even if the program has a low expected payoff compared with its costs.

Implications for Commercial R&D Programs

The implications of the political analysis of the environment in which R&D programs must operate can usefully be discussed in two parts. The first is the initial political decision to undertake a commercial R&D project, and the second is the sequence of later decisions about its management, including decisions whether to continue, terminate, or reorient the program. Because the problems encountered in the second set of decisions are predictable and can therefore be anticipated by political decisionmakers, they are likely to affect the first decision, so that logically it makes sense to start with the management issues.

Suppose that, for whatever reason, a commercial R&D project is under way. In the initial phases most of the work is relatively inexpensive research and engineering design to explore a range of technical issues before committing to a particular technological option. During this phase the distributive consequences of expenditures on the program will probably fall below the threshold of political importance, because the program is small in scale and its expenditures more dispersed than they will be when it enters the stage of prototype development and testing. The main particularistic political pressure relating to the program, if one exists at all, will come from the target industry and will reflect its judgment on the ultimate private value of the technology, ignoring, of course, the government's share of the cost and spillover benefits. But even the target industry will not have a large stake in the program when it is still small and research intensive. As results from the early research come in, therefore, the political evaluation of the program is likely to be based largely on its technical success and promise.

There are two main difficulties in the research stage. First, information about the success of the program will be filtered through the agency and the industry, which for different reasons tend to be overoptimistic. And second, political officials are likely to be impatient and risk averse, which suggests they will favor early termination of research and adoption of technical options that are too conservative. To some degree these two problems are offsetting, with the optimism of the advocates about radical technologies facing the natural desire of the politicians to be more incremental. But in any case, regardless of the technical design ultimately adopted, the research phase will tend to be subject to premature termination, and thus the ultimate performance of the new technology will be hampered.

Once the more expensive bricks-and-mortar phases of prototype construction, scale models, and tests begin, distributive politics increase in importance. As more money is spent on fewer technical alternatives, the jobs and contracts eventually become politically salient to at least some members of Congress. When they do, the project will begin to exhibit the familiar characteristics of the federal pork barrel. Its continuation will become more likely to reflect the political desire to avoid the short-term political costs of cancellation. If so, the relevant budgetary allocation process will start to show logrolling and the norm of unanimity among members of Congress, factors that enable legislators to provide one another with politically attractive, but economically questionable, locally oriented programs. As the importance of distributive effects rises, of course, how the project performs will become less relevant in congressional decisions whether to continue or reorient it. Once the pork barrel phase begins, technical news will have to be very bad indeed to overcome the political inertia that prolongs a program for distributive reasons. The president could lose interest in the program and even succeed in curtailing it. More likely, however, the program will be used in bargaining with Congress to obtain other presidential objectives.

Because of the asymmetry in distributive consequences of cuts and increases in a program, stretching out one phase of an R&D program will be somewhat easier than returning to an earlier, less expensive phase or killing the program altogether. Stretchouts of, say, prototype experiments preserve the current level of employment and contracts. Nevertheless, impatience among political leaders militates against a stretchout. Impatience demands that higher standards be set for continuing R&D at current levels in the face of bad technical news than under optimally efficient management. Once again, the unwillingness to "go back to the drawing board" causes the ultimate performance of the project to be less satisfactory than otherwise could be expected.

Decisions at the point of transition from one phase to another are therefore complex. Whereas distributive concerns and impatience militate against reductions (returning to a less expensive, more research-oriented phase), distributive politics are less important in deciding between continuing the current phase or moving on to the next one. Consequently, the life of a project will contain a few windows when performance is more important than distributional effects in influencing political decisions, corresponding to the points in the schedule at which decisions are made to increase the scale of the program by moving closer to a commitment to commercial demonstration. Because of impatience, combined with the optimism of the agency and its clients, the government will tend to be too willing, relative to an efficiency standard, to move toward commercial development.

One last complexity involves distributive political liabilities. During the life of the project the structure of the industry or the nature of the most promising technologies may change, either because of outside events or because of the results of the early stages of the program. If so, organized interests may emerge that will be harmed by the technology, altering the basis for its political support. For example, the industry may become more competitive so that not all its participants can be brought under the tent of the program. Or an unanticipated technological advance may open new potential applications of a developing technology that pose a new threat to an established industry. Or some of the technical approaches in a parallel

R&D strategy may fail so badly that they become unattractive to the firms exploring them, even considering their subsidy. For the most part, these negative distributional aspects are a result of success, in that only a project which has identified serious technical opportunities for producing a commercially attractive technology will threaten an established interest. At best, this can be a benefit in evaluating programs, because the distributive effects of cancellation and continuation may be more in balance once a threatened interest emerges. But if the losers far outnumber the winners, the effect can be the cancellation of a program *because* it is successful enough to pose a threat.

Presumably the initial decision to adopt a program will be taken with full realization of the problems the program will encounter later. Hence decisionmakers will realize that R&D projects are susceptible to cost overruns and performance underruns because the proponents will be overoptimistic and because those who make the subsequent decisions will be impatient. Some bad programs will be continued if the political costs of killing them are too high, and some good programs will be curtailed if they become a threat to a politically important constituency. Consequently, political leaders have even more cause to be reluctant to adopt such programs than for the reasons given at the beginning of the chapter. Besides being long-term and risky, R&D projects are likely to create political problems later on because of the managerial incentives at work in the governmental decision process.

The preceding analysis suggests that the political attractiveness of R&D projects will vary depending on several characteristics of the project and the industry it seeks to serve, holding constant the expected economic returns of the program.

First, the government is more likely to be willing to undertake programs oriented toward a concentrated industry than a competitive one. It is usually easier to provide programmatic benefits to a few firms than to a large number, and hence to avoid running the risk of disadvantaging politically relevant groups at some stage of the process. Of course, this tendency is not consistent with the view that atomistically competitive industries are most likely to underinvest in R&D.

Second, R&D projects will be more attractive if they address a broadly salient national political issue, so that they plausibly constitute an effective response to a concern of the citizenry at large. Undertaking a commercial R&D program because a specific industry is in decline is not likely to seem particularly attractive to political officials other than those who count the industry among their support constituencies.

Third, an R&D program will be more attractive if it has a short time horizon and does not entail a radical change in the technological base of an industry. These characteristics comport with the impatience of the political process and the desire to avoid risks of damaging distributional consequences because the project either might fail or might succeed in a way that undermines established interests. But they are inconsistent with the view that market failures in R&D are most likely for risky, long-term projects.

Fourth, the net benefits of a program are likely to play an important role early in the history of a program, simply because there are only weak political reasons to undertake a program unless it is economically attractive. But as time goes on, the relative importance of the expected net benefits of a program will decline as a factor affecting its political success. Indeed, once the high-cost demonstration phase is entered, its performance must be especially poor to overcome the distributive political costs of terminating it.

Fifth, programs that can be fragmented into many, largely independent components are usually more attractive politically than programs that can be implemented only if they are centralized. Fragmentation allows greater flexibility for bringing more firms into the support coalition of the program. Moreover, it is more likely to keep the distributive aspects of the program beneath the politically relevant threshold for members of Congress. Usually legislators are assumed to seek programs that have politically important distributive effects; however, given the risks of R&D programs, the likelihood that they will be tied to salient national political issues, and the chance that their distributive consequences will end up being politically disadvantageous, members of Congress would probably prefer not to use commercial R&D as an element of the political pork barrel. Once a program is under way, it may assume pork barrel characteristics, but before the program assumes distributive importance, it is not attractive as a prospective means for delivering distributive benefits. A member of Congress, viewing the uncertainties of the distributive consequences of a program, should prefer to take his or her share of the pig in a more prosaic and safer form, such as federal construction projects. If so, legislators representing areas that might receive significant funds from an R&D project in the future would be somewhat negatively inclined to support it, especially if its principal motivation came from a government agency rather than the target industry.

Sixth, commercial R&D projects may tend to arrive in waves or cycles. Proponents of unattractive ongoing projects (usually members of Congress representing contractors) will often seek logrolls with advocates of new programs, thereby achieving majority support and presidential consent for the entire package.

An interesting aspect of these conclusions is how they correspond to the market failure justifications for undertaking commercial R&D projects in the first place. First, a project which will produce knowledge that is not appropriable (namely, that is easily copied by people and organizations not involved in its creation) is not likely to raise concerns that it could unfairly advantage a subset of an industry, and is likely to attract industrial cooperation. Hence this market failure does not create a political problem for political leaders who might seek to solve it. Second, the more fundamental projects that will take a long time and pose high technical risks will appear politically unattractive. Political leaders are not apt to want to correct problems arising from high discount rates and excessive risk aversion in private industry, because they too share these characteristics. Third, competitive industries, which may be prone to the greatest departure of private R&D from the economic optimum, are usually not attractive targets for a federal R&D project. The exception is the case where the program can be fragmented; however, if that is possible, there is more reason to believe that individual firms will engage in research on their own. The greatest problems for a competitive industry emerge when the minimum efficient scale of research is large compared to the resources of a single firm, yet these are the cases that present the greatest political risk to an elected official.

Chapter 5

The Structure of the Case Studies

LINDA R. COHEN, ROGER G. NOLL
& WILLIAM M. PEGRAM

THE PRINCIPAL criticism of detailed case studies is that they provide a shaky basis for generalizations about the class of policies from which they are drawn. Obviously, one datum does not constitute an adequate sample for establishing a trend. But undertaking several related case studies to generate more data is also dangerous because the richness and uniqueness of the individual cases may be lost in a relatively superficial treatment of each that, perhaps forcedly, focuses only on the commonalities.

The structure of this book reflects a compromise between richness of detail and a diversity of observations for the purpose of generalization. Six cases provide sufficient material for plausible generalizations while still permitting each case to contain a reasonably detailed history of a program. Even so, six cases, while better than one, are insufficient to establish as scientific fact the validity of the conclusions about the selection, management, and performance of R&D commercialization projects that were developed in chapter 4. Moreover, to accommodate the page constraint for even a lengthy book, the cases were not designed to be comprehensive histories. Each case could easily fill an entire book by itself.

The six cases do not exhaust the interesting, important commercial research and development (R&D) projects undertaken in the postwar era, and they were not selected according to systematic criteria that would ensure variance among them in all the relevant factors that might affect the success of an R&D program. We did select cases to minimize overlap with two previous collections of case studies, even though neither collection dealt comprehensively with the politics of commercial R&D.[1] Thus our study expands the set of programs for which a moderately detailed economic history exists. Obviously, this selection criterion limited our choice of cases. Nevertheless, the six cases represent a spectrum of differences in the underlying technology, the implementation and management strategy, the target industry, and ultimate performance.

Only one of the cases, the communications satellite program of the 1960s, was an unambiguous success. At least three—the supersonic transport (SST), the Clinch River Breeder Reactor, and nearly all the synthetic fuels projects—were unambiguous failures. The space shuttle, though a commercial disaster, remains a near-miraculous technical achievement that expands the feasible range of national space projects. Depending on how one values its unique capabilities for manned space flight, a plausible case can be made that it was worth developing. The photovoltaics program failed to achieve commercial success by the originally scheduled date of 1985, but it made remarkable progress. By 1990 the technology had become commercially feasible for some uses, and could become an economically attractive source of power for electric utilities sometime in the 1990s. One synthetic fuels project, the "Cool Water" combined-cycle coal gasification project, is also on the verge of commercial feasibility.

The programs also differ according to the structure of the target industry. The SST, breeder reactor, and space shuttle programs directly involved highly concentrated industries, aerospace and nuclear generation systems. Synthetic fuels and photovoltaics were directed at two industries, hydrocarbon fuels and microelectronics, that are structurally competitive, though the former contains extremely large firms. Communications satellites are intermediate, having been virtually monopolized by Hughes early in the

1. One of the two previous studies was undertaken by a team at the Rand Corporation in the early 1970s. For the most part, it dealt with programs that were smaller than the ones studied in this book. The Rand study is summarized in Walter S. Baer, Leland L. Johnson, and Edward W. Merrow, *Analysis of Federally Funded Demonstration Projects: Executive Summary*, Research Report R-1925-DOC, prepared for the Experimental Technology Incentives Program, Department of Commerce (Santa Monica, Calif.: Rand Corporation, April 1976). The other is a compendium of very detailed economic and technical histories of programs that are more like the ones studied in this book: agricultural research, the development of computers and semiconductors, and federal support for the development of airframes for commercial aircraft. Although this study does discuss some of the political factors affecting these programs, its primary focus is on their economics. These case studies, and an interpretive introduction and conclusion, are published in Richard R. Nelson, ed., *Government and Technical Progress: A Cross-Industry Analysis* (Pergamon Press, 1982).

history of the industry, but having become reasonably competitive by the early 1970s.

All six programs dealt with user groups for the new technology that were either regulated or nationalized throughout the world. Communications satellites, which were the most important source of commercial business for the space shuttle, serve regulated or nationalized telecommunications firms, and were especially heavily controlled by government during the 1960s and early 1970s when the two programs were initiated. Hydrocarbon fuels, though increasingly less regulated since the late 1970s, were also regulated or nationalized when the big push to commercialize synthetic fuels technology began. The breeder reactor and, to a lesser extent, photovoltaics were intended to develop new generation technology for the electric utility industry, which is regulated in most of the United States and nationalized in the Tennessee Valley and Pacific Northwest as well as almost everywhere outside the United States. The SST was expected to be adopted by commercial airlines, which in the 1960s, when the program was implemented, were all either regulated or nationalized.

The management structure of the programs also differed. The SST, photovoltaics, and synthetic fuels programs were established as multi-source, parallel R&D programs, initially including just about anyone in the industry who wanted to participate. The SST program was structured to develop a single design that would be pushed to commercialization, and so meant direct competition among the original participants to become the ultimate single prime contractor. The synthetic fuels program did not have an automatic, built-in competitive aspect but did affect a large number of firms. The photovoltaics program was an intermediate case in that poorly performing projects were eliminated if they could not meet benchmark standards, which led to a steady attrition of the participants. But at no stage was a dramatic reduction in the number of participants designed into the program.

The programs also varied in their duration and scale. At one extreme, the communications satellite program consisted of a sequence of projects. Each had a time horizon of two or three years and was relatively inexpensive. Both the space shuttle and synthetic fuels were, from the beginning, expected to be multibillion-dollar programs and to take approximately a decade to prove commercial feasibility. They differed in that the synthetic fuels project explored many technologies and was expected to bring several of them to commercial feasibility. The SST and photovoltaics programs were also expected to take a decade or more to achieve commercial success but were expected to cost less than a billion dollars. The breeder reactor was transformed during its life from a relatively inexpensive program that would take about ten years for development to a multibillion-dollar project that would not be commercially interesting for several decades.

Finally, the programs differed in the range of feasible technical options that were seriously considered. At one extreme, the space shuttle was compared only with existing expendable launch vehicles. The lead agency, the National Aeronautics and Space Administration, actively suppressed consideration of many alternative means for performing the commercial functions of the space shuttle in order to promote a technology that would serve other goals of the agency that were not shared with any enthusiasm by the president or Congress. Likewise, the breeder reactor program selected a particular technology for commercialization—the sodium-cooled fast breeder—without much of a technical rationale and subsequently, because of cost overruns, forced large reductions in research on alternatives. Synthetic fuels and photovoltaics pursued a variety of technological options but foreclosed some on purely political grounds: synthetic fuels focused on technologies for converting Eastern coal to liquids, and in the early years the entire solar energy program emphasized "distributive" systems that would enable consumers to produce their own energy, independent of electric and gas utilities. Finally, the SST and the communications satellite programs considered only a few technical options; however, there is no indication that any relevant technical paths were intentionally excluded. Indeed, the communications satellite program might have survived longer had NASA not pursued options that proved politically unacceptable.

In summary, the six projects demonstrate interesting characteristics that may have been relevant in determining technical and political success. Each was justified initially on the basis of claims that its commercial benefits would exceed its costs. Though varying in quality and comprehensiveness, benefit-cost analyses were performed by proponents of each program, and every study concluded that the expected net benefits were positive. Moreover, each analysis developed a market failure explanation for the unwillingness of the private sector to develop the technology as rapidly as was desirable.

Methods and Scope of the Case Studies

Each of the six case studies adopts essentially the same approach. The central objective is to identify the key decision points in the history of the project, characterize the information available about its progress and likely success at each point, and explain the causes of its performance record as well as the forces that drove the decisions made about the future of the program. The last requires investigating the technical, managerial, economic, and political factors that affected the performance of the program and its ultimate fate.

Objectives and Performance

In each case, an understanding of the economic and political history of the program requires some appreciation of the project's principal technical problems and objectives. Hence each case addresses the scientific and engineering issues that had to be resolved if the project were to succeed, and the basis for the belief by technologists that the program was worth pursuing. In addition, the commercial success of a project depends on the economic environment of the target industry, which can change dramatically over the decade or more of an R&D project. Each study therefore considers external shocks to the program arising from changed economic circumstances.

To address the question of the overall performance of the program, each case study provides a sequence of estimates of its expected prospective net benefits at the principal decision points in its history. Hindsight enables us to correct mistakes in even the best economic evaluations from the past. Though this exercise is interesting and useful, it does not accurately account for why people made the decisions they did. Hence the analysis of project benefits and costs focuses primarily on the information available to analysts when decisions were made. That is especially important when expectations about the future strongly influence calculations of expected net benefits. For example, evaluations of energy R&D projects are highly sensitive to future energy prices, requiring analysts to forecast supply, demand, and prices decades into the future. To examine reasons for past decisions that depended on such forecasts, it makes sense to use what were then presumed to be a realistic range of expectations about the future, rather than what subsequently happened.

Because all of the projects were affected by benefits and costs over many years, a key part of the economic evaluations was the choice of a discount rate. The appropriate discount rate for public expenditure programs is a long-standing controversy not only among government officials but among economists as well. As with future energy prices, the most sensible approach for the case studies is to mimic the practice by government officials at the time, rather than to impose the current view of the profession on studies that are a decade or more old. Generally speaking, the discount rates used for evaluating the programs ranged from 7 percent in the 1960s to 10 percent in the 1970s.

Detailed updates of benefit-cost studies as a program proceeds are usually not necessary to track the performance of a program. The final success of a program usually hinges on a few key technical objectives and baseline economic assumptions about demand or the cost of alternative technologies, or both. The results of the research that addressed the key technical issues, and realizations a few years after that program was started of the key unknown economic parameters, typically made the likely success of a project very clear. An important element of each case study is to identify when in the history of a project information is revealed that determines its success, and whether the information is provided in a timely fashion to political decisionmakers.

In all the cases examined here, the initial economic justification for the program, though certainly not beyond question, was also not implausible. On balance, a reasonable person who was called upon to decide whether to support the program could rationally have decided to do so on the basis of the information then available about the expected net benefits of the program. Indeed, in most instances the arguments against the project in the beginning tended to be based more on ideology and national priorities than on the intrinsic merits of the project, although that is not uniformly true. In any event, the conclusion that a project is economically worthwhile is based on key assumptions about technical possibilities and the economic future, so that it is possible, as information is gained through the life of the project, to determine how a person who accepted these key assumptions would revise expectations about the merits of the project through time. For failed programs, which in either economic or political terms applies to some degree to all the cases considered, the sequence of updated benefit-cost analyses identifies when, if ever, the program turns sour, in the sense that, considering the original conceptualization of the program, it is no longer succeeding on its own terms. This approach also separates three independent causes of a program's failure: bad technical luck (a technical objective is not reached), bad economic luck (expectations about future prices, demand, and costs of alternatives are unfulfilled), or bad political luck (for reasons unrelated to the performance of the program, its base of political support disappears or new, unexpected sources of political opposition emerge).

Each case study also tries to determine the extent to which unsatisfactory outcomes were due to bad luck or bad decisions. In some sense that cannot be determined, for the initial assumptions and future expectations underpinning a program can always be disputed. Here we resolve the benefit of the doubt in favor of the initial advocates of the program, reflecting the perspective that the case for any of these programs was plausible at the beginning. Bad luck, then, consists of future events that produce disappointing information relative to the initial assumptions of the program. Bad decisions are responses to new information that are superficially irrational from the perspective of economic efficiency. Examples are a decision by a proponent to turn against a program when the information is good, or to continue to support the program after it is certain to be a commercial failure. Here rationality is defined in terms of the overall objectives of the program, rather than the actual, perhaps perverse, incentives acting upon a political decisionmaker at the time. Surely voting for a program that is a failure but that brings large contracts to constituencies is hardly an irrational act for a member of Congress.

Factors Influencing Key Decisions

The next step in each case study is to explain the decisions that were made about the program as it proceeded to conclusion or termination. Here the objective is to separate and detect the effects of all the factors that might enter into such decisions. One is the overall performance of the project, another is its disruptive consequences, and a third is its consistency with the reigning political values of the time.

The vehicle for this assessment is an analysis of the budgetary history of the programs. One approach to this issue is to examine the actual appropriations of Congress in relation to the original expectations when the program was designed and to the proposals from the administration. Was the program roughly on target with respect to costs, and who seemed at the forefront of its support—Congress or the administration? The expectation is that as the reasons for a program become increasingly distributional and decreasingly based on its likely success, its appropriation should exceed both initial plans and administration budget proposals, revealing a greater willingness to spend on the program by the more distributionally sensitive Congress.

The second approach is to examine the sources of political support in Congress. Here the principal method of analysis is a statistical examination of the voting behavior of legislators on authorizations and appropriations for the program. The basic idea is to use parallel specifications for estimating equations in all of the case studies so that the extent to which their politics are consistent can be observed. In practice, data limitations preclude using precisely the same approach in all studies. Most important, not all R&D commercialization projects are the subject of specific votes that deal only with the project. For communications satellites, and, after its initial R&D period, for the space shuttle, the closest Congress ever came to voting on the program was in the votes for the budget of the entire agency (in both cases the National Aeronautics and Space Administration). Because in its later stages the space shuttle was the driving force behind the NASA budget, one can plausibly use these votes as indicators of support for the shuttle, but one cannot do this for communications satellites. During the heyday of the satellite program, NASA's primary activity was the Apollo Project to land Americans on the moon, so that one cannot plausibly believe that votes for NASA's budget were in any meaningful way related to the satellite program. A similar problem arises in the photovoltaics program. Early in its history, a coalition emerged of supporters of all renewable energy sources, so that congressional votes typically were on combinations of programs.

Tests to determine the importance of the various factors in influencing the political success of a program in Congress are based on two procedures. One is the significance and signs of coefficients in equations explaining voting behavior in Congress, and the other is examination of whether votes on various budgetary bills can be pooled. The latter tests the effects of changes in the presidency. The methods used to analyze roll call votes differ from the approaches taken in the rather extensive literature on the subject and so require some explanation.

In every case the statistical model employed is a conditional LOGIT model, which uses the logarithm of the odds favoring an event (in this case, voting for a program) as the dependent variable. The research literature is not consistent in the choice of models, which is unfortunate since it makes comparisons of the results of different studies problematic. The first generation of roll call studies used either simple correlation analysis or ordinary least squares (OLS), but since the mid-1970s most studies have used either PROBIT or LOGIT. The more complicated estimation techniques take into account that the probability of an event can never exceed unity or fall below zero. Failure to account for this fact can lead to serious problems of inefficiency and bias in estimating the relationships between voting behavior and the factors hypothesized to influence it. The choice between PROBIT and LOGIT is less clear than the decision to use something more sophisticated than OLS. Nevertheless, we use LOGIT in this study for two reasons.

First, the characteristic logistic (S-shaped) curve that it estimates comports with the theoretical expectations, explained in chapter 4, that many independent variables may have important threshold-like effects. That is, at the extreme values of variables, such as distributive effects, the effect of a change in the independent variable may be small compared with its effect in the midrange, whereas choice of an appropriate functional form for the independent variables allows any method to approximate any hypothesis about the nature of the estimated relationship. LOGIT accomplishes our objectives without adding complexity to the estimation and without reducing the degrees of freedom in the estimation.

Second, one difference between PROBIT and LOGIT is the assumption about the nature of the tails of the distribution. In LOGIT the estimated probability of an event is implicitly never zero or one, whereas PROBIT assumes a distribution of probabilities with truncated tails. We believe the former assumption is a more accurate description of our congressional votes, and especially of the capabilities of political theory to explain those votes.

A novel feature of the statistical analysis used in some of the case studies is the test for the appropriateness of pooling votes on different bills. Many previous studies have tried to analyze a large number of votes, but typically one of two approaches has been taken. The first is to estimate each voting equation separately and then to make qualitative observations about their similarities and differences.[2] The second is to adopt a technique for combining several votes into a single measure of the voting behavior of

2. See, for example, James B. Kau and Paul H. Rubin, "Voting on Minimum Wages: A Time-Series Analysis," *Journal of Political Economy*, vol. 86 (April 1978), pt. 1, pp. 337–42; "Public Interest Lobbies: Membership and Influence," *Public Choice*, vol. 34, no. 1 (1979), pp. 45–54; and "Self-Interest, Ideology, and Logrolling in Congressional Voting," *Journal of Law and Economics*, vol. 22 (October 1979), pp. 365–84; Gregory B. Markus, "Electoral Coalitions and Senate Roll Call Behavior: An Ecological Analysis," *American Journal of Political Science*, vol. 43 (August 1974), pp. 595–607; Henry W. Chappell, Jr., "Conflict of Interest and Congressional Voting: A Note," *Public Choice*, vol. 37, no. 3 (1981), pp. 331–36, and "Campaign Contributions and Congressional Voting: A Simultaneous Probit-Tobit Model," *Review of Economics and Statistics*, vol. 64 (February 1982), pp. 77–83; James B. Kau, Donald Keenan, and Paul H. Rubin, "A General Equilibrium Model of Congressional Voting," *Quarterly Journal of Economics*, vol. 97 (May 1982), pp. 271–93; and John R. Wright, "PACs, Contributions and Roll Calls: An Organizational Perspective," *American Political Science Review*, vol. 79 (June 1985), pp. 400–14.

a member of Congress. One example is the use of an index of general support for a particular type of program. Peltzman used an index of economic liberalism constructed from Senate votes on economic policy; Riddlesperger and King constructed an index of the support of senators for energy producers; Markus used four general indexes of Senate voting; Lopreato and Smoller measured energy policy positions by a member's voting conformity with the majority of the Texas delegation; Kenski and Kenski used the score of members by the League of Conservation Voters as a measure of voting on environmental bills; Wayman used an index of voting on defense bills; Kalt counted the number of times in thirty-six Senate votes on oil price controls that a member voted to increase the wealth of oil producers; and Kalt and Zupan used the frequency of votes favorable to strip miners.[3] The other approach is to construct a Guttman scale for adding up votes, assigning different weights to different actions (votes, pairs, abstentions) and on different bills.[4]

The problem with simply estimating separate equations is that it entails a loss of power in estimation if, in fact, different votes legitimately do represent separate trials of the same fundamental process. Whether there is behavioral consistency across different votes, and, if so, the importance

3. Sam Peltzman, "An Economic Interpretation of the History of Congressional Voting in the Twentieth Century," *American Economic Review*, vol. 75 (September 1985), pp. 656–75; James W. Riddlesperger, Jr., and James D. King, "Energy Votes in the U.S. Senate," *Journal of Politics*, vol. 44 (August 1982), pp. 838–47; Markus, "Electoral Coalitions"; Sally C. Lopreato and Fred Smoller, *Explaining Energy Votes in the Ninety-Fourth Congress*, University of Texas at Austin, Center for Energy Studies, June 1978; Henry C. Kenski and Margaret C. Kenski, "Partnership, Ideology, and Constituency Differences in Environmental Issues in the U.S. House of Representatives, 1973–1978," *Policy Studies Journal*, vol. 9 (Winter 1980), pp. 325–35; Frank Whelon Wayman, "Arms Control and Strategic Arms Voting in the U.S. Senate: Patterns of Change, 1967–1983," *Journal of Conflict Resolution*, vol. 29 (June 1985), pp. 225–51; Joseph P. Kalt, *The Economics and Politics of Oil Price Regulation: Federal Policy in the Post-Embargo Era* (MIT Press, 1981), and "Oil and Ideology in the United States Senate," *Energy Journal*, vol. 3 (April 1982), pp. 141–66; and Joseph P. Kalt and Mark A. Zupan, "Capture and Ideology in the Economic Theory of Politics," *American Economic Review*, vol. 74 (June 1984), pp. 279–300.

4. See, for example, John E. Jackson, "Statistical Models of Senate Roll Call Voting," *American Political Science Review*, vol. 65 (June 1971), pp. 451–70; Robert A. Bernstein and William W. Anthony, "The ABM Issue in the Senate, 1968–1970: The Importance of Ideology," *American Political Science Review*, vol. 68 (September 1974), pp. 1198–1206; and Otto A. Davis and John E. Jackson, "Representative Assemblies and Demands for Redistribution: The Case of Senate Voting on the Family Assistance Plan," in Harold M. Hochman and George E. Peterson, eds., *Redistribution through Public Choice* (Columbia University Press, 1974).

and magnitude of the consistent effects, remain unestimated and perhaps even undetected. The problem is not solved by simply constructing an index of voting behavior across several bills. That assumes the validity of the method of constructing a voting index, rather than allowing the statistical model to provide information about the validity of regarding the votes as representing the same underlying causal model. Moreover, if each vote really is an independent trial from the same behavioral model, combining several votes into a single index throws away valuable information in the separate data and leads to loss of power in the estimation procedure.

The approach taken here, followed also by Weingast and Moran,[5] is to estimate a pooled regression in which each vote on each bill is regarded as a separate observation from the same underlying process. In addition, separate equations are estimated for each vote, as are equations that pool only some of the votes. The validity of pooling is examined by testing whether each step in the pooling process—from adjacent votes in time through pooling the entire sample—causes a significant loss in the explanatory power of the regressions. This constitutes a formal test of whether the basis of political support for a program changed during its history. For example, pooling within a presidency, but not between presidents, provides a basis for examining the influence of the president on Congress.

The selection of independent variables for the regressions was designed to test a wide range of alternative hypotheses about the source of support for the programs. Consequently, variables were constructed to capture traditional political influences, the economic performance of the program, and the program's distributive effects. The more traditional political variables are party, ideology, and membership on the oversight committees and subcommittees for the program. Economic performance was typically measured by one or two simple measures of a program's success that dominated its expected net benefits. The distributive effects were measured by spending and contracts within a member's constituency, and variables concerning the constituency that indicate whether it would enjoy benefits from the new technology were it to prove successful.

Variables indicating membership on relevant committees are included to capture the importance of the structure of Congress in explaining voting behavior. All R&D programs are reviewed by two sets of committees. First, authorizing committees propose legislation establishing the objectives of

5. Barry R. Weingast and Mark J. Moran, "Bureaucratic Discretion or Congressional Control? Regulatory Policymaking by the Federal Trade Commission," *Journal of Political Economy*, vol. 91 (October 1983), pp. 765–800.

a program and authorizing expenditures to carry it out. Second, the appropriations committees in each body provide annual funds to carry out the program, but not necessarily the full amount that is authorized. Usually the committees of Congress are further separated into subcommittees, but in some cases committees have no subcommittees.

For several reasons, legislators might be more inclined to support legislation pertaining to subjects within their committee's jurisdiction than other members.

First, committee assignments in the House reflect the preferences of the members. Members of Congress choose committees on which to serve, and, although they do not always receive their first choice of committee assignments, they usually serve on committees whose jurisdictions are of particular relevance to them and their constituents.[6] Logically, members can be expected to seek membership on a committee that is especially important to their constituents. To some degree, this is measured by the other independent variables included in the analysis. Members of the Committee on Science and Technology were more likely than other House members to have aerospace companies in their districts, which received shuttle contracts; members of the Joint Committee on Atomic Energy were more likely to have nuclear facilities (both related and unrelated to the breeder program) within their districts than nonmembers. This pattern tends to hold for both authorization committees and the relevant subcommittee of the Appropriations Committee.[7]

Second, support for programs by committee members is also related to factors not measured by the other independent variables. Committee members may seek membership because, besides constituency interests, they have some other reason to be particularly drawn to the subject area: solar energy, space, commercial nuclear power, and so on. These reasons may be related to electoral politics (for example, campaign contributions), or may reflect a member's special expertise (for example, John Glenn and the space program). As discussed earlier, for bureaucrats interest typically accompanies proponency. If so, committee members might support the bills to a greater degree than would be predicted by the other characteristics of the committee members.

Third, relatively small, expert committees, composed of members

6. Kenneth A. Shepsle, *The Giant Jigsaw Puzzle: Democratic Committee Assignments in the Modern House* (University of Chicago Press, 1973).

7. See Douglas R. Arnold, *Congress and the Bureaucracy: A Theory of Influence* (Yale University Press, 1979).

with atypically strong interest in the programs within the committee's jurisdiction, provide an excellent opportunity for negotiating differences, working out compromises, and logrolling with other bills or programs. Thus committees should exhibit greater voting solidarity on the bills they propose than would otherwise be implied by their personal policy preferences and constituency interests.

Finally, for purely statistical reasons we would expect committee members to be more likely to support the bills they propose, since the bills are almost always subject to approval within the committee before they are subject to a floor vote. Because we observe only bills that are brought to a floor vote, unmeasured and random effects on voting are likely to be distributed differently between the committee and the entire legislature. By controlling for committee membership, we correct for potentially biased coefficients for other variables correlated with committee membership.

In some cases, strong correlation between membership and distributive variables creates a specification problem. Including the committee variable may well result in underestimating the impact of distributive variables, particularly before expenditures for these programs become large. Committee members with, for example, nuclear facilities within their districts may be more confident of receiving major contracts in the breeder program because of their relationship with the bureaucracy.[8] Consequently, they may be more willing to support the program for its pork barrel attributes before appropriations for construction are passed, whereas other eventual recipients of the government's largesse await actual expenditures. The extent to which collinearity problems arise, and the proper interpretation of results, depends on the particulars of each case, and so is discussed in the subsequent chapters.

Our interest in committee support for bills relates as well to committee power in Congress. Some students of Congress argue that the organization and behavioral norms of Congress give committees disproportionate influence over subjects within their jurisdictions.[9] The committee variable

8. This argument was made by Weingast and Moran. "Bureaucratic Discretion," about a different set of policies.

9. An excellent summary and discussion of the literature is contained in Keith Krehbiel and Douglas Rivers, "The Analysis of Committee Power: An Application to Senate Voting on the Minimum Wage," *American Journal of Political Science*, vol. 32 (November 1988), pp. 1151–74. See also Barry R. Weingast, "Floor Behavior in Congress: Committee Power under the Open Rule," Working Papers in Political Science P-88-2 (Stanford University, Hoover Institution, 1988); Kenneth A. Shepsle and Barry R. Weingast, "The Institutional

detects—with the possible problems just discussed—internal agreements within these groups to support the product of the committee on the floor of the larger body. To the extent that committees are disproportionately influential over the subjects within their jurisdictions, agreement within the committee on a program may retard changes in the program.

Suppose, for example, that the authorizing committee disproportionately (compared with the entire legislature) supports a project that has been in existence for several years. One source of committee power is "gatekeeping": the power to prevent new legislation within the committee's jurisdiction from reaching the floor.[10] If once such legislation is brought to the floor, it is subject to amendment, then even if the committee would like some program modifications—say, in the event of changed economic projections—it may prefer the status quo to the final results if a committee bill were brought to the floor.[11] If a committee lacks internal agreement, we expect that the program is more subject to political and economic vagaries than one that has strong institutional support.

A considerable literature has developed on measuring the effects of party and ideology on roll call voting in Congress, and on the actual meaning of these variables. In this book, ideology is measured by the legislator's score according to the ratings system of the Americans for Constitutional Action (ACA). Each year the ACA chooses about twenty-five key roll call votes in the House and the Senate and assigns each member of Congress a score equal to the percentage of votes on which he or she voted in agreement with the ACA. Legislators with high scores generally fit the conservative label; those with low scores vote in a typically liberal fashion.[12]

Foundations of Committee Power," *American Political Science Review*, vol. 81 (March 1987), pp. 85–104; and Keith Krehbiel, "Sophisticated Committees and Structure-Induced Equilibrium in Congress," in Mathew D. McCubbins and Terry Sullivan, eds., *Congress: Structure and Policy* (Cambridge University Press, 1987).

10. Of course, the powers of a committee are not absolute and can reasonably be regarded as a source of bias in legislative outcomes away from the position the entire body would otherwise adopt. See Krehbiel and Rivers, "Analysis of Committee Power"; Shepsle, *The Giant Jigsaw Puzzle*; Thomas W. Gilligan and Keith Krehbiel, "Organization of Informative Committees by a Rational Legislature," Stanford University, Graduate School of Business, June 1989; and Arthur Denzau and Robert Mackey, "Gatekeeping and Monopoly Power of Committees: An Analysis of Sincere and Sophisticated Behavior," *American Journal of Political Science*, vol. 27 (November 1983), pp. 740–61.

11. See John A. Ferejohn and Charles Shipan, "Congress and Telecommunications Policy," in Paula R. Newberg, ed., *New Directions in Telecommunications Policy* (Duke University Press, 1989).

12. The ACA score was selected rather than the principal alternative, the ratings by Americans for Democratic Action (ADA), primarily because of the differences in the way

Virtually all studies find that ideological scores are statistically significant in explaining roll call votes, but disagree on what that means. Because ideological scores are constructed on the basis of actual votes, they represent a composite of all the factors affecting voting behavior, including party membership and constituency characteristics.[13] Kalt and Zupan, and Peltzman, each undertook exhaustive statistical studies of the relationships between ideology scores (Americans for Democratic Action) and constituency characteristics and concluded that such characteristics do an excellent job of explaining the scores.[14] But they disagreed about the conclusion implied by their results. Much unexplained variance remains, and, as a predictor of voting behavior, these and other authors find that ADA scores significantly increase the explanatory power of an equation even when constituency and party variables are included. Peltzman concluded that proper specification would cause the significance of ideological scores to disappear entirely; Kalt and Zupan disagreed, believing it measures the independent personal attributes of a representative.

The significant independent explanatory power of ideological scores is not surprising because the constituency characteristics that are included in the regressions on voting behavior incompletely measure the relevant interests as seen by the legislator. Typically the constituency characteristics are socioeconomic and demographic variables rather than direct indicators of voter preferences. Even if these characteristics did determine voter preferences exactly, specification error in the voting behavior equation could cause an ideological index to be significant. The latter could be interpreted as detecting unmeasured constituency effects or correcting for errors in the functional form of the equation.

A more fundamental problem is that the constituency variables all measure the average or median characteristics of the entire district rather than simply the part of the electorate that supports a member of the

---

the two organizations treat a failure to vote on one of the issues included in their ratings. The ACA considers only yes-no votes that are cast or paired, whereas ADA records an abstention or a failure to vote as equivalent to a vote against the ADA's interests. In practice, the choice between the two scores is not very important, for they are extremely highly correlated, even though the ADA does occasionally give liberals a low (conservative) score because of poor participation.

13. Morris P. Fiorina, *Representatives, Roll Calls, and Constituencies* (Lexington, Mass.: Lexington Books, 1974).

14. Joseph P. Kalt and Mark A. Zupan, "The Apparent Ideological Behavior of Legislators: Testing for Principal-Agent Slack in Political Institutions," *Journal of Law and Economics*, vol. 33 (April 1990), pp. 103–31; and Sam Peltzman, "Constituent Interests and Congressional Voting," *Journal of Law and Economics*, vol. 27 (April 1984), pp. 181–210.

legislature. Members of Congress need only succeed in capturing the support of more than half their voting constituents, not 100 percent of the total population in the districts. In addition, the means of obtaining majority support includes actions that please not only their support constituency but also potential financial contributors from outside the district.

Studies that attempt to explain voting behavior from ideological scores imply that issues that concern constituents can be modeled as having a single dimension. That is, several different factors may determine voting behavior, but all issues perfectly correlate along a single dimension—for example, a strong environmentalist position usually goes along with an antidefense spending posture, big budgets for social programs, and fiscally liberal government programs. To some degree, this holds, but it holds poorly enough to make us reject the single dimensional view of politics. If it held, we would expect legislators to support the median position within their constituency. In a single-dimensional world, the median voter's position maximizes the probability of winning an election; hence successful candidates would probably hold that position.[15] In a multidimensional world, a median voter is unlikely to exist, so that positioning oneself at the median in each dimension is not likely to be a successful campaign strategy. Consequently, candidates put together some coalition of supporters that may or may not reflect a median along any dimension, and one that is typically not stable over time. Indeed, we expect it not to be stable: successful candidates (that is, elected representatives) from the same district represent, at different times, different coalitions of supporters.[16] Thus to explain how much voting behavior reflects the legislator's constituency, we need much more detailed information about the composition of the supporters of the legislator than are contained in all empirical studies of floor votes in Congress. Differences in voting behavior of representatives from district median positions may represent shirking, as has been claimed by Kalt and Zupan;[17] however, it could represent a perfect representation of different coalitions in the same constituency. Markus, in investigating the empirical importance of the structure of coalition support, constructed an index of a candidate's support constituency and found it did a better job

15. Duncan Black, *The Theory of Committees and Elections* (Cambridge University Press, 1958).

16. For an excellent exposition of the implications of social choice theory for public policymaking, see William H. Riker, *Liberalism against Populism: A Confrontation between the Theory of Democracy and the Theory of Social Choice* (San Francisco: Freeman, 1982).

17. Kalt and Zupan, "Capture and Ideology."

of explaining indexes of roll call voting of senators than did the standard list of state socioeconomic and demographic variables.[18]

The resolution of this argument is relevant to our interpretation of the coefficients in the regressions that follow. First, we expect that the coefficient of the ACA score will reflect in part economic characteristics of constituents that are imperfectly measured by other variables included in the regressions. The other variables, then, will pick up the extent to which these bills differ in distributive consequences from those chosen by the ACA to make up the scores, for example, particular benefits or costs to the representatives' districts in excess of those that make up the ACA's bundle of relevant votes. Consequently, the conservative statistical strategy is to include the ACA variable to determine whether other variables of interest independently explain additional variance in voting. Insofar as R&D projects differ from other programs, separate measures of the performance and distributional effects can be used to detect the importance of these departures from the norm. Alternatively, we run the risk of overinterpreting the importance of ideology; as with other collinearity problems, we resolve this case by case.

For the purposes of the case studies in this book, two ideological factors are directly applicable. One is a legislator's position on the principle that government has a legitimate role to subsidize the development of new technology for private industry. The second is the legislator's position on the principle that a particular type of technology, or technology in general, is an appropriate way to resolve a social issue. These two factors are not collinear. The studies reported in subsequent chapters suggest that, though the second factor tends to correlate with conservatism, as measured by the ACA, the first does not. In particular, the most "conservative" Democrats (who normally have ACA scores of about 50) tend to be comfortable with both principles, while the most liberal Republicans (also with ACA scores of about 50) are comfortable with neither. Thus we usually obtain opposite and significant coefficients for party and ACA scores. Indeed, in most of the cases the two ideological principles are in conflict with the ACA score, so that the party coefficient is typically of the "wrong" sign. We interpret the party coefficient as correcting in part for measurement errors in ideology and in part as reflecting party solidarity and partisan support for the issue. In the studies that follow we take several tacks to disentangle the effects. In some cases independent information is available on party positions; in

18. Markus, "Electoral Coalitions."

others that span multiple presidential administrations we use pooling techniques to test for changes in party coefficients that are distinct from changes in the ACA coefficients.

ACA scores are based on a wide variety of issues. Some are domestic and some international, and some are budgetary while others have no budgetary significance. Moreover, the votes that are included in the ratings differ from year to year, so that the score may not be a meaningful indicator of shifts in the position of a legislator over time or of changes in the overall ideological complexion of Congress. As a result, the use of ideological scores in a time-series regression may fail to detect ideological phenomena even when they are present. One advantage of the pooling tests just described is that the validity of a time-series use of ideology scores can be tested statistically, rather than simply assumed or rejected on a priori grounds.

In pooled regressions including votes in different years, the performance of the program can be incorporated directly into the regression analysis. Although performance could be measured by the prospective net benefits of the program, it is most easily monitored by tracking a few key variables that reveal the ongoing technical success of the project or the external conditions that will affect its attractiveness in the market. The significance of these variables can be used to test the importance of performance on the support for a program.

The distributive effects of a program are captured by measures of the expenditures of the program in the constituency and factors that may cause program benefits to be unevenly dispersed throughout the nation. In general, past attempts to detect distributive effects have produced mixed results. Bernstein and Anthony found that Senate votes on the antiballistic missile issue from 1968 to 1970 were not significantly affected by the amount of contract expenditures in the state (as a fraction of state income). Subsequently Wayman found that an index of voting on defense issues was not greatly influenced by the net gains from defense contracts (where the tax burden of defense was also included as an independent variable), a result that was robust to several alternative specifications. Davis and Jackson found that support for President Richard M. Nixon's welfare reform (the Family Assistance Plan) was generally consistent with the expected distribution of benefits, but there were some notable anomalies. Silberman and Durden, and Kau and Rubin, found that support for minimum wage legislation generally reflected constituency interests. Finally, several studies dealt with the relationship between votes on energy bills and constituency interests regarding energy resources. Though the results are mixed, the most detailed studies usually find important constituency effects. For example, Kalt found that votes on oil price controls were influenced by the extent to which the senator represented an oil-producing state. Riddlesperger and King found that net energy dependence (the ratio of production to consumption within a state) was significant in explaining an index of Senate votes on a variety of energy issues, although it was less important in explaining variance than party and region.[19] Overall, the distribution of expenditures and program benefits usually, but not always, can be detected as a factor influencing voting behavior.

From the perspective of the analysis of chapter 4, these equivocal results are unsurprising. Distributive effects ought to be difficult to detect if, as argued in that chapter, they have a strong threshold effect, are asymmetric between cuts and increases, and are susceptible to excessive discounting of future authorizations. Testing for these more complex effects is conceptually possible. For example, expenditures can be divided between current and prospective spending and further divided into categories according to their magnitude. Unfortunately, such divisions can cause a proliferation of independent variables that produces an equation that cannot be estimated reliably using an appropriate limited dependent variable model.

When the data permit, we examine several alternative specifications of the effect of distributive aspects of a program. These include the separation of current from future expenditures, as well as separate estimation of coefficients for different ranges of the level of expenditures. For Senate votes, expenditures per capita were used as an indicator of the importance of the program. In some cases, data on the geographic distribution of expenditures were not available, so dummy variables indicating important contractors were used as a surrogate for expenditures.

Where data were sufficient, we undertook tests to determine whether one party was more sensitive to distributive effects than the other. Because the decision to support a program has overall policy significance, party, ideology, and performance together can make a member of one party very likely to vote for a program regardless of its distributive effects. Hence distributive effects may be observable only for conflicted members whose

19. Bernstein and Anthony, "The ABM Issue"; Wayman, "Arms Control"; Davis and Jackson, "Representative Assemblies"; Jonathan I. Silberman and Garey C. Durden, "Determining Legislative Preferences on the Minimum Wage: An Economic Approach," *Journal of Political Economy*, vol. 84 (April 1976), pp. 317–29; Kau and Rubin, "Voting on Minimum Wages"; Kalt, *Economics and Politics*; and Riddlesperger and King, "Energy Votes."

constituents receive distributive benefits but whose party is opposed to the program. This is an especially interesting possibility if the legislator is of the same party as the president, and the president actively seeks a vote against the distributive interests of the legislator's constituents.

The effects of an administration may be detected not only by interacting distributive variables with party but also through attempts to pool votes over time. The relevant test is whether votes within an administration are more likely to pool than votes across an administration.

General Assessment

Each case study concludes with an assessment of the program that it examines. The ultimate causes of the technical, economic, and political fate of the program are examined, as well as the quality of the managerial strategy. The assessment of management quality includes whether the program was structured to maximize its chance for technical success and whether along the way changes in the program reflected bad decisions or bad luck.

Each case study assesses the future of the technology that the corresponding program attempted to advance. Every program examined in this book, including the two that died in the early 1970s (the SST and communications satellites), has successor programs that are under way or that are pending before Congress. Hence the lessons from the case studies are useful in assessing the likely performance of the follow-up projects.

Chapter 6

The American Supersonic Transport

SUSAN A. EDELMAN

There is no such thing as quick-and-easy development of a new transport carrier; to ignore this truism is to court disaster.

Donald W. Douglas, Jr., President,
Douglas Aircraft Company, May 1960

IN THE LATE 1950s the British, French, and Soviets began developing civilian supersonic aircraft. The federal government of the United States, a nation priding itself then as now on technological leadership, did not want to fall behind; hence it embarked, cautiously at first, on a program to develop an American supersonic transport (SST). As time wore on and the perceived threat of European aviation dominance loomed larger, caution was abandoned. The program forged ahead in the face of technological setbacks and mounting evidence that the financial and business communities lacked faith in the aircraft's commercial prospects. After nearly ten years and almost $920 million of government money, the SST program sputtered to a halt on the floor of Congress.

The key question about the SST is why the program was permitted to be such a costly and dismal failure, especially since sound management and decisionmaking were stressed so strongly at the start. Furthermore, earlier civilian efforts by the National Advisory Committee on Aeronautics (NACA, the predecessor to NASA) are regarded as successes. In large part the answer lies in the differences between the guiding principles of NACA

The source of the epigraph is *Supersonic Air Transports*, Hearings before the House Committee on Science and Astronautics, 86 Cong. 2 sess. (Government Printing Office, 1960), p. 171.

and the task undertaken by the Federal Aviation Agency (FAA), the administrator of the SST program.[1] During the 1920s and 1930s NACA developed testing facilities, undertook basic research important to progress in aeronautics, and pursued applied research on structural aspects of aircraft design such as retractable landing gear, engine positioning, engine design, and fuels. This was in essence "generic" research from what was in name and fact an advisory committee.[2] NACA did no work to support any particular commercial transport.[3]

In the late 1940s proponents lobbied for NACA's support in the development of commercial jet aircraft, arguing that because the British government was helping its industry, the United States was in danger of losing its leadership in aviation. Congress denied the requests for such a program. Less than twenty years later, Congress accepted the same arguments for the SST. As a consequence, the country pursued a program that persisted despite its poor performance.

History of the American SST

Ever since the Wright brothers' first flights, aviation research has sought to increase the speed of flight. Exceeding the speed of sound long captured the imagination, as demonstrated by the excitement felt when Chuck Yeager first broke the sound barrier in 1947 and by the recent interest in a program to develop a hypersonic spaceplane.

Congress first confronted the possibility of a civilian supersonic transport in 1960, concluding that such an aircraft would be technically feasible.[4] In March 1961 President John F. Kennedy charged Najeeb E. Halaby, administrator of the Federal Aviation Agency, with preparing "a statement of national aviation goals for the period between now and 1970."[5] One

1. In April 1967 the Federal Aviation Agency was made a part of the new Department of Transportation, and its name was changed to the Federal Aviation Administration.

2. David C. Mowery and Nathan Rosenberg, "The Commercial Aircraft Industry," in Richard R. Nelson, ed., *Government and Technical Progress: A Cross-Industry Analysis* (Pergamon Press, 1982), p. 128.

3. George Eads and Richard R. Nelson, "Governmental Support of Advanced Civilian Technology: Power Reactors and the Supersonic Transport," *Public Policy*, vol. 19 (1971), pp. 405–08; Mowery and Rosenberg, "The Commercial Aircraft Industry," pp. 128–29.

4. *Supersonic Air Transports*, Hearings before the Special Investigating Subcommittee of the House Committee on Science and Astronautics, 86 Cong. 2 sess. (Government Printing Office, 1960), p. 165.

5. Federal Aviation Agency, *Report on the Task Force on National Aviation Goals: Project Horizon* (1961), p. iii.

month later, before the statement was completed, Halaby requested funds for SST feasibility studies, claiming that American leadership and security dictated "that we proceed immediately with the development of a commercially feasible supersonic transport." He proclaimed the inevitability of civilian supersonic transportation and candidly admitted "that to be honest with you, we want to be there ahead of potential competitors."[6]

Not surprisingly, when the task force report on a national aviation goal was released in September 1961, it strongly endorsed the development of the SST and emphasized the importance of developing the first such aircraft.[7] The necessity of being first was surely dubious, for after NACA decided not to help develop the first jets Britain did indeed beat the United States, yet the American aviation industry easily became dominant in commercial jet aircraft shortly thereafter. Nevertheless, Congress granted the FAA $11 million in fiscal 1962 and $20 million in fiscal 1963 for technical and economic feasibility studies. These studies were meant to enable the government to make the correct decision on whether and how to proceed with SST development.[8]

The preoccupation with being first was spurred by the French, British, and Soviet programs. In April 1963 Kennedy asked the FAA to hasten its feasibility studies so that foreign competition would not have too big a head start. The development of an American SST was virtually ensured when, on June 4, 1963, Pan American World Airways placed an order for six Anglo-French Concordes. The next day Kennedy reversed the historical role of the federal government in aviation by announcing that the U.S. govenment would help the American aviation industry develop an SST.

Kennedy's guidelines for the program incorporated suggestions from the Department of Defense, NASA, and the FAA's *Commercial Supersonic Transport Aircraft Report* of 1961. These called for the government to exploit the market's competitive forces, to maintain rational standards for the selection of technology, and to define cost and time schedules for the project.[9] In addition, Kennedy proposed that manufacturers pay for 25 percent of the project, that "in no event" would government's contribution exceed $750 million, and that the entire program would be reconsidered if

6. *Independent Offices Appropriations for 1962,* Hearings before a Subcommittee of the House Committee on Appropriations, 87 Cong. 1 sess. (GPO, 1961), pp. 64, 83.

7. *Report on the Task Force on National Aviation Goals*, pp. 76, 80.

8. *Independent Offices Appropriations for 1963,* Hearings before a Subcommittee of the House Committee on Appropriations, 87 Cong. 2 sess. (GPO, 1962), p. 654.

9. *Contemporary and Future Aeronautical Research,* Hearings before the House Committee on Science and Astronautics, 87 Cong. 1 sess. (GPO, 1961), p. 111.

it ever appeared that the aircraft could not be operated economically or if industry were unwilling to take adequate risks with its own capital.[10]

Early Stages of the Program

In October 1963 Halaby presented the Senate Committee on Government Research with a timetable for the program (table 6-1). Phase I, the initial design competition, would begin in August 1963, and phase V, initial production, would be entered in 1970.[11] If phase I produced a clearly superior combination of airframe and engine, then phase II, the detailed design competition, would be skipped. The winners of the design competition would progress to phase III, development and construction of the prototypes, and then to phase IV, certification of the aircraft and production development. Halaby also proposed the first deviation from Kennedy's guidelines when he stated in October 1963 that the program should proceed even if the manufacturers proved unwilling to pay 25 percent of the development costs "because we are in a race with the British and French."[12]

For the aircraft to reach the desired performance specifications, several technical obstacles had to be overcome. As identified in 1960 and 1961, the major obstacles were providing for efficient flight both at subsonic and supersonic speeds, ensuring that the aircraft would withstand the heat produced during supersonic flight, and dealing with the adverse effects of sonic boom.[13] It was believed that variable sweep wing geometry would be the route to maximum efficiency;[14] such a design was chosen but it proved

10. "Presidential Report," *Congressional Quarterly Weekly Report*, June 21, 1963, p. 1035. (Hereafter *CQ*.)

11. *Independent Offices Appropriations, 1964*, Hearings before the Senate Committee on Appropriations, 88 Cong. 1 sess. (GPO, 1963), pt. 2, pp. 1983, 2037; and *Federal Research and Development Programs*, Hearings before the House Select Committee on Government Research, 88 Cong. 1 sess. (GPO, 1963), pt. 1, p. 146.

12. *Independent Offices Appropriations, 1964*, Senate Hearings, pt. 2, p. 2247.

13. *Contemporary and Future Aeronautical Research*, House Hearings, p. 95; *Supersonic Air Transports*, House Hearings, p. 139; and *Report on the Task Force on National Aviation Goals*, p. 78.

14. Maximum flight efficiency requires the wings to be more perpendicular to the fuselage at subsonic than at supersonic speeds. (Though the SST would cruise at supersonic speeds, it would have to take off and land subsonically.) Variable sweep wings can be adjusted so that the aircraft maintains the most efficient configuration at all times. No civilian transport uses variable sweep wings.

Table 6-1. *Schedules for SST Development, 1963, 1965, 1967, 1969*

Phase	*1963*[a]	*July 1965*[b]	*1967*[c]	*1969*[d]
Phase I: preliminary design competition	August—Dec. 1963	. . .	. . .	. . .
Phase II: detailed design competition	Mid 1964–mid 1965	June 1964–Dec. 1966	. . .	. . .
Phase III: prototype development	Mid 1965–mid-end 1968	Early 1967–end 1970	Jan. 1967–mid 1971	Jan. 1967–mid 1973
Phase IV: certification and production development	. . .	Early 1970–mid 1974	Mid 1970–end 1974	Mid 1973–early 1978
Phase V: initial production	1970 and later	Early 1970–74 and later	Mid 1970–75 and later	Early 1974–78 and later

a. *Independent Offices Appropriations, 1964,* Hearings before the Senate Committee on Appropriations, 88 Cong. 1 sess. (GPO, 1963), pt. 2, pp. 1983, 2037; and *Federal Research and Development Programs,* Hearings before the House Select Committee on Government Research, 88 Cong. 1 sess. (GPO, 1963), pt. 1, p. 146.

b. *Independent Offices Appropriations for Fiscal Year 1967,* Hearings before a Subcommittee of the Senate Committee on Appropriations, 89 Cong. 2 sess. (GPO, 1966), pt. 1, p. 366.

c. *Department of Transportation Appropriations for 1968,* Hearings before a Subcommittee of the House Committee on Appropriations, 90 Cong. 1 sess. (GPO, 1967), p. 287; and *Department of Transportation Appropriations for Fiscal Year 1968,* Hearings before a Subcommittee of the Senate Committee on Appropriations, 90 Cong. 1 sess. (GPO, 1967), p. 397.

d. *Department of Transporation Appropriations for Fiscal Year 1970,* Hearings before a Subcommittee of the Senate Committee on Appropriations, 91 Cong. 1 sess. (GPO, 1969), pp. 602, 615.

unsatisfactory. The temperature limit of aluminum, the standard material for aircraft shells, is too low for the speed desired from the SST. Titanium alloy was regarded as the most promising material, but a suitable process for fabricating it had not yet been found when the program was terminated. Likewise, the sonic boom problem was never solved, so that eventually it was conceded that SSTs could not be permitted to fly over land at supersonic speed.

Kennedy asked Eugene R. Black, a former president of the World Bank, and Stanley de J. Osborne, board chairman of the Olin Matheson Chemical Company, to prepare a report on the potential for private financing and management of the SST. Their report, submitted to President Lyndon B. Johnson on December 19, 1963, recommended that the race with the Concorde be abandoned.[15] The Black-Osborne report was not well received and was made public on March 2, 1964, only under pressure. Two months later, before phase I was completed, let alone phase II begun, the FAA requested an appropriation for phase III on the grounds that this would enable prototype construction to begin as soon as the winning design was

15. "Around the Capitol," *CQ,* March 6, 1964, pp. 467–68; and "Around the Capitol," *CQ,* February 21, 1964, p. 360.

chosen, preventing "a delay in the competition with the British and French."[16]

In January 1964 Boeing, Lockheed, North American–Rockwell, General Electric, Pratt and Whitney, and Curtiss-Wright submitted designs for the initial competition, the first three for the airframe and the latter three for the engine.[17] Although these plans were submitted according to Halaby's original schedule, none of the airframe designs met the standards for range, payload, or economical operation.[18] At this key decision point, President Johnson announced a delay and turned to his recently formed President's Advisory Committee on Supersonic Transport, chaired by Secretary of Defense Robert McNamara, for advice. In May both the committee and the FAA recommended that Boeing, Lockheed, General Electric, and Pratt and Whitney be chosen for the detailed design competition, even though none of the airframe designs met the specifications. On July 1, 1965, Johnson announced that phase II would continue through 1966, eighteen months past its original deadline (table 6-1, column 2) so that financial and development risks could be minimized.

Developing a Prototype

On September 6, 1966, the manufacturers submitted their phase III proposals. All proposed that the manufacturers pay for only 10 percent of development costs, not the hoped-for 25 percent.[19] The proposals were evaluated by representatives from the FAA, NASA, the Department of Defense, the Civil Aeronautics Board, and thirty-one airlines. Boeing and GE won the competition, and phase III began on schedule. The contracts for the prototype stage specified that by July 1, 1968, the manufacturers had to submit proposals for private financing of the project past the development stages; failure to produce acceptable plans would permit the

16. *Independent Offices Appropriations for 1965*, Hearings before a Subcommittee of the House Committee on Appropriations, 88 Cong. 2 sess. (GPO, 1964), pt. 1, p. 1172.

17. Federal Aviation Administration, "United States Supersonic Transport Milestones," *News*, April 1970, p. 3, located at the Washington National Records Center, Reference Branch, Suitland, Md. CC. No. 72A-6174, Box No. 25 of 151, File No. 7; and *Independent Offices Appropriations for 1965*, House Hearings, pt. 1, p. 1160.

18. "Around the Capitol," *CQ*, May 1, 1964, pp. 867–68. The range of an aircraft declines as its payload increases. Though it is possible to meet one of these technical standards by compromising on the other, it is the ability to carry many passengers over a long range that would be necessary to make the aircraft attractive to airlines.

19. Federal Aviation Administration, "United States Supersonic Transport Milestones," p. 5; and "Around the Capitol," *CQ*, September 9, 1966, p. 1965.

government to cancel the contracts for nonperformance. That cancellation would occur, however, seemed doubtful, for the FAA admitted that the government might have to pay for the production of the SST.[20]

By February 1968 it was apparent that Boeing's lauded variable sweep wing design would not be able to meet the range-payload specifications, and on February 21 Johnson announced that the program would be delayed again, perhaps for up to a year. The government gave Boeing until January 15, 1969, to submit revised plans.

President Nixon entered office without a public position on the SST. He formed the SST Ad Hoc Review Committee to study the program, and seven of the eleven members recommended that spending on the SST prototype cease until more research had been completed.[21] Furthermore, Lee A. DuBridge, the president's science adviser as well as a committee member, stated that "the government should not be subsidizing a device which had neither commercial attractiveness nor public acceptance." Nixon next "apparently" asked DuBridge to provide a report in which scientists evaluated the SST. The group was chaired by Richard L. Garwin, a former member of the President's Science Advisory Committee. Their report was kept secret, suggesting that it was not favorable; a member of the House reported that the scientists concluded that the SST was "economically wasteful and environmentally harmful." Nevertheless, on September 23, 1969, Nixon, flanked by a delegation from Washington, Boeing's home state, announced his unqualified support for the SST program as a means of maintaining American aviation leadership.[22]

Earlier plans had called for overlapping phases III, IV, and V, but this was abandoned (table 6-1, columns 3 and 4) when the airframe was redesigned to enable completion of the prototype before certification and production began. This change would reduce the technical risk of the project, making it more attractive to private financiers. Amid fears that the Concorde would allow Britain to "lead the world clearly and beyond challenge in a field the long term benefits of which are almost beyond

20. *Department of Transportation Appropriations for 1968*, Hearings before a Subcommittee of the House Committee on Appropriations, 90 Cong. 1 sess. (GPO, 1967), pp. 289, 937.

21. *Department of Transportation and Related Agencies Appropriations for Fiscal Year 1970*, Hearings before a Subcommittee of the Senate Committee on Appropriations, 91 Cong. 1 sess. (GPO, 1969), p. 747.

22. Congressional Quarterly, *Congress and the Nation*, vol. 3: *1969–1972* (Washington, 1973), p. 159.

calculation,"[23] there was now talk of reinstating the overlap,[24] which, if all went well, would have permitted the SST to enter commercial service about a year earlier.

As a result of the transition from a wartime to a peacetime economy, the aircraft industry was not faring well in November 1969, with employment down 10 percent and Washington State's unemployment rate at 7.5 percent because of layoffs at Boeing.[25] Nine months later employment in the aerospace industry was down 12.5 percent for the year, and SST supporters emphasized that both blue- and white-collar jobs had been lost.[26] The SST program was now credited not only with maintaining American aeronautical supremacy but also with the ability "to preserve the health of one of America's greatest industries."[27]

The original phase III proposals had called for government-backed bonds to finance production, with the government thus assuming the risk. Boeing had already received an extension on its plan for production financing until December 31, 1969, and in 1970 received another one, until June 30, 1972. In light of these difficulties, James Beggs, the administrator of NASA, suggested that Congress pay for production outright, the better to maintain control of and review over the program. Furthermore, if private financing were not forthcoming after the successful completion of phase III, he would flat out ask Congress for funds to pay for production so that the government could recover its development costs.[28] Beggs thus adopted the view that the SST would succeed in the aircraft market even though it lacked private financing.

By 1970 it seemed likely that the SST would lose its race with the Concorde by about five years.[29] The FAA contended that the first Concordes

23. *Department of Transportation and Related Agencies Appropriations for 1970*, Hearings before a Subcommittee of the House Committee on Appropriations, 91 Cong. 1 sess. (GPO, 1969), pt. 3, p. 190.

24. *Department of Transportation and Related Agencies Appropriations for Fiscal Year 1970*, Senate Hearings, p. 602.

25. *Department of Transportation and Related Agencies Appropriations for Fiscal Year 1970*, Senate Hearings, pp. 748, 609.

26. *Department of Transportation and Related Agencies Appropriations for Fiscal Year 1971*, Hearings before a Subcommittee of the Senate Committee on Appropriations, 91 Cong. 2 sess. (GPO, 1970), pt. 2, p. 1352.

27. "On the Floor," *CQ*, January 2, 1970, p. 47.

28. *Department of Transportation and Related Agencies Appropriations for 1971*, Hearings before a Subcommittee of the House Committee on Appropriations, 91 Cong. 2 sess. (GPO, 1970), pt. 3 pp. 555, 613.

29. In 1967 the estimated date Concorde would enter commercial service was 1971; in 1968 it was 1972; in 1969 it was early 1973; in early 1970 it was 1973 or 1974; in early 1970

would be paid off within four years, so that when the clearly superior SST was introduced in 1978, the original purchasers would be ready to retire almost half these planes. The early Concordes would then be snapped up by smaller airlines at less than half their original purchase price. Why the smaller airlines would want aircraft able to fly efficiently only on overwater routes remains unanswered.

In 1970 the FAA abandoned the goal that the SST prototype would meet the aircraft's production objectives, characterizing this goal as a "waste of time and money."[30] The FAA stated that the prototype would be a success if it permitted tests on the design of systems for use on the production aircraft, provided information about requirements for building the production aircraft, and demonstrated that the production objectives could be met at the time of production certification. The production plane was expected to weigh 115,000 pounds more than the prototype, not the 40,000 pounds originally planned.[31]

The road to construction of an SST prototype began in 1964 with airframe proposals that failed to meet desired technical and economic specifications, continued on to a winning design that failed to meet range-payload standards, and finally progressed to plans for a prototype that would be unable to test all production systems and would weigh 15 percent less than the production aircraft. Yet this dismal record, for a plane that Garwin told Congress was inferior to the one originally endorsed by Congress,[32] did not spell doom for the SST.

Political Battle over the SST

Until December 1970 the closest Congress ever came to canceling the SST program was when it was proposed. The SST obtained its original

it was early 1974. *Department of Transportation Appropriations for 1968*, House Hearings, p. 294; "Around the Capitol," *CQ*, November 1, 1968, p. 3067; *Department of Transportation and Related Agencies Appropriations for 1970*, House Hearings, pt. 3, p. 50; *Economic Analysis and Efficiency of Government*, Report of the Subcommittee on Economy and Government of the Joint Economic Committee, 91 Cong. 2 sess. (GPO, 1970), p. 921; and *Department of Transportation and Related Agencies Appropriations for Fiscal Year 1971*, Senate Hearings, pt. 2, p. 1379.

30. *Department of Transportation and Related Agencies Appropriations for 1971*, House Hearings, pt. 3, pp. 614–16.

31. *Department of Transportation and Related Agencies Appropriations for 1971*, House Hearings, pt. 3, p. 621; and *Department of Transportation Appropriations for 1968*, House Hearings, pp. 295–96.

32. *Economic Analysis and Efficiency of Government*, Joint Economic Committee Report, pp. 904–05.

appropriation for feasibility studies in 1961 when a Senate roll call vote to delete it failed by a tie vote (35 to 35). This vote was called by Senator Stuart Symington (Democrat of Missouri) to "protest subsidizing a commercial airliner" while at the same time not speeding up the development of the supersonic B-70 bomber.[33] The later Senate votes were all called by Senator William Proxmire (Democrat of Wisconsin) as protests against a program he deemed a failure. Between this first roll call vote and December 3, 1970, SST backers in the Senate defeated three attempts to reduce or delete the aircraft's appropriations by margins of no less than twenty-four votes.

In December 1970 the Senate passed an amendment to the Department of Transportation appropriations bill to delete the SST's appropriation for the remainder of the fiscal year (52 to 41). Nevertheless, in the conference committee four of the seven Senate conferees voted in favor of the SST, causing a disgruntled Proxmire to note that this violated Senate rules calling for conferees to reflect the wishes of the Senate majority.[34] On December 15, the House approved the conference report, which provided $210.2 million for the SST, but Senate opponents began a filibuster on the 17th. Two attempts to invoke cloture failed. Meanwhile, Senate Majority Leader Mike Mansfield (Democrat of Montana) proposed a continuing resolution for the Department of Transportation, and thus for the SST, at fiscal 1970 levels through the third quarter of fiscal 1971. This resolution was passed by the House on December 31 and by the Senate on January 2, 1971, with Proxmire threatening to continue the filibuster if he did not receive an on-the-record assurance of a vote solely on the SST in March.[35]

On March 18 the House voted 215–204 to delete the SST from the Department of Transportation's fourth-quarter continuing resolution, and on March 24 the Senate defeated an amendment to restore SST funds 46–51. The government thus canceled its contracts with Boeing and GE. The battle, however, had not quite come to a close.

When the House Appropriations Committee recommended $85.33 million in termination fees to repay Boeing and GE for their contributions, Representative Edward P. Boland (Democrat of Massachusetts) brought forth an amendment that would instead allow these funds to be used for

33. "Floor Action," *CQ*, July 28, 1961, p. 1342.

34. "On the Floor," *CQ*, December 25, 1970, p. 3047. There is no explicit rule on this, only the expectation that conferees will do so.

35. *Congress and the Nation*, vol. 3, p. 160.

continued prototype construction.[36] The amendment was defeated in the first roll call, but then House Minority Leader Gerald R. Ford (Republican of Michigan) summoned six Republicans to change their nay votes to present, and the amendment passed.

Nixon, who had called the defeat of the SST a "mortal blow for our aerospace industry for years to come," favored continued prototype development, arguing that it would be cheaper than paying termination fees and public assistance to laid-off aerospace employees.[37] Support for reviving the SST collapsed, though, when Boeing's chairman announced that the company would resume prototype development only if it were to receive between $500 million and $1 billion over the original contract. Since the original Boeing contract had been for considerably less than $1 billion, Boeing now clearly believed that its original development estimates had been wildly incorrect. On May 19 the Senate voted to delete the funds the House had reallocated just a week earlier. The battle was over.

Technical Concerns over the SST

To meet performance goals that would make the SST commercially viable, certain technical characteristics first had to be achieved. The most salient change in the SST's technical specifications over time (table 6-2) is its weight gain between 1961 and 1970. Hauling around the extra weight would mean higher direct operating costs and thus a poorer commercial prognosis for the craft. Boeing's original design (column 3) compared favorably with the FAA's desires (column 2), but it had to be scrapped because the heavy, 20,000–25,000 pound, variable sweep wing assembly made the range-payload standards unattainable.[38] Boeing's final design (column 4) weighed in at 35 percent more than the 1965 goal.[39] Prospects had deteriorated even more by 1970, when the FAA revealed that the

36. *Second Supplemental Appropriations Bill, 1971*, Hearings before the House Committee on Appropriations, 92 Cong. 1 sess. (GPO, 1971), p. 695; and "On the Floor," *CQ*, May 14, 1971, p. 1066.

37. *Congress and the Nation*, vol. 3, p. 159; and "Executive Branch," *CQ*, May 21, 1971, p. 1096, and "On the Floor," p. 1100.

38. C. M. Plattner, "Boeing SST Emphasizes Economy, Growth," *Aviation Week and Space Technology*, August 15, 1966, p. 38; William H. Gregory, "Joint SST Effort Proposed Again," ibid., May 6, 1968, p. 37; Harold D. Watkins, "Boeing Shifts Emphasis on SST," ibid., July 15, 1968, p. 25; and Harold D. Watkins, "Boeing Approaches Critical SST Decision," ibid., July 22, 1968, p. 32.

39. Harold D. Watkins, "Boeing to Brief Airlines on SST Design," *Aviation Week and Space Technology*, October 14, 1968, p. 29.

Table 6-2. *Proposed SST Designs, 1961, 1965, 1967, 1969*

Item	*Commercial supersonic transport report, 1961*[a] (1)	*Federal Aviation Administration, 1965*[b] (2)	*Boeing's original design, 1967*[c] (3)	*Boeing's revised design, 1969*[d] (4)
Cruise speed (mach)	3	2.7	2.7	2.7
Range	3,500 n.m.	4,000 m.	4,000 m.	4,000 m.
Gross takeoff weight (pounds)	400,000	500,000	500,000	675,000 maximum[e]
Payload (pounds)	30,000	40,000	75,000 maximum, 58,600 over 4,000 miles[f]	65,580, 75,000 over 3,700 miles, 65,580 over 4,000 miles[f]
Capacity (seats)	120	200–50	280	262–98

a. *Contemporary and Future Aeronautical Research*, Hearings before the House Committee on Science and Astronautics, 87 Cong. 1 sess. (GPO, 1961), pp. 92–117.

b. *Supplemental Appropriation Bill, 1966*, Hearings before the House Committee on Appropriations, 89 Cong. 1 sess. (GPO, 1965), p. 8.

c. *Department of Transportation Appropriations for 1968*, House Hearings, pp. 288, 295, 296.

d. *Department of Transportation and Related Agencies Appropriations for 1970*, Hearings before a Subcommittee of the House Committee on Appropriations, 91 Cong. 1 sess. (GPO, 1969), pt. 3, pp. 10, 29, 30, 57, 140.

e. Even though the specifications were unchanged, by 1970 the production aircraft was expected to weigh 750,000 lbs. *Department of Transportation and Related Agencies Appropriations for 1971*, Hearings before a Subcommittee of the House Committee on Appropriations, 91 Cong. 2 sess. (GPO, 1970), pt. 3, p. 621.

f. The testimony is inconsistent on whether these are maximum payloads, or those attainable at the specified range of the aircraft. *Department of Transportation Appropriations for 1968*, House Hearings, p. 296; and *Department of Transportation and Related Agencies Appropriations for 1970*, House Hearings, pt. 3, p. 30.

production aircraft would weigh 750,000 pounds, nearly twice the original goal.[40]

By 1970 the prototype's noise levels were still about twice as high as permitted in the contract specifications. The SST's fuel inlet performance was still unsatisfactory, a fuel sealant that would not degrade in the hot temperatures of flight had not yet been found, and the work undertaken on fabricating a titanium shell that could withstand high temperatures was not promising.[41]

In May 1970, Richard Garwin of the president's ad hoc review committee

40. *Department of Transportation and Related Agencies Appropriations for 1971*, House Hearings, pt. 3, p. 621.

41. *Economic Analysis and Efficiency of Government*, Joint Economic Committee Report, 1970, p. 900; *Department of Transportation and Related Agencies Appropriations for 1970*, House Hearings, pt. 3, pp. 313, 314; and *Department of Transportation and Related Agencies Appropriations for 1971*, House Hearings, pt. 3, p. 525.

stated that the aircraft now being developed was inferior to the original proposal, and that it "has not fulfilled its contract requirements; it is too heavy, too noisy."[42] Throughout the lifetime of the project, though, the SST's supporters claimed that it would achieve the contract specifications.

Economic Evaluations of the SST

The commercial success of the SST required that airlines be able to operate the plane at a profit. From 1963 through 1969 the FAA received a stream of optimistic studies on the feasibility of the SST as a commercial venture. The few pessimistic studies were dismissed with the claims that they used assumptions that would not be used "by anyone knowledgeable in the industry of the air transport market," and that they were based on historical industry figures that were inappropriate for the SST.[43] In 1969 Boeing completed a study that predicted sales of 515 aircraft, within the FAA's prediction of 500–1,200 (the maximum applied if there were no overland restrictions).[44] The Boeing study, which reflected such real-world situations as actual routes and time-zone differentials, was hailed by the FAA as "outstanding." The one real-world likelihood not included was a fare surcharge, which many believed would be necessary for the aircraft to prove profitable for the airlines.[45]

In the early 1960s, while deciding whether to enter a design in the preliminary competition, the Douglas Aircraft Company concluded that the SST would weigh and cost twice what the FAA hoped for. Furthermore, under the assumptions that the project would proceed for ten years before cancellation and that the company could borrow money for its contribution

42. *Economic Analysis and Efficiency of Government*, Joint Economic Committee Report, 1970, pp. 904–05; and *Congress and the Nation*, vol. 3, p. 159.

43. *Department of Transportation and Related Agencies Appropriations for 1971*, House Hearings, pt. 3, p. 577; and *Department of Transportation and Related Agencies Appropriations for 1970*, House Hearings, pt. 3, p. 223.

44. *Department of Transportation and Related Agencies Appropriations for 1970*, House Hearings, pt. 3, p. 125; and *Department of Transportation Appropriations for 1968*, House Hearings, p. 293.

45. *Economic Analysis and Efficiency of Government*, Report of the Subcommittee on Economy and Government of the Joint Economic Committee, 91 Cong. 1 sess. (GPO, 1969), pt. 3, p. 674. How Boeing concluded that the surcharge would not be necessary is perplexing, given the study's estimated 50 percent higher direct operating cost for the SST over subsonics. Richard S. Shevell, "Technological Development of Transport Aircraft—Past and Future," *Journal of Aircraft*, vol. 17 (February 1980), pp. 67–80. The operating cost estimate was not reported to Congress. Perhaps Congress relied on the phenomenal estimated returns on investment noted later in the text.

at 6 percent without ever having to repay the principal, the interest payments alone would still exceed the company's net worth. Citing Douglas's commitment to the DC-9, the company did not submit a design.[46]

The economic studies also estimated the rate of return on investment (ROI) that airlines, the customers of the SST, could expect to receive. The estimates were 20 to 50 percent in 1965, 30 percent in 1967, and 21.2 percent in 1969.[47] By contrast, the 1960–69 average ROI over international routes was 9.63 percent, with a peak of 14.67 percent in 1965 that fell steadily to 5.46 percent in 1969.[48] Thus if the studies had proved correct, the SST would indeed have been attractive to airlines.

In 1969 the president's SST ad hoc review committee tendered an unfavorable report. The Department of Transportation provided a blow-by-blow response of 120 pages.[49] Hendrik S. Houthakker, a member of the Council of Economic Advisers and the ad hoc committee, recommended more research into technical and economic characteristics of the SST before continuing with the prototype. He also mentioned the committee's concern over the conflicts of interest that could arise because the airplane's principal booster, the FAA, was also in charge of all SST studies as well as the certification of the craft's airworthiness.[50]

In 1970 chinks also began to mar the claim that the SST would provide employment benefits. Arnold R. Weber, assistant secretary for manpower in the Department of Labor and also a member of the ad hoc committee, reported a labor shortage in the aircraft industry and stated that the SST program could provide no benefits to the hard-core unemployed because it required such highly skilled labor.[51]

46. Interview with Richard Shevell, professor of aeronautical engineering, Stanford University, formerly director of aerodynamics and later of commercial advanced design, Douglas Aircraft Company, Atherton, California, June 10, 1986.

47. *Continuing Appropriations, 1966,* Hearings before the Senate Committee on Appropriations, 89 Cong. 1 sess. (GPO, 1965), p. 30; *Supplemental Appropriation Bill, 1966,* Hearings before the House Committee on Appropriations, 89 Cong. 1 sess. (GPO, 1965), p. 8; "U.S. SST Seen More Economical Than Concorde, Some Subsonics," *Aviation Week and Space Technology,* March 20, 1967, p. 42; and *Economic Analysis and Efficiency of Government,* Joint Economic Committee Report, 1969, pt. 3, p. 66.

48. *Department of Transportation and Related Agencies Appropriations for Fiscal Year 1971,* Senate Hearings, pt. 2, p. 1391.

49. *Department of Transportation and Related Agencies Appropriations for Fiscal Year 1970,* Senate Hearings, pp. 633–783.

50. *Department of Transportation and Related Agencies Appropriations for 1970,* House Hearings, pt. 3, pp. 331–32.

51. *Department of Transportation and Related Agencies Appropriations for 1970,* House Hearings, pt. 3, pp. 85–87, 332.

Sonic Boom and Environmental Effects

From the beginning, the acceptability of the SST was known to depend on the magnitude and significance of its sonic boom, the pressure wave with a harsh thunderlike sound sensed as "a kind of explosion" that is caused by a projectile traveling faster than the speed of sound.[52] The sonic boom issue, however, never did achieve the saliency of the potential environmental impact of the SST that was raised in the summer of 1970.

In an attempt to ascertain the effects of the SST booming residential areas, the FAA and Air Force carried out sonic boom tests in Oklahoma City in 1964. Although many residents felt that sonic booms degraded living conditions, the FAA concluded that people would learn to live with it.[53] The Citizens League against the Sonic Boom worked with Proxmire to convince legislators to oppose the SST, but they were not successful until environmental groups joined with them to form the Coalition against the SST in early 1970, which had the support of fifteen national and fourteen state and local groups.[54]

Environmental groups had remained silent until late 1969 for fear of alienating Senator Henry Jackson (Democrat of Washington), chairman of the Senate Interior Committee.[55] When the environmental issues finally were raised, two effects were stressed. The first was the increase in water vapor in the stratosphere; however, analysts could not decide whether the effect would be a decrease in temperature and an ice age or an increase that would melt the polar ice caps and result in flooding. The second issue was destruction of ozone by the aircraft's exhaust, which would allow increased ultraviolet radiation to hit the earth.[56] This radiation is absorbed by DNA and can damage or destroy human cells. It may also harm crops and kill organisms living at the surface of the ocean.

Studies by the National Center for Atmospheric Research and the Science Policy Research Division of the Library of Congress found the

52. *Report on the Task Force on National Aviation Goals*, p. 78; and *Independent Offices Appropriations for 1965*, House Hearings, pt. 1, p. 1177.

53. "Decision Awaited on Supersonic Transport Contract," *CQ*, October 14, 1966, p. 2492.

54. Joshua Rosenbloom, "The Politics of the American SST Programme: Origin, Opposition and Termination," *Social Studies of Science*, vol. 11 (1981), pp. 403–23; and "Lobby Report," *CQ*, March 26, 1971, p. 718.

55. Rosenbloom, "Politics of the American SST Programme," p. 413.

56. *Economic Analysis and Efficiency of Government*, Joint Economic Committee Report, 1970, pp. 898, 899.

effect of the SST on ozone depletion to be trivial.[57] MIT's 1970 *Study of Critical Environmental Problems* was not so sanguine, recommending that the environmental consequences be clarified before an even larger commitment to the SST program was made.[58] This recommendation echoed the conclusions of the ad hoc committee's report. Russell E. Train, under secretary of the Department of the Interior and ad hoc committee member, reiterated the committee's conclusions while also noting that incomplete combustion of fuel during subsonic portions of SST flights could increase air pollution levels. Furthermore, Train, speaking for his department, stated that the SST probably would have adverse effects on the environment and that he did not think the program should continue in the absence of "overwhelming" benefits, which, quite obviously, he did not find.[59]

Cost-Benefit Analyses of the Program

The cost-benefit analyses presented here simulate those that a conscientious economist working in Congress or the Executive Office of the President could have performed during the lifetime of the SST program to provide guidance for policy decisions. Cost-benefit analyses are performed for all years, and forward-looking (net of sunk costs) analyses are reported for the years the benefit-cost ratio dips below one. In keeping with the studies performed in the 1960s, only the direct costs of the program are considered. Throughout its history the SST was known to be noisy on takeoff and to produce sonic booms. Later, it was claimed that the SST would cause serious environmental problems. Both sources of external costs, as well as indirect costs, are neglected in the following analyses. In addition, much of the data for these calculations are from favorable studies completed for the FAA, and as such are biased in favor of the SST. Furthermore, as the probability of attaining the desired specifications became smaller and smaller, the studies continued to assume they would be met with a probability of one. Except for operating costs, this is done here, too, which further biases the conclusions in favor of the program.

All monetary values are in constant 1967 dollars, and a rate of 7 percent is used to discount all costs and benefits to 1967. The present discounted value of development costs, production costs, revenue from sales of the

57. "Lobby Report," *CQ*, March 26, 1971, p. 718.

58. Rosenbloom, "Politics of the American SST Programme," p. 417.

59. *Department of Transportation and Related Agencies Appropriations for 1970*, House Hearings, pt. 3, pp. 324–26, 335, 340.

aircraft, and the value of time saved by travelers using the SST are calculated for each year of the program from the data most recently available in each year. It is assumed that when a new value was not reported, the belief was that in now-current dollars the old estimate was still correct.

The results of these calculations are shown in table 6-3. (The underlying data are reported in appendix table 6A-1.) Calculating the present discounted value of development costs in a given year proceeds as follows. First, the funds already spent through this year on the program, evaluated in the dollars for the year in which they were spent, are subtracted from the current estimate of development costs, in current year dollars. The rationale for using current dollars at this stage is the belief that as the cost estimate was revised over time the projected expenses were just added to the actual money already spent. The resulting difference is then divided equally among the years left for development, from the coming year through the estimated first flight date. These are then converted to 1967 dollars, discounted, and summed, producing the prospective costs of continuing the program. Full development costs are estimated by adding the prospective costs to the present discounted value of the funds already spent.

The components of costs that are included in the analysis are the appropriations from the government, the $17 million paid by manufacturers during the phase II competition in 1966, the cost sharing by Boeing and GE during phase III (equal to one-ninth of the government's payments), and the $60 million in risk capital put up by the airlines.[60] Fifty-two million dollars of this risk capital was in by 1967 and is attributed to spending that year; the remainder was profferred by late 1969 and is divided equally between 1968 and 1969.[61]

To these development costs must be added the projected costs associated with certifying the aircraft. The first explicit statement of certification costs was in 1969, and it was clearly separated from development and production costs.[62] Using the phase III cost estimates presented in the same year, the phase IV costs are between 51 and 57 percent of the former. So for the years before 1969, certification costs are estimated by subtracting phase II

60. *Congress and the Nation*, vol. 3, p. 168; *Department of Transportation and Related Agencies Appropriations for Fiscal Year 1971*, Senate Hearings, pt. 2, p. 1363; and *Department of Transportation Appropriations for 1968*, House Hearings, p. 345.

61. "Floor Action," *CQ*, July 21, 1967, p. 1219; and *Department of Transportation and Related Agencies Appropriations for 1970*, House Hearings, pt. 3, p. 617.

62. *Department of Transportation and Related Agencies Appropriations for Fiscal Year 1970*, Senate Hearings, p. 615.

Table 6-3. *Summary of Cost-Benefit Analysis of SST, 1961–71*

Billions of 1967 dollars

Year	*Development, certification, and production costs, with no operating cost differential* (1)	*Value of time saved plus revenue from sale of craft, with no operating cost differential* (2)	*Development, certification, and production costs, with operating cost differentials per the text*[a] (3)	*Net value of time saved plus craft sales revenue, with operating cost differentials* (4)	*Column 1 minus costs already sunk (entries only if total benefit-cost ratio is less than 1)*[b] (5)
1961	1.01–2.10	13.28–18.73	. . .	. . .	. . .
1962	0.99–2.11	13.08–19.79	. . .	. . .	. . .
1963	1.07–2.24	14.40–22.00	1.05–2.20	4.33–5.02	. . .
1964	1.06–2.20	16.38–20.82	1.03–2.16	5.12–7.29	. . .
1965	.84–2.43	15.58–26.73	1.75–2.42	5.02–14.85	. . .
1966	5.51–9.98	13.92–23.48	3.15–9.59	4.51–13.06	3.14–9.29
1967	5.27–9.15	14.14–20.72	2.88–4.37	2.74–5.04	2.23–3.72
1968	6.60–6.80	13.24–19.33	3.32–3.52	2.54–4.67	2.55–2.75
1969	6.03–6.15	14.54–19.50	5.12–5.26[c] 3.13–3.25	9.32–13.29[c] 1.85–3.59	2.35–2.47
1970	5.97–6.04	14.54–17.38	5.13–5.19 3.23–3.29	9.32–11.59 1.85–2.85	2.37–2.45
1971	5.77–5.82	14.54–17.38	4.95–5.01 3.14–3.19	9.32–11.59 1.85–2.85	2.16–2.21

Sources: From raw data and sources in table 6A-1, manipulated as described in the text.

a. Percent by which SST's total operation costs exceed those of advanced subsonics: 6–22 percent for 1963–66, 32 percent for 1967–71, and also 10 percent for 1969–71.

b. Operating cost differential is 32 percent for these entries.

c. For this and all subsequent entries in this column, the first row is with a 10 percent operating cost differential and the second is with a 32 percent differential.

costs from the total development costs, all in current year dollars, and taking 54 percent of this as the certification cost.[63] That amount is then divided evenly among the years after the estimated first flight through the estimated service date of the aircraft. This stream is then converted to 1967 dollars and discounted. The previous calculation of development costs plus this calculation of certification costs constitutes the estimated present discounted value of total development costs of the SST.

The estimates of production costs are transformed into 1967 dollars, evenly divided among the years from the start of service date through 1990, the end of the official studies' horizon, and discounted.[64] The only estimate of production cost available from 1961 through 1965 is for the total program including development costs, so the present discounted value of phases I through III costs is subtracted from the present discounted value of "production" costs. The estimated phase IV cost is not subtracted, because in later years the certification costs seemed to be considered a thing apart, and thus separate, from development and production costs; this notion is followed here.

The revenue stream from sales of the aircraft is generated by dividing the current market estimate by the number of years from service date through 1990, as is done for production costs. This result is multiplied by the estimate of the SST's selling price, in 1967 dollars, and the stream is discounted.

The calculation of the value of travelers' time saved with the SST starts with estimates of passenger miles of travel and their expected rate of growth. The two passenger-mile figures available are 81 billion for 1962 and 250 billion for 1968.[65] The former is used for all calculations before 1969, and the latter from 1969 on, since the 1968 figure was entered into testimony in 1969. The annual rate of growth used is 10 percent. In 1963 it was estimated that 43 percent of the total passenger miles would be

63. Preliminary research and feasibility costs are neglected, since it is not clear that they were always considered part of development costs and are comparatively small, and Phase I expenses are neglected because they were not government financed and were not reported to Congress.

64. *Panel on Science and Technology, Sixth Meeting,* Proceedings before the House Committee on Science and Astronautics, 89 Cong. 1 sess. (GPO, 1965), p. 46; and *Independent Offices Appropriations for 1967*, Hearings before the House Committee on Appropriations, 89 Cong. 1 sess. (GPO, 1966), pt. 1, p. 1412.

65. *United States Commercial Supersonic Aircraft Development Program,* Hearings before the Senate Committee on Commerce, 88 Cong. 1 sess. (GPO, 1963), p. 62; and *Economic Analysis and Efficiency of Government*, Joint Economic Committee Hearings, 91 Cong. 1 sess. (GPO, 1969), pt. 3, p. 676.

suitable for SST travel; this figure is used through the 1966 calculation. In 1967 the FAA announced that henceforth only overwater routes would be used in economic feasibility studies, which obviously reflected the likelihood that the aircraft would be so restricted; this convention is followed here.[66] It seems that the shortest overwater route the SST could fly is between New York and London, a distance of somewhat more than 3,400 miles. A chart of air traffic by mileage blocks, as estimated in 1963 for 1975,[67] is used to figure that trips of at least 3,400 miles would be expected to constitute 30 percent of air traffic. Thirty percent of total passenger miles are therefore considered available to the SST from 1967 on. This figure overstates the passenger miles the SST would cover if a sizable portion of the trips over 3,400 miles were not overwater routes, and so would bias the analyses in favor of support for the program.

It was expected that advanced subsonic jets, and later the Boeing 747, would be the SST's competitors. The 747's cruise speed of 580 mph is used to calculate the total minimum travel time at subsonic speeds. The SST was expected to cruise at mach 2.7 or 1,780 mph, making it three times as fast as the 747.[68] The upper bound of the percent of travel time saved is therefore 67 percent, and the lower bound is 50 percent. The 1965 estimate that to the 12 to 14 percent of Americans who fly the average hour is worth $7–$8, or $7.40–$8.50 in 1967 dollars, is used to arrive at the value of $8 per hour of time saved.[69] The value of time saved in each year of service is then discounted, and for each year's analysis the sum is computed for the years from the date for entering service through 1990.

The benefit-cost ratios resulting from the analysis just presented are shown in table 6-4. The net benefit of the program is always positive, but the analysis assumes that the desired specifications of the aircraft would be met. After the failure of Boeing's original design in 1968 it should have been realized that the probability of meeting the specifications had fallen and caused the expected net benefit also to fall. Subsequent failures to accomplish goals in a timely manner should have resulted in the probability

66. *Department of Transportation Appropriations for 1968*, House Hearings, p. 318.

67. *Independent Offices Appropriations, 1964*, Senate Hearings, pt. 2, p. 2024.

68. *Supplemental Appropriations Bill, 1966*, House Hearings, p. 8; and *Department of Transportation and Related Agencies Appropriations for 1970*, House Hearings, pt. 3, p. 140. Though mach 1, the speed of sound, is usually quoted as 760 mph at sea level, at the altitude the SST would have flown it is only 660 mph. "On Supersonic Transport," *CQ*, November 8, 1963, p. 1908.

69. *Independent Offices Appropriations for 1966*, Hearings before the House Committee on Appropriations, 89 Cong. 1 sess. (GPO, 1965), pt. 1, p. 1191.

Table 6-4. *Benefit-Cost Ratios of SST, 1961–71*

Year	*Benefit-cost ratio, with no operating cost differential* (1)	*Benefit-cost ratio, given operating cost differentials in text*[a] (2)	*Forward-looking benefit-cost ratio; costs are net of sunk costs (entries only if total ratio less than 1)*[b] (3)
1961	6.32–18.54	. . .	. . .
1962	6.20–19.91	. . .	. . .
1963	6.43–20.56	1.97–4.97	. . .
1964	7.45–19.64	2.37–7.08	. . .
1965	6.41–14.53	2.07–8.49	. . .
1966	1.39–4.26	0.47–4.15	0.49–4.16
1967	1.55–3.93	0.63–1.75	0.74–2.26
1968	1.95–2.93	0.72–1.41	0.92–1.83
1969	2.36–3.23	{1.77–2.60[c] / 0.57–1.15}	0.75–1.53
1970	2.41–2.91	{1.80–2.26 / 0.56–0.88}	0.76–1.19
1971	2.50–3.01	{1.86–2.34 / 0.58–0.91}	0.84–1.32

a. See table 6-3, note a.
b. See table 6-3, note b.
c. See table 6-3, note c.

falling even further. In addition, this analysis assumes that there would be no surcharge on SST fares, which implies operating costs per passenger mile comparable to those of a subsonic jet. Although this assumption was used in many studies up through 1969, many believed it would prove incorrect. A second analysis is therefore performed, with SST operating costs greater than those of subsonic jets.

The second analysis assumes that passengers would be forced to pay for higher operating costs, which would decrease the mileage flown on SSTs as well as the net value of time saved. First, the total operating cost for the SST's competition is calculated, using 3.11 cents per passenger mile from 1963 through 1966 and 2.03 cents from 1967 through 1971.[70] In 1963 it was reported that the SST's total operating cost for the top 43 percent of the traffic distribution was expected to be 6 to 22 percent greater for the SST than for advanced subsonic jets. In 1967 the Office of SST Development

70. *Independent Offices Appropriations, 1964*, Senate Hearings, pt. 2, p. 2035 (with the figure changed to 1967 dollars); *Department of Transportation Appropriations for 1968*, House Hearings, p. 349; and *Economic Analysis and Efficiency of Government*, Joint Economic Committee Report, 1969, pt. 3, p. 680.

noted that SST fares were expected to be 32 percent greater than comparable subsonic fares.[71] Although no reason was given, it seems reasonable to link the higher fares to higher operating costs.[72]

In 1969 the updated feasibility study performed by the FAA predicted that the SST's operating costs would be only slightly higher than the 747's, but allowed that to meet sonic boom restrictions the SST's total operating cost might be 10 percent greater.[73] Therefore, for 1963–66 the SST's total operating costs are calculated at 6 to 22 percent above subsonics', for 1967–71 at 32 percent above the 747's, and also at 10 percent above the 747's for 1969–71.[74] The speed-fare preference curve indicates that at fare surcharges of the same percent that the SST's operating costs surpass subsonic jets', the SST would garner available passengers and, presumably, passenger miles, as follows: 95 percent for 6 percent higher operating costs, 80 percent for 10 percent higher, 40 percent for 22 percent higher, and 35 percent for 32 percent higher.[75] It is assumed that the number of aircraft produced, production costs, and aircraft sales revenue would therefore be these percentages of their values when no surcharge is imposed. The value of travel time saved is reduced similarly, and since it is assumed that passengers, not airlines, would bear the operating cost differential, this value of travel time saved is further reduced by the fare surcharge, equal to the extra operating costs of the SST over subsonic jets. Development costs are, of course, unchanged.

The benefit-cost ratios for this analysis are in table 6-4, column 2. Through 1965 the net benefit is clearly positive. In 1966 the minimum ratio is less than one, but the maximum is large enough so that one could credibly conclude that the expected net benefit is positive. Calculations with the 10 percent differential in operating costs maintain positive net benefits

71. *Independent Offices Appropriations, 1964*, Senate Hearings, pt. 2, text p. 2035, and graph p. 2024; and *Department of Transportation Appropriations for 1968*, House Hearings, p. 332.

72. Indeed, the same Boeing study that assumed no fare surcharge estimated direct operating costs per passenger mile 50 percent above that for the 747. Considering indirect operating costs also, that would have meant SST fares about 31 percent greater than those for the wide-body subsonics. Shevell, "Technological Development of Transport Carriers," p. 73. The Boeing operating cost estimate was not reported to Congress.

73. *Economic Analysis and Efficiency of Government*, Joint Economic Committee Report, 1969, pt. 3, p. 680.

74. The 10 percent rate is included because it was presented to Congress as an updated, and thus presumably better, figure. In light of the 1981 study mentioned in footnote 54, however, it seems much less legitimate than the 32 percent rate.

75. *Economic Analysis and Efficiency of Government*, Joint Economic Committee Report, 1969, pt. 3, p. 677.

throughout the life of the project; however, this result is not credible, because it is at odds with the 1967 estimates from both the Office of SST Development and Boeing. Calculations using the 32 percent operating cost differential show that by 1969 the range of the benefit-cost ratio was roughly symmetric around unity and so casts doubt on the desirability of the program. By 1970 the net benefit is clearly negative. Furthermore, these results assume that the contract specifications for the SST would be met, even though evidence to the contrary had been accumulating since the end of phase I.

Prospective incremental benefit-cost ratios are reported in table 6-4, column 3, for the case of an operating cost differential of 32 percent. For this analysis the funds already sunk into development are subtracted from the total cost of the project. It again must be emphasized that these results assume that the SST would meet specifications, a feat never accomplished during the program.

From 1966 to 1968 these prospective benefit-cost ratios indicate that the project remains justifiable. From 1969 on, the picture is not so rosy. Although the maximum benefit-cost ratio never dips below one, the range includes ratios substantially less than one. The results would be more pessimistic if the probability of achieving the touted specifications was less than one. That would mean smaller expected benefits, costs at least as great, and overall expected benefit-cost ratios less than those in column 3.

Retrospectively, even under the assumption that the plane would meet its specifications, the economic case for the SST was weaker than contemporary studies suggested. During the life of the program, the greater fuel intensity and lesser labor intensity of the SST as compared with subsonic jets was regarded as an advantage because the rate of inflation on fuel was expected to be less than that on labor.[76] The drastic fuel price increases of the middle and late 1970s shattered this expectation. As a result, the Concorde, the aircraft the SST was intended to preempt, has total operating costs about twice those of subsonic jets.[77]

A major concern, at least among legislators, was the recovery of government capital. The contracts with Boeing and GE provided for 100 percent recovery only upon the sale of 300 aircraft. A positive net benefit is not sufficient to fulfill this objective. The cost-benefit analysis with a 32 percent operating cost differential suggests that only 175 aircraft would

76. *Department of Transportation and Related Agencies Appropriations for 1970*, House Hearings, pt. 3, p. 235.

77. Shevell, "Technological Development of Transport Carriers," p. 73.

have been sold (table 6A-1, column 3). So while the net benefit of the entire program could still be construed as positive, it nevertheless implies a government subsidy to private industry.

Balance-of-payments benefits were also claimed for the SST, with estimates ranging from one-third to two-thirds of the production going to foreign airlines.[78] This is not an economic benefit beyond that from the net sales of the aircraft, but it is a political benefit. The SST was not expected to displace American subsonics, because the introduction of the Concorde was expected to cause subsonic planes to lose those sales anyway.

The Institute of Defense Analysis claimed that the net loss in the tourism balance-of-payments account would offset the gains from the sale of the airplanes.[79] Because SST fares were never envisaged as being less than those for subsonics, the SST could increase the number of American travelers abroad only if it made a previously prohibitively long journey possible, a plausible idea only for business travel. But even that is dubious, for there is no reason to believe that American businessmen would go abroad more frequently than their foreign business associates would come to the United States.

In conclusion, the conscientious staff economist, taking account of the risk that the plane would not perform to specifications, might still have concluded through 1966 that the SST would prove socially beneficial but would have had to express doubts in 1967. By 1969, after Boeing had redesigned its airframe, the analyst would have had to report that a positive net benefit from the SST program was doubtful even if based on incremental benefit-cost analysis. In 1970, to the extent that the environmental issues raised that summer were regarded as significant, the direct net benefits would have been lower still.

The turning point in the cost-benefit analyses between 1967 and 1969 is confirmed by the history of the airlines' support for the SST. Success was contingent on airlines purchasing the aircraft; their support ebbed and flowed with their confidence in the project. The $60 million provided by the airlines through 1969 for SST development demonstrated strong

78. *Independent Offices Appropriations for 1962,* House Hearings, p. 295; *Department of Transportation Appropriations for 1968,* House Hearings, pp. 294, 299; *Economic Analysis and Efficiency of Government,* Joint Economic Committee Report, 1969, pt. 3, pp. 672–73; and *Department of Transportation and Related Agencies Appropriations for 1970,* House Hearings, pt. 3, p. 86.

79. *Economic Analysis and Efficiency of Government,* Joint Economic Committee Report, 1969, pt. 3, p. 687.

Table 6-5. *Reservations for SST Delivery Position, 1963–68*

Date	Total number of reservations to date
October 1963	27
December 1963	45
February 1964	72
April 1964	84
January 1965	96
End 1966	113
July 19, 1967	129
March 1968	122

Sources: For 1963–64, *Supplemental Appropriations Bill, 1966, Hearings*, House, p. 42; for 1965, ibid., pp. 42–43; for 1966 and 1967, "Second-Round Orders Surge for Boeing 2707," *Aviation Week and Space Technology*, July 24, 1967, p. 30; for 1968, Harold E. Watkins, "SST Faces Drastic Cut in Weight," ibid., March 11, 1968, p. 29.

support.[80] Reservations for the 129 delivery positions purchased from 1963 through 1967 did too (table 6-5), especially the last 16, which were placed within 48 hours after Boeing began accepting orders on July 17, 1967. Whereas the first 113 positions cost a refundable $200,000, the last 16 went for a nonrefundable $750,000.[81] In 1968 the number of purchased delivery positions declined to 122.[82] As best as can be reconstructed from available information, one of the cancellations came from the original 113 orders and 6 from the later 16.[83] There were no further cancellations or orders.

The SST in Congress

The budget history of the SST is shown in table 6-6, which presents the annual budget requests of the president and the ultimate appropriations by Congress from 1962 through 1971. In six of the ten years the president

80. *Department of Transportation and Related Agencies Appropriations for 1970*, House Hearings, pt. 3, p. 96.

81. "Second-Round Orders Surge for Boeing 2707," *Aviation Week and Space Technology*, July 24, 1967, p. 30; and *Second Supplemental Appropriations Bill, 1971*, House Hearings, p. 705.

82. Harold D. Watkins, "SST Faces Drastic Cut in Weight," *Aviation Week and Space Technology*, March 11, 1968, p. 29.

83. *Department of Transportation and Related Agencies Appropriations for 1970*, House Hearings, pt. 3, p. 99; *Independent Offices Appropriations, 1964*, Senate Hearings, pt. 2, pp. 389, 415; "Second-Round Orders Surge for Boeing 2707," *Aviation Week and Space Technology*, July 24, 1967, p. 30; and *Second Supplemental Appropriations Bill, 1971*, House Hearings, p. 705.

Table 6-6. *SST Budget Requests and Congressional Appropriations, Fiscal Years 1962–71*
Millions of current dollars

Year	*Request*	*Appropriation*
1962	12	11
1963	25	20
1964	60	60
1965	0	0
1966	140	140
1967	280	280
1968	198	142.375
1969	223	−30
1970	95.958	85
1971	289.965	210.2[a]

Source: *Department of Transportation and Related Agencies Appropriations for Fiscal Year 1971*, Hearings before a Subcommittee of the Senate Committee on Appropriations, 91 Cong. 1 sess. (GPO, 1970), p. 1217.
a. Congressional Quarterly, *Congress and the Nation*, vol. 3 (Washington, 1973), p. 168. This sum was appropriated through a continuing resolution, though not all of it had been spent at the program's termination.

requested more than Congress was willing to appropriate, and in one of the remaining four years, 1965, no appropriations were considered because the design phase of the program had already been fully financed.

Throughout the life of the program, the president was a strong advocate of the SST. Indeed, Presidents Lyndon B. Johnson and Richard M. Nixon rejected independent studies they had commissioned that reached negative conclusions about the prospects for the aircraft. The combination of the technological optimism in the FAA, support from the industry, and the presence of a technology race with both Russians and Western Europeans led several administrations to be persistently strong advocates of the program. As the budgetary figures indicate, Congress was not so enthusiastic, which, for reasons discussed in chapter 4, gave the legislature a strategically advantageous position.

Analysis of congressional votes on the SST provide an opportunity to examine how presidential advocacy, program performance, and distributive politics influenced the viability of the program. Unfortunately, most of this analysis must be confined to the Senate, for the House cast roll call votes on the SST only when the program was being killed.

Voting in the Senate

The Senate took seven roll call votes on the SST program. The first, in July 1961, was to delete $12 million for research and development. This vote is unlike subsequent ones because it was taken before the program even began, when the Senate had no information on its technical prospects. Because the program then had no established stakeholders, the importance of distributive politics should be relatively minor for this vote. Opponents protested aid to a private commercial venture amidst refusal to accelerate the development of the B-70 bomber,[84] and that could have caused strategic voting intended to protect or otherwise affect the B-70. Evidence that this vote differs from the others is that it is the only vote in which, on average, Republican and conservative Democratic senators were more likely to vote against the SST.

The second vote, in August 1966, was to reduce the fiscal 1967 appropriation from $280 million to $80 million.[85] At the time this amendment was considered, its passage was expected to delay the beginning of phase III, prototype construction, from January to July 1967. Because the prototype phase had not yet started, this vote should also be comparatively free of distributive influences, though not entirely so, since the likely candidates for phase III contracts had been narrowed to two each for the engines and the aircraft. As it turned out, authorization for the phase III contracts was delayed until late April 1967. Had the amendment passed, it would have only briefly extended the delay, not ended the program.

The October 1967 vote was to reduce the fiscal 1968 appropriation from $142.375 million to $1 million.[86] By this time the phase III contracts had been awarded, so that the stakeholders in prototype construction were identified. Consequently, this vote and those that follow should be more sensitive to distributive issues. Besides, the worst news about performance had not yet appeared, so that the program should not yet have become as controversial as it would be in later years.

The December 1969 and December 1970 votes were on amendments to delete the entire SST appropriations for fiscal 1970 and fiscal 1971, respectively.[87] The passage of either amendment would have killed the SST program. By the time of the 1969 vote, the bad news about performance had been made public, but the environmental issues had not yet been effectively raised. The 1969 amendment was easily defeated, but the 1970

84. "Floor Action," *CQ*, July 28, 1961, p. 1342.

85. "Floor Action," *CQ*, August 12, 1966, p. 1725.

86. "Senate Rejects $198-Million Cut in Poverty Authorization; Refuses to Reduce Funds for SST Development, Air Safety," *CQ*, October 13, 1967, p. 2101.

87. "Senate Votes $21.4 Billion for Labor, HEW Departments; Approves $2.2 Billion for Transportation Department," *CQ*, December 26, 1969, p. 2718; and "Senate Deletes Funds for SST; Approves $66.4 Billion for Defense, $17.7 Billion for HUD Department," *CQ*, December 11, 1970, p. 2978.

amendment passed. At the time of the latter vote, funds for the SST for the first half of fiscal 1971 had already been appropriated through a continuing resolution. The 1970 amendment deleted what remained of the $289.9 million budget request. Once that amendment passed, the SST program would be ended if the House agreed.

On March 18, 1971, the House did agree. On March 24, 1971, the Senate rejected the restoration of funds, and the government canceled its contracts for the SST. The House voted on May 12 to use the program's termination funds for continued prototype development, but on the 19th the Senate adopted the final Proxmire amendment on the SST, striking this $85.3 million from the Second Supplemental Appropriations Bill for fiscal 1971. The American SST was dead.

Two logistic models are estimated here, one an annual model for each Senate vote and the other a time-series model to test the effects of the program's performance and environmental threats on its support. Precise definitions of variables are found in figure 6-1. The 1966 and 1967 votes are pooled in the cross-sectional regressions, as are the 1970 and two 1971 votes.[88]

The statistical information used to examine the voting behavior of senators is found in tables 6A-2 through 6A-5. The regression results for subsamples that pass pooling tests are shown in table 6-7. The time-series regressions for larger samples that do not pool are shown in table 6-8.

PARTY AND IDEOLOGY. The party and to some extent the ideology variable can be an instrument for either differences in the constituencies of the legislative parties or party loyalty to the president. During the history of the SST the president, whether Democrat or Republican, always supported the program.

The difference between the coefficients on *PARTY* for 1966–67 and 1969 (table 6-7) provides some support for the hypothesis that when a Republican president supported the program Republican senators were more likely to do so too. Despite Nixon's firm support for the aircraft, the coefficient on *PARTY* (table 6-7) decreased significantly after the environmental concerns were raised in 1970. This suggests that once the SST conflicted with a

88. The validity of the 1966–67 pooling is confirmed at the 95 percent level by a likelihood ratio test with χ^2 (16) = 24.60, and for the 1970–71 pooling by χ^2 (16) = 0.78 for the two 1971 votes and χ^2 (16) = 6.88 for the addition of the 1970 vote. Adding the 1969 vote to the 1966 and 1967 votes results in χ^2 (16) = 27.20, which is significant at the 95 percent level.

Figure 6-1. *Definition of Variables, Senate Analysis*

VOTE: 1 if voted or paired in favor of supporting the SST; 0 if voted or paired against supporting the SST.

PARTY: 1 if Republican; 0 if Democrat.

ACA: Member's score on the Americans for Constitutional Action index. For the 1970–71 regression, *ACA* is replaced with *ACALE20, ACA2180,* and *ACAGE81*. These variables equal a senator's *ACA* score when it is no greater than 20, between 21 and 80 inclusive, and 81 and above; they are 0 otherwise.

SUBAPP: 1 if a member of the Independent Offices Subcommittee (fiscal 1967 and earlier) or the Transportation Subcommittee (fiscal 1968 and later) of the Appropriations Committee; 0 otherwise.

NSUBAPP: 1 if a member of the Appropriations Committee, but not one of the subcommittees above; 0 otherwise.

AUTH: 1 if member of the Committee on Aeronautical and Space Sciences, the authorization committee for the SST; 0 otherwise.

Distributive Variables

PRIME: 1 if represent a state with a prime, or potential prime, contractor; 0 otherwise

COMPET: per capita dollars flowing into the state for contracts with Boeing, Lockheed, General Electric, and Pratt and Whitney for phase II, the detailed design competition; 0 if not from a home state of the above contractors.

SPENT1: per capita dollars accruing to the state for phase III subcontracts, if the total is above zero and less than or equal to $0.2612 (this represents the lower quartile among the states receiving these subcontracts); 0 otherwise.

SPENT2: per capita subcontracting dollars, if greater than $0.2612 and no greater than $0.55811 (this represents the middle two quartiles); 0 otherwise.

SPENT3: per capita subcontracting dollars, if greater than $0.55811 and no greater than $14.262, the maximum (this is the top quartile); 0 otherwise.

SPENTZER: 1 if no subcontracting dollars flowed into the state; 0 otherwise.

FUTURE1: predicted per capita dollars of benefits that would accrue to a state from the SST program subsequent to *SPENT*, if they are over zero and no greater than $0.69858 (this represents the lowest quartile among the states that were "promised" these funds); 0 otherwise.

FUTURE2: predicted per capita dollars of benefits that would accrue subsequent to *SPENT*, provided they are greater than $0.69858 and do not exceed $42.13386 (this represents the middle two quartiles); 0 otherwise.

FUTURE3: predicted per capita dollars of benefits promised after *SPENT*, if they are above $42.13386 and no greater than $207.84, the maximum (this is the top quartile); 0 otherwise.

FUTURZER: 1 if it was predicted that the state stood to gain no production benefits from the SST after *SPENT*; 0 otherwise.

Longitudinal Variables

FLY: estimated flight date of the SST prototype. The year is broken into thirds: early, middle, and late. Mid-1966, the baseline, equals one.

ENVIR: 1 from the 1970 vote on; 0 otherwise.

Table 6-7. *Regression Results for the Annual Model, Senate*[a]

Variable	*1961*	*1966 and 1967*	*1969*	*1970, March 1971, and May 1971*
Constant	2.351*	2.470*	0.280	−0.692
	(2.65)	(4.03)	(0.373)	(−1.41)
PARTY	0.564	1.896*	3.013*	0.5286
	(0.688)	(3.32)	(3.41)	(1.41)
ACA	−0.052*	−0.035*	0.011	. . .
	(−3.49)	(−3.80)	(1.04)	
ACALE20	. . .	. . .	. . .	−0.101*
				(−2.21)
ACA2180	. . .	. . .	. . .	0.015*
				(1.74)
ACAGE81	. . .	. . .	. . .	0.011*
				(1.71)
SUBAPP	2.277*	0.981	0.354	0.206
	(2.14)	(1.55)	(0.368)	(0.445)
NSUBAPP	0.307	0.923	1.413	1.640*
	(0.284)	(1.37)	(1.31)	(3.09)
AUTH	−0.410	0.314	−0.492	0.890*
	(−0.414)	(0.590)	(−0.579)	(1.81)
PRIME	−1.147	−0.818	−0.059	1.247
	(−0.961)	(−0.553)	(−0.0347)	(1.03)
COMPET	0.0932	0.0814	0.156*	0.075
	(0.988)	(1.06)	(1.82)	(1.33)
SPENTZER	−2.418	−1.226	−2.045*	−0.565
	(−1.64)	(−1.57)	(−1.74)	(−0.856)
FUTURZER	3.159*	0.254	3.394*	0.478
	(1.76)	(0.303)	(2.24)	(0.675)
SPENT1	−88.42	−47.06	33.77	−42.64
	(−1.15)	(−1.10)	(0.515)	(−1.26)
SPENT2	0.822	−3.365*	−4.298*	1.877
	(0.334)	(−1.73)	(−1.83)	(1.08)
SPENT3	0.402	1.190*	0.198	1.248*
	(0.819)	(1.80)	(0.516)	(3.39)
FUTURE1	2.047	0.985	0.121	0.887
	(0.904)	(1.25)	(0.158)	(1.38)
FUTURE2	−0.126*	−0.109*	−0.085*	−0.203*
	(−2.21)	(−2.94)	(−1.85)	(−3.05)
FUTURE3	−0.010	−0.032*	−0.020*	−0.037*
	(−0.742)	(−2.28)	(−1.68)	(−3.63)
Number	82	171	96	291
χ^2	41.86	42.80	42.88	125.10
df for χ^2	15	15	15	17
log *L*	−35.88	−86.74	−38.18	−136.80

Sources: See figure 6A-1.

* In a one-tailed *t*-test, significantly not equal to zero at a significance level of 5 percent.

a. Numbers in parentheses are *t*-statistics.

Table 6-8. *Regression Results for the Time-Series Model, Senate*[a]

Variable	*All votes*	*Without 1961*	*1969 and 1970*[b]
	Regressions on all variables		
FLY	0.051*	0.022	. . .
	(2.79)	(0.573)	
ENVIR	−1.491*	−1.451*	−1.482*
	(−5.87)	(−4.65)	(−3.83)
	Regressions on only FLY and ENVIR		
FLY	0.043*	0.021	. . .
	(2.65)	(0.600)	
ENVIR	−1.234*	−1.122*	−1.021*
	(−5.44)	(−4.15)	(−3.38)

Sources: See figure 6A-1.
* In a one-tailed *t*-test, significantly not equal to zero at a significance level of 5 percent.
a. If *FLY* and *ENVIR* were orthogonal to the other variables, the results would be the same with and without the other variables. These results indicate little spurious correlation with the other variables, validating these poolings for this purpose. The numbers in parentheses are *t*-statistics.
b. See text for explanation.

salient national political issue, Republicans abandoned supporting their president.

A conservative senator might oppose government subsidies to override private business decisions, and yet might also support a program touted as raising American prestige and increasing the country's technological capabilities beyond those of other nations. The coefficient on *ACA* before the failure of Boeing's first design is significantly negative. The coefficient supports the view that conservatives were concerned about government involvement in commercial ventures.

Votes in later years show that conservative senators supported the program after a conservative president endorsed it. The *ACA* coefficient changed in sign in the 1969 vote and is significantly positive for the top two categories in the 1970–71 regression.[89]

COMMITTEE BLOCK VOTING. The committee variables permit a test of the hypothesis of block voting by the relevant committees.[90] The

89. In the 1970–71 regression, the *ACA* variable is replaced by three variables measuring *ACA* score, as described in figure 6-1. A likelihood ratio test indicates that this regression provides significantly more explanatory power, at the 99.5 percent level, than one with only *ACA*. The highest confidence level for which this is true in the other regressions is 50 percent.

90. Since senators try to obtain committee assignments that can benefit their constituents, it would be plausible for *SPENT* and *FUTURE* to be higher for members of these committees than for other senators. This is not always the case, and when it is the difference is never significant.

significantly positive coefficients on *AUTH* and *NSUBAPP* for the 1970–71 votes support the hypothesis that when the SST was under heaviest pressure, supporters were able to arrange mutual accommodation and logrolling in the oversight committees. While the impact of block voting in the full Appropriations Committee is largest in 1970, the joint hypothesis that this was a factor in the support for the SST from 1961 on is easily accepted. Whereas in 1966–67 and 1969 the percent of members of the Appropriations subcommittee supporting the SST is significantly greater than for senators who were not members of the relevant committees, in 1970–71 that was true for members of all three committee groupings (table 6A-5). Other evidence on the difference between committee members and other senators is somewhat less conclusive. Between the 1969 and 1970–71 votes, support from senators not on these committees fell significantly, while among committee members that was true only for the Appropriations subcommittee. However, in both regressions the subcommittee's support was insignificantly different from that of members of the full committee.

DISTRIBUTIVE BENEFITS. The distributive aspects of the SST program are measured by four variables: *PRIME*, *COMPET*, *SPENT*, and *FUTURE*. Benefits measured by these variables accrued to partially overlapping sets of states at different times and so permit tests of both the importance of distributive benefits and the relevance of their timing.

In 1961 the potential prime contractors consisted of the manufacturers of commercial airframes and engines: Boeing (Wash.), Lockheed (Calif.), North American–Rockwell (Calif.), Convair (Calif.), McDonnell (Mo.), Douglas (Calif.), General Electric (Ohio), Pratt and Whitney (Conn.), and Curtiss-Wright (N.J.). All nine companies received requests for proposals at the start of the design competition. By 1966 potential prime contractors had been narrowed to Boeing, Lockheed, General Electric, and Pratt and Whitney. From 1967 on, Boeing and GE were the prime contractors. The presumption is that the potential windfall that would accrue to a state with a prime contractor is so large that harboring a prime or even a potential prime contractor increases the probability of voting in favor of SST funding.

The *COMPET* variable measures per capita contract dollars from the phase II competition. This money was being spent during the 1966 vote. The phase III funds represented by *SPENT* were contracted from April 1967, the start of phase III, through September 1970 by GE and through June 1970 by Boeing. Boeing, because of its design problems, did not actually start its subcontracting until 1969.

The theory of retrospective voting holds that constituents give electoral credit to incumbents for recently received economic benefits, but give little credit for promises of future benefits. It is therefore expected that *COMPET* would influence voting for the SST most favorably in 1966 and 1967 and that the positive influence of *SPENT* would increase over time as these funds progressed from potential to actual spending. Moreover, the probability that a senator's reelection constituency contains program beneficiaries, and that the continuation of the program will therefore be a politically salient issue in the election, increases with per capita spending until it becomes large enough to make it virtually certain a senator will vote for a program. Separating the spending variable into three ranges (*SPENT1*, *SPENT2*, and *SPENT3*) is useful for detecting the inflection points in the voting function.

The states that obtain no contracts receive only the public benefits of the program. Hence, if distributive benefits were an important aspect of the SST program, senators representing states with no contracts should be more likely to oppose the program. The expected coefficient of the *SPENTZER* variable is therefore nonpositive.

FUTURE represents potential production benefits from the SST. If constituents behave as if dollars speak louder than promises, the importance of *FUTURE* is expected to be smaller than that of *SPENT* and perhaps even insignificant. Like the spending variable, future production promises can usefully be separated into three ranges (*FUTURE1*, *FUTURE2*, and *FUTURE3*) to detect inflection points in the voting function. States receiving no future benefits are represented by the *FUTURZER* dummy coefficient. Because their senators should be less likely to support the SST, the expected sign of this coefficient is nonpositive.

The significance of the positive coefficients on *COMPET* after 1969 but not before supports the hypothesis that senators try to protect their constituents' distributive benefits once they are being received, but less when they are prospective. Even though the coefficient is insignificant in 1961 and 1966–67, the joint hypothesis that it is positive from 1961 through 1971 is easily accepted. The pattern of coefficients on *PRIME* supports the same view, although all the coefficients are insignificant, perhaps because of the collinearity of *PRIME* and *COMPET*.

Within each party *SPENT* is significantly greater for supporters than opponents (table 6A-3, column 3). The significantly positive coefficient on *SPENT3* for the 1966–67 and 1970–71 votes, coupled with the insignificant or incorrectly signed coefficients on *SPENT1* and *SPENT2*, support the hypothesis that a threshold must be reached before the benefits become salient. The coefficients on *SPENTZER* from 1961 through 1969 are jointly significantly negative, suggesting that the nondistributive net benefits were perceived as negative.

The coefficients on *FUTURE* are always either insignificant or significant with the wrong sign, even though the magnitude of *FUTURE* is considerably larger than *SPENT*. This supports the hypothesis of retrospective voting and implies that legislators can safely ignore future distributive benefits. That is further shown in the 1970–71 votes, when opponents of the SST would have received, on average, significantly more in future benefits than supporters, but received significantly less in phase III benefits than supporters did (table 6A-3, columns 3 and 4). Among Democrats the mean of *FUTURE* distributional benefits is significantly higher for opponents than supporters (table 6A-3, column 4), whereas the difference for Republicans is in the same direction but is not significant.

LONGITUDINAL VARIABLES. Two time-series measures of project performance are the date of first flight and the degree of environmental concern. Movement of the first flight further into the future indicates that the program was not living up to expectations. *FLY* is a measure of but one program performance characteristic, but it is one that is inextricably linked to the performance of the entire project.

In May and August of 1970 the Joint Economic Committee heard considerable testimony on the SST's detrimental environmental effects. By 1970 environmental policy had become a salient national political issue, and major reforms in environmental policy were being adopted and implemented. Raising environmental questions about the SST made it more controversial and more visible to the general public. Consequently, by 1970 senators were more likely to believe that the environmental harm from the SST might outweigh its economic benefits and that supporting the SST might be politically costly.

Three time-series regressions were performed (table 6-8), over all the votes (column 1), the votes after 1961 (column 2), and only the 1969 and 1970 votes (column 3).[91] The justification for the second regression is that in 1961 the projected first-flight date for the SST must be construed as mere hope, since research into the technical feasibility of a civilian SST had

91. The only coefficients in the time series model that can be interpreted are for *FLY* and *ENVIR*. This is legitimate even though the pooling is invalid as long as *FLY* and *ENVIR* are independent of the other variables, and spurious correlation is rejected in the bottom half of table 6-8.

not yet even begun. The third regression is an attempt to pinpoint whether *FLY* and *ENVIR* finally turned the Senate against the SST.

The insignificant coefficient on *FLY* from 1966 on affirms the hypothesis that once the near-term distributional benefits were being put in place, support for the SST program did not flag as a result of its deterioration in performance. The significantly positive coefficient in the first regression can confidently be discounted because of the aforementioned ephemeral nature of *FLY* in 1961.

The significantly negative and large coefficient on *ENVIR* supports the hypothesis that the environmental protection issue was an important factor in killing the SST, a hypothesis also backed up by the constant term becoming negative in the 1970–71 estimation of the annual model (table 6-8). These findings show that the program was mortally wounded when opponents raised a previously overlooked negative externality.

Because *FLY* and *ENVIR* are perfectly collinear in 1969 and 1970, only one parameter can be estimated in the third regression. The coefficient is significantly negative and large, and in light of the previous result it seems proper to interpret it as the effect of *ENVIR*. This result is again consistent with the hypothesis that the SST's journey from firm support in 1969 to defeat in 1970 was fueled by concern over the aircraft's environmental effects.

Voting Behavior in the House

Glancing only at the record of votes in the House gives the impression that the SST was not nearly as contentious an issue in the House as in the Senate. A cursory perusal of the hearings before the House Committee on Appropriations quickly dispels this notion. The ensuing absence of votes in the House during the 1960s most likely reflects this chamber's more stringent rules and procedures; in the Senate, amendments and debating time are unrestricted, whereas in the House a fixed period is allowed for debate and the number and nature of amendments are determined in advance by the Rules Committee.

The analysis of House votes is based on a pooled logit regression. Even though the SST lost the March roll call by 13 votes and won the May vote by 4, the differences between the votes are so slight that the statistical propriety of pooling is not in doubt.[92] Since *FLY* and *ENVIR* did not change between March and May of 1971, only a cross-sectional model is estimated.

92. χ^2 (12) = 1.82.

Figure 6-2. *Selected Variable Definitions, House Analysis*

SUBAPP: 1 if member of Transportation Subcommittee of the Appropriations Committee; 0 otherwise.

NSUBAPP: 1 if member of the Appropriations Committee, but not the Transportation Subcommittee; 0 otherwise.

SUBAUTH: 1 if member of Advanced Research and Technology Subcommittee of the Committee on Science and Astronautics, the authorization subcommittee for the SST; 0 otherwise.

NSUBAUTH: 1 if member of the Committee on Science and Astronautics but not the Advanced Research and Technology Subcommittee; 0 otherwise.

COMPET: hundreds of thousands of dollars flowing into the district for detailed design competition contracts in phase II; 0 if not from a home district of the four contractors in the detailed design competition.

SPENT1: hundreds of thousands of dollars accruing to the district for phase III contracts if the total is above zero and less than or equal to $1.3400 (this represents the lower quartile of districts receiving these subcontracts); 0 otherwise.

SPENT2: hundreds of thousands of subcontracting dollars if greater than $1.3400 and no greater than $14.333 (this represents the middle two quartiles); 0 otherwise.

SPENT3: hundreds of thousands of subcontracting dollars if greater than $14.333 and no greater than $320.44, the maximum (this is the top quartile); 0 otherwise.

For the House analysis, the variables *VOTE*, *PARTY*, *ACA*, and *SPENTZER* are defined as they were in the Senate model.[93] The *ACA* variable is not decomposed into different ranges, because the improvement in the regression is not judged significant enough to warrant it. Precise definitions of all other variables are found in figure 6-2.

The statistical information used to test hypotheses about voting behavior in the House is found in appendix tables 6A-6 through 6A-9. Table 6-9 reports the results of the regression.

PARTY AND IDEOLOGY. The significantly negative regression coefficient on *PARTY* reinforces the finding in the Senate that by 1971 the SST could not ride on the Republican president's coattails. The highly significant *ACA* coefficient again lends credence to the hypothesis that in 1971 support for the SST was in accord with conservative, nationalistic precepts. Evaluating the effect of *ACA* at the party means mitigates the effect of *PARTY* to such an extent that the average Republican, independent of all other characteristics, was less likely to oppose the SST than the average Democrat.

COMMITTEE SUPPORT. The coefficients on *SUBAPP*, *NSUBAPP*,

93. *PRIME* is not in this model because its extreme collinearity with *COMPET* prevented the iterative estimation procedure from converging. Data necessary to construct the *FUTURE* variables by congressional district are unavailable.

Table 6-9. *Regression Results for the Annual Model, House*

Variable	*Coefficient*[a]
Constant	−2.446*
	(−8.19)
PARTY	−1.824*
	(−7.12)
ACA	0.046*
	(11.26)
SUBAPP	1.023*
	(1.70)
NSUBAPP	0.077
	(0.299)
SUBAUTH	−0.115
	(−0.178)
NSUBAUTH	0.636
	(1.64)
COMPET	0.005*
	(2.18)
SPENTZER	0.731*
	(2.63)
SPENT1	0.700
	(1.40)
SPENT2	0.137*
	(2.93)
SPENT3	0.026*
	(3.46)
Number	849
χ^2	160.60
df for χ^2	11
log *L*	−478.13

Sources: See figure 6A-1.
* In a one-tailed *t*-test, significantly not equal to zero at a significance level of 5 percent.
a. The numbers in parentheses are *t*-statistics.

SUBAUTH, and *NSUBAUTH* provide little support for block voting. Whereas all committees supported the program by a greater margin than members who were on no committee, the only committee group for which support was statistically significantly greater was authorization committee members who were not members of the relevant authorization subcommittee (table 6A-9). This difference is not significantly different from support in the subcommittees, or any other committee group.

DISTRIBUTIVE BENEFITS. As in the Senate, *SPENT* is either smaller for committee members or insignificantly greater. The coefficients on *SPENT* reinforce the Senate finding that in 1971 legislators voted to protect their constituents' distributional benefits if they exceeded a minimum threshold. Further support for this hypothesis is that, as in the Senate, supporters received significantly larger phase III benefits than opponents (table 6A-7, column 3). The mean *SPENT* for Democratic opponents is significantly smaller than for supporters (table 6A-7, column 3), the reverse of the Senate case, and supports the hypothesis that representatives believe more strongly than senators that their constituents will hold them accountable for the loss of their distributional benefits.

The party disparity in the mean of *SPENT* is insignificant in the House (table 6A-8, column 3), whereas it is significant in the Senate (table 6A-4, column 3). This difference supports the hypothesis that for a project to be politically viable in the House, benefits must be spread more equally among constituencies than in the Senate, and is consistent with representatives' having greater need for bringing something home to the constituents.

The significantly positive coefficient on *SPENTZER*, considering that the average size of *SPENT1* is near unity, further supports the hypothesis of a threshold effect of distributional support because the coefficients on these variables are essentially identical. The product of the sample mean and the coefficient is much larger for *SPENT2* and *SPENT3*, showing that members in those two categories are more likely to support the program than members in the first two categories.[94] These results suggest that the coefficients on *SPENTZER* and *SPENT1* reflect specification error.

Assessment of the Program

When Halaby asked Congress for the first SST appropriation, he emphasized the aircraft's importance for American leadership in commer-

94. The strong support for the SST among Republicans suggests that permitting the effects of the independent variables to differ by party would improve the explanatory power of the regressions. Allowing the *ACA* and financial variables to vary by party did not significantly improve the House regressions. In the Senate, statistical improvements exist, though the problems with using *ACA* scores as explanatory variables argue against making much of these regression results. In the 1970 and 1971 regressions, Senate Republicans appear more sensitive to current distributive benefits than Democrats. This result has two possible interpretations: first, that on average Republican senators are more likely to have within their constituency base those interests that benefit from expenditures on the SST program; and second, that the Republican senators were less cohesive as a group in their voting on the SST than were Democrats, so that additional factors were more likely to influence their votes.

cial aviation and the maintenance of American prestige. To meet these goals, the craft would have to be "safe, reliable, and economically competitive."[95]

The FAA's 1963 report on the SST development program included five "decision points" at which the program could be reviewed and decisions rendered on whether the appropriate action would be the continuation, redirection, or termination of the SST. These decision points were the following: (1) after phase I, if technical or economic characteristics were unacceptable; (2) after phase II, for the same reasons; (3) if manufacturers were unwilling to contribute 25 percent of development costs; (4) if a significant number of reservation orders had not been placed within six months of the end of phase II; and (5) if after 50 hours of flight tests the prototypes could not demonstrate the required characteristics.[96] The fifth decision point was never reached, although not even the paper plane could meet the required characteristics. It can be persuasively argued that the fourth was passed satisfactorily.

None of the phase I airframe proposals met the desired standards, and the decision on how to proceed was postponed, as decision theory would recommend. A blue-ribbon panel chose two of the three designs to progress to the detailed design competition. The proposals had not changed in the intervening month, but it was nonetheless decided that the best course of action was to invest in further research along two competing designs for both aircraft and engines in the hope that one would produce a feasible SST. This decision can also be defended on decision-theoretic grounds.

The second decision point was reached in 1968, when Boeing's winning design was found to be unsuitable. Allowing more research at that point, in an attempt to develop an aircraft that could meet standards, was a defensible decision. As this research progressed and prototype construction began, the prospects for the SST did not improve. The FAA did not rethink its support of the plane or seek further extension of the research phase. That was not a defensible decision.

The first inkling that the third decision point would come into play was in 1963, when manufacturers balked at the idea of a 25 percent cost share.[97] The phase III contracts, negotiated in 1967, provided for a 10 percent share before cost overruns.[98] This split was recommended by the Black-Osborne

95. *Independent Offices Appropriations for 1962*, House Hearings, pp. 64, 65.
96. *Independent Offices Appropriations, 1964*, Senate Hearings, pt. 2, pp. 2013–14.
97. "On Supersonic Transport," *CQ*, November 8, 1963, p. 1909.
98. *Department of Transportation Appropriations for 1968*, House Hearings, p. 345.

report,[99] and since the report was critical of program management, one can reasonably conclude that 25 percent was too optimistic. The violation of the original principle of private production financing, perhaps even with direct government control, is a powerful indicator that management was willing to compromise on the commercial nature of the project, although it continued to claim that the SST was a joint venture and not a subsidy.[100]

Once the project began running into unanticipated technical delays, extra pains were taken to stress the employment aspects of the SST program. To help make the point more forcefully, a map showing the total subcontracting dollars that would flow into each state was presented to Congress in 1969.[101] In addition, Congress was informed that "subcontractor teams have been established throughout the country,"[102] so as to drive home the argument that many voters received distributional benefits from the project.

Once the program encountered its fiercest political trouble, Secretary of Transportation John A. Volpe stressed that, having spent almost $900 million, "we have gone too far . . . to let this all go down the drain with no tangible results," and Senator Jackson wondered whether it would be "responsible" to quit after we had sunk so much into the effort.[103] Proceeding with a project for the sake of purportedly salvaging sunk costs is never a reasonable move.

Within the span of ten years the justification for the SST program had progressed from one of developing the fastest, safest, most economical supersonic transport to the economically groundless desire of not wasting sunk costs. The framework so carefully laid out to ensure sound management proved unable to check the excesses its creators sought to guard against. The executive branch was too wedded to the program, and Congress to its distributive economic benefits, for the SST to be canceled because of its dwindling prospects as a feasible civilian transport.

In the summer of 1970 Senator Proxmire created a highly visible, public controversy in hearings of the Joint Economic Committee over the SST's

99. "Around the Capitol," *CQ*, March 6, 1964, p. 468.

100. *Department of Transportation and Related Agencies Appropriations for 1970*, House Hearings, pt. 3, p. 290.

101. *Department of Transportation and Related Agencies Appropriations for 1970*, House Hearings, pt. 3, p. 93.

102. *Department of Transportation Appropriations for Fiscal Year 1970*, Senate Hearings, p. 583.

103. "In Committee," *CQ*, March 5, 1971, p. 517; and "In Committee," *CQ*, March 19, 1971, p. 606.

detrimental environment impact. Unlike the sonic boom issue that had arisen a decade earlier, the issue of ozone depletion arose at the peak of the environmental movement's political influence. Soon after the issue was raised, the tide turned against the SST. Even though the severity of the SST's environmental effects remained controversial, the environmental problems cost the program more votes than the distributional and nationalistic benefits could produce, and by early 1971 the program was dead.

Table 6A-1. *Data Used in Cost-Benefit Analyses, 1961–71*[a]
Current dollars unless otherwise noted

Year	*Development costs (billions of dollars)*	*Production costs (billions of dollars)*	*Projected fleet through 1990*	*Projected selling price per aircraft (millions of dollars)*
1961	0.5[d]–0.7[e]	2.00–4.18[f]	145–220[f]	12.5–20[e]
1962	0.5–0.6[i]	. . .	. . .	10–30[j]
1963	0.70–0.94[k]	. . .	185–325[l]	12–25[m]
1964	. . .	. . .	200[p]	30[q]
1965	1.2–1.6[r]	. . .	400–800[s]	23–32[t]
1966	. . .	10–20[v]	. . .	. . .
1967	1.278[y]–1.455[z]	. . .	279–479[a′]	40[b′]
1968	. . .	14.9[e′]	. . .	. . .
1969	1.339[h′]–1.515[i′]	15.0[j′]	500[k′]	37–50[l′]
1970	1.7–1.8[o′]	. . .	. . .	37[p′]
1971	. . .	. . .	. . .	. . .

a. Absence of an entry means no new value was estimated that year.
b. Decisions made with the consideration that spending occurs in fiscal years, whereas scheduling was in calendar years.
c. Congressional Quarterly, *Congress and the Nation*, vol. 3: *1969–1972* (1973), p. 168.
d. Federal Aviation Agency, *Report on the Task Force on National Aviation Goals: Project Horizon* (1961), p. 82; and *Contemporary and Future Aeronautical Research*, Hearings before the House Committee on Science and Astronautics, 87 Cong. 1 sess. (GPO, 1961), p. 20.
e. *Contemporary and Future Aeronautical Research*, House Hearings, p. 14.
f. *Contemporary and Future Aeronautical Research*, House Hearings, p. 102.
g. *Contemporary and Future Aeronautical Research*, House Hearings, p. 96.
h. *Contemporary and Future Aeronautical Research*, House Hearings, p. 21.
i. *Independent Offices Appropriations for 1963*, Hearings before a Subcommittee of the House Committee on Appropriations, 87 Cong. 2 sess. (GPO, 1962), p. 961.
j. *Independent Offices Appropriations, 1963*, Hearings before the Senate Committee on Appropriations, 87 Cong. 2 sess. (GPO, 1962), p. 528.
k. *Federal Research and Development Programs*, Hearings before the House Select Committee on Government Research, 88 Cong. 1 sess. (GPO, 1963), pt. 1, p. 147.
l. *Independent Offices Appropriations, 1964*, Hearings before the Senate Committee on Appropriations, 88 Cong. 1 sess. (GPO, 1963), pt. 2, p. 1992.
m. *United States Commercial Supersonic Aircraft Development Program*, Hearings before the Aviation Subcommittee of the Senate Committee on Commerce, 88 Cong. 1 sess. (GPO, 1963), p. 214.
n. *Independent Offices Appropriations, 1964*, Senate Hearings, pt. 2, p. 2034.
o. *Independent Offices Appropriations for 1964*, Hearings before a Subcommittee of the House Committee on Appropriations, 88 Cong. 1 sess. (GPO, 1963), pt. 2, p. 88.
p. *Independent Offices Appropriations for 1965*, Hearings before a Subcommittee of the House Committee on Appropriations, 88 Cong. 2 sess. (GPO, 1964), pt. 1, p. 1174.
q. *Independent Offices Appropriations for 1965*, House Hearings, pt. 1, p. 1174.
r. *Continuing Appropriations, 1966*, Hearings before the Senate Committee on Appropriations, 89 Cong. 1 sess. (GPO, 1965), p. 21.
s. *Independent Offices Appropriations for 1966*, Hearings before a Subcommittee of the House Committee on Appropriations, 89 Cong. 1 sess. (GPO, 1965), pt. 1, p. 14.
t. 1964 dollars. *Continuing Appropriations, 1966*, Senate Hearings, p. 21,
u. *Panel on Science and Technology, Sixth Meeting*, Proceedings before the House Committee on Science and Astronautics, 89 Cong. 1 sess. (GPO, 1965), p. 45.

Table 6A-1 *(continued)*

Projected date of prototype's first flight	*First flight date used in the analyses*[b]	*Predicted date SST enters commercial service*	*Spending by government (millions of dollars)*[c]
late 1964–early 1968[g]	1966	1970[h]	. . .
. . .	. . .	. . .	11
late 1968[n]	1969	1968[o]	20
. . .	. . .	. . .	60
. . .	. . .	1972[u]	. . .
late 1969[w]	1970	1974[x]	140
late 1970[c']	1971	1975[d']	280
. . .	. . .	1976[f']	112.375[g']
late 1972[m']	1973	1978[n']	. . .
early 1973[q']	1973	. . .	8.5
. . .	. . .	. . .	210.2

v. *Independent Offices Appropriations, 1967*, Senate Hearings, pt. 1, p. 349.
w. *Independent Offices Appropriations for 1967*, Hearings before a Subcommittee of the House Committee on Appropriations, 89 Cong. 2 sess. (GPO, 1966), pt. 1, pp. 1387–88.
x. *Independent Offices Appropriations for 1967*, House Hearings, pt. 1, p. 1387.
y. *Department of Transportation Appropriations for 1968*, Hearings before a Subcommittee of the House Committee on Appropriations, 90 Cong. 1 sess. (GPO, 1967), p. 407.
z. *Economic Analysis and Efficiency of Government*, Report of the Subcommittee on Economy and Government of the Joint Economic Committee, 91 Cong. 1 sess. (GPO, 1969), pt. 3, p. 683.
a'. *Department of Transportation Appropriations for 1968*, House Hearings, pp. 943, 299.
b'. *Department of Transportation Appropriations for 1968*, House Hearings, p. 294.
c'. *Department of Transportation Appropriations for 1968*, House Hearings, p. 397.
d'. With overlap of phases III–V. *Department of Transportation Appropriations for 1968*, House Hearings, p. 287.
e'. *Department of Transportation and Related Agencies Appropriations for 1970*, Hearings before a Subcommittee of the House Committee on Appropriations, 91 Cong. 1 sess. (GPO, 1969), pt. 3, p. 222.
f'. "Around the Capitol," *CQ*, November 1, 1968, p. 3067.
g'. $142.375 million was appropriated, but $30 million that remained unspent because of Boeing's delays was rescinded in fiscal 1969.
h'. *Department of Transportation Appropriations for Fiscal Year 1970*, Senate Hearings, p. 617.
i'. *Economic Analysis and Efficiency of Government*, Joint Economic Committee Report, 1969, p. 683.
j'. *Department of Transportation and Related Agencies Appropriations for 1970*, House Hearings, pt. 3, p. 220.
k'. *Department of Transportation and Related Agencies Appropriations for 1970*, House Hearings, pt. 3, p. 86.
l'. 1967 dollars. *Economic Analysis and Efficiency of Government*, Joint Economic Committee Hearings, 1969, pt. 3, p. 670 (price decrease due to lower manufacturing cost of fixed-wing aircraft).
m'. *Department of Transportation Appropriations for Fiscal Year 1970*, Senate Hearings, p. 602.
n'. No overlap. *Department of Transportation and Related Agencies Appropriations for 1970*, House Hearings, pt. 3, p. 602.
o'. *Federal Transportation Expenditures*, Hearings before the Joint Economic Committee, 91 Cong. 2 sess. (GPO, 1970), p. 18.
p'. 1967 dollars. *Department of Transportation and Related Agencies Appropriations for 1971*, House Hearings, pt. 3, p. 588.
q'. *Department of Transportation and Related Agencies Appropriations for 1971*, House Hearings, pt. 3, p. 507.

Figure 6A-1. *Regression Data*

VOTE and *PARTY*: *Congressional Quarterly Weekly Report,* August 4, 1961, p. 1373, Senate vote no. 109; August 12, 1966, p. 1758, Senate vote no. 138; October 13, 1967, p. 2101, Senate vote no. 215; December 26, 1969, p. 2718, Senate vote no. 224; December 11, 1970, p. 2978, Senate vote no. 375; March 26, 1971, p. 727, Senate vote no. 23; May 28, 1971, p. 1195, Senate vote no. 54; March 26, 1971, pp. 728–29, House vote no. 15; May 25, 1971, pp. 1144–45, House vote no. 52.

ACA: Voting Scores for Members of the United States Congress, 1945–1978 (ICPSR 7645). The data used in this study were made available by the Inter-university Consortium for Political and Social Research. The data for voting scores for members of Congress were originally collected by Congressional Quarterly, Inc., and published in *Congressional Quarterly Almanacs*. Neither the collector of the original data nor the consortium bears any responsibility for the analyses or interpretations presented here.

AUTH, SUBAUTH, APP, and *SUBAPP:* Commerce Clearinghouse, *Congressional Index* (Chicago, various years).

PRIME: Independent Offices Appropriations for 1965, Hearings before the House Committee on Appropriations, 88 Cong. 2 sess. (Government Printing Office, 1964), pt. 1, p. 1166; and Federal Aviation Administration, "United States Supersonic Transport Milestones," *News* (April 1980), pp. 4, 5.

COMPET: Continuing Appropriations, 1966, Hearings before the Senate Committee on Appropriations, 89 Cong. 1 sess. (GPO, 1965), p. 303; and *Independent Offices Appropriations, 1967*, Hearings before the Senate Committee on Appropriations, 89 Cong. 2 sess. (GPO, 1966), pt. 1, p. 337.

SPENT: General Accounting Office, Project Office Data, *National Impact of SST Prototype Program*. Location: Washington National Records Center, Reference branch, Suitland, MD, CC. No. 72A-6174, Box No. 25 of 151, File No. 7.

FUTURE: Department of Transportation and Related Agencies Appropriations for 1970, Hearings before the House Committee on Appropriations, 91 Cong. 1 sess. (GPO, 1969), pt. 3, p. 93; and *SPENT*.

FLY: Contemporary and Future Aeronautical Research, Hearings before the House Committee on Science and Astronautics, 87 Cong. 1 sess. (GPO, 1961), p. 96; *Independent Offices Appropriations for 1967*, Hearings before the House Committee on Appropriations, 89 Cong. 1 sess. (GPO, 1966), pt. 1, pp. 1387–88; *Department of Transportation Appropriations for 1968*, Hearings before the House Committee on Appropriations, 90 Cong. 1 sess. (GPO, 1967), p. 287; *Department of Transportation Appropriations for Fiscal Year 1968*, Hearings before the Senate Committee on Appropriations, 90 Cong. 1 sess. (GPO, 1967), p. 397; *Department of Transportation Appropriations for Fiscal Year 1970*, Hearings before the Senate Committee on Appropriations, 91 Cong. 1 sess. (GPO, 1969), p. 602; and *Department of Transportation and Related Agencies Appropriations for 1971*, Hearings before the House Committee on Appropriations, 91 Cong. 2 sess. (GPO, 1970), pt. 3, p. 507.

ENVIR: Economic Analysis and the Efficiency of Government, Hearings before the Joint Economic Committee, 91 Cong. 2 sess. (GPO, 1970); and *Federal Transportation Expenditures*, Hearings before the Subcommittee on Economy in Government of the Joint Economic Committee, 91 Cong. 2 sess. (GPO, 1970).

Table 6A-2. *Means of Variables in Senate Regressions*[a]

Variable	*1961*	*1966 and 1967*	*1969*	*1970, March 1971, and May 1971*
PARTY	0.35	0.36	0.44	0.44
	(0.48)	(0.48)	(0.50)	(0.50)
ACA	39.1	41.3	41.9	43.9
	(34.9)	(32.8)	(31.9)	(31.5)
SUBAPP	0.15	0.15	0.14	0.12
	(0.36)	(0.35)	(0.34)	(0.32)
NSUBAPP	0.10	0.11	0.11	0.14
	(0.30)	(0.31)	(0.32)	(0.34)
AUTH	0.15	0.18	0.16	0.12
	(0.36)	(0.38)	(0.37)	(0.33)
PRIME	0.13	0.07	0.04	0.04
	(0.34)	(0.26)	(0.20)	(0.20)
COMPET	1.15	1.12	1.11	1.10
	(4.72)	(4.42)	(4.43)	(4.40)
SPENTZER	0.18	0.21	0.21	0.20
	(0.39)	(0.41)	(0.41)	(0.40)
FUTURZER	0.11	0.13	0.13	0.12
	(0.31)	(0.34)	(0.33)	(0.33)
SPENT	0.85	0.78	0.95	0.95
	(2.12)	(2.16)	(2.42)	(2.40)
FUTURE	24.83	22.75	23.53	24.00
	(44.91)	(42.98)	(43.68)	(43.39)
Number	82	171	96	291

a. The numbers in parentheses are standard deviations.

Table 6A-3. *Means of Selected Variables, by Party and Stance, Senate*[a]

Votes	*PARTY*	*ACA*	*SPENT*	*FUTURE*	*SPENTZER*	*FUTURZER*	*Number*
1961							
Pro							
Total	0.21 (0.42)	22.8 (24.7)	0.62 (1.49)	29.62 (53.12)	0.19 (0.40)	0.17 (0.38)	42
Republican	. . .	44.2 (19.1)	0.77 (1.66)	50.76 (65.78)	0.11 (0.33)	0.11 (0.33)	9
Democrat	. . .	17.0 (22.9)	0.58 (1.47)	23.85 (48.72)	0.21 (0.42)	0.18 (0.39)	33
Con							
Total	0.50 (0.51)	56.3 (36.1)	1.09 (2.62)	19.81 (34.22)	0.18 (0.38)	0.05 (0.22)	40
Republican	. . .	75.7 (20.5)	1.86 (3.53)	23.47 (41.42)	0.10 (0.31)	0.00 (0.00)	20
Democrat	. . .	36.9 (38.3)	0.32 (0.54)	16.15 (25.69)	0.25 (0.44)	0.10 (0.31)	20
1966 and 1967							
Pro							
Total	0.39 (0.49)	40.2 (33.2)	0.93 (2.56)	21.12 (46.29)	0.22 (0.41)	0.14 (0.35)	115
Republican	. . .	70.2 (23.56)	1.68 (3.80)	31.38 (60.55)	0.18 (0.39)	0.09 (0.29)	45
Democrat	. . .	20.9 (22.3)	0.45 (1.03)	14.51 (33.01)	0.24 (0.43)	0.17 (0.38)	70
Con							
Total	0.29 (0.46)	43.5 (32.2)	0.47 (0.76)	26.10 (35.37)	0.20 (0.40)	0.11 (0.31)	56
Republican	. . .	67.9 (26.7)	0.43 (0.60)	14.06 (16.57)	0.19 (0.40)	0.13 (0.34)	16
Democrat	. . .	33.8 (29.0)	0.48 (0.82)	30.92 (39.69)	0.20 (0.41)	0.10 (0.30)	40

1969							
Pro							
Total	0.55 (0.51)	48.9 (31.5)	1.00 (2.77)	17.7 (38.98)	0.23 (0.42)	0.17 (0.38)	66
Republican	...	58.3 (27.3)	1.68 (3.60)	31.38 (42.87)	0.18 (0.38)	0.09 (0.32)	36
Democrat	...	37.6 (32.8)	0.19 (0.57)	9.40 (32.52)	0.30 (0.47)	0.23 (0.43)	30
Con							
Total	0.20 (0.41)	26.5 (27.5)	0.84 (1.38)	36.4 (50.95)	0.17 (0.38)	0.03 (0.18)	30
Republican	...	28.5 (24.7)	0.48 (0.35)	31.41 (28.78)	0.17 (0.41)	0.00 (0.00)	6
Democrat	...	26.0 (28.7)	0.92 (1.53)	37.62 (55.5)	0.17 (0.38)	0.04 (0.20)	24
1970, March 1971, May 1971							
Pro							
Total	0.56 (0.50)	57.1 (29.4)	1.39 (3.36)	15.89 (35.36)	0.24 (0.43)	0.17 (0.37)	127
Republican	...	69.4 (24.2)	2.41 (4.21)	24.12 (40.61)	0.15 (0.36)	0.11 (0.32)	71
Democrat	...	41.5 (28.1)	0.10 (0.40)	5.45 (23.82)	0.34 (0.48)	0.23 (0.43)	56
Con							
Total	0.34 (0.48)	33.7 (29.3)	0.60 (1.11)	30.27 (47.88)	0.16 (0.37)	0.09 (0.29)	164
Republican	...	51.9 (27.6)	0.52 (0.72)	31.16 (39.04)	0.13 (0.33)	0.07 (0.26)	56
Democrat	...	24.2 (25.5)	0.64 (1.27)	29.80 (52.04)	0.19 (0.39)	0.10 (0.30)	108

a. The numbers in parentheses are standard deviations.

Table 6A-4. *Means of Selected Variables, by Party, Senate*[a]

Votes	*ACA*	*COMPET*	*SPENT*	*FUTURE*	*SPENTZER*	*FUTURZER*	*Percent pro SST*	*Number*
1961								
Republican	65.9	0.86	1.53	31.94	0.10	0.03	31.0	29
	(24.7)	(3.94)	(3.09)	(50.65)	(0.31)	(0.19)	(47.1)	
Democrat	24.5	1.31	0.48	20.95	0.23	0.15	62.3	53
	(30.9)	(5.13)	(1.21)	(41.43)	(0.42)	(0.36)	(48.9)	
1966 and 1967								
Republican	69.6	0.25	1.36	26.84	0.18	0.10	73.8	61
	(24.2)	(0.96)	(3.31)	(53.07)	(0.39)	(0.30)	(44.4)	
Democrat	25.6	1.60	0.46	20.48	0.23	0.15	63.6	110
	(25.6)	(5.41)	(0.96)	(36.28)	(0.42)	(0.35)	(48.3)	
1969								
Republican	54.0	0.23	1.51	25.58	0.17	0.10	85.7	42
	(28.7)	(1.08)	(3.36)	(40.93)	(0.38)	(0.30)	(35.4)	
Democrat	32.4	1.79	0.51	21.95	0.24	0.15	55.6	54
	(31.3)	(5.77)	(1.15)	(46.02)	(0.43)	(0.36)	(50.2)	
1970, March 1971, May 1971								
Republican	61.7	0.70	1.58	27.23	0.14	0.09	55.9	127
	(27.1)	(2.86)	(3.31)	(39.92)	(0.35)	(0.29)	(49.8)	
Democrat	30.1	1.48	0.45	21.49	0.24	0.15	34.1	164
	(27.6)	(5.27)	(1.09)	(45.86)	(0.43)	(0.35)	(47.6)	

a. The numbers in parentheses are standard deviations.

Table 6A-5. *Committee Membership Variable Means, Senate*[a]

Committee membership	*1961*	*1966 and 1967*	*1969*	*1970, March 1971, and May 1971*
Authorization[b]				
Number	12	30	15	36
Percent pro SST	50.0	73.3	73.3	66.7
	(52.2)	(45.0)	(0.46)	(47.8)
SPENT	0.36	0.42	1.40	1.14
	(0.84)	(0.85)	(2.10)	(2.05)
FUTURE	24.84	20.75	31.16	29.15
	(50.16)	(39.64)	(51.57)	(52.37)
Appropriations subcommittee				
Number	12	25	13	34
Percent pro SST	67.0	80.0	84.6	52.9
	(49.2)	(40.8)	(37.6)	(50.7)
SPENT	1.13	0.15	1.18	0.57
	(2.23)	(0.25)	(3.93)	(2.43)
FUTURE	24.71	8.11	3.54	10.59
	(58.73)	(18.52)	(13.02)	(19.16)
Appropriations, but not subcommittee				
Number	8	18	11	40
Percent pro SST	50.0	77.8	81.8	65.0
	(53.5)	(42.8)	(40.5)	(48.3)
SPENT	0.12	0.64	0.08	0.14
	(0.20)	(1.65)	(0.10)	(0.30)
FUTURE	18.62	34.63	9.90	6.41
	(35.03)	(66.34)	(19.84)	(13.67)
None of above				
Number	55	107	62	192
Percent pro SST	47.3	61.7	61.3	34.4
	(50.4)	(48.8)	(49.1)	(47.6)
SPENT	0.93	1.00	0.88	1.09
	(2.33)	(2.58)	(2.21)	(2.60)
FUTURE	23.76	23.09	26.63	27.89
	(41.10)	(41.31)	(46.43)	(46.50)

a. The numbers in parentheses are standard deviations.
b. Committee on Aeronautical and Space Sciences.

Table 6A-6. *Means of Variables in House Regression*

Variable	*Mean*[a]	*Variable*	*Mean*[a]
PARTY	0.41 (0.49)	*NSUBAUTH*	0.04 (0.21)
ACA	50.5 (31.9)	*COMPET*	6.60 (52.58)
SUBAPP	0.02 (0.13)	*SPENTZER*	0.65 (0.48)
NSUBAPP	0.11 (0.31)	*SPENT*	5.99 (23.91)
SUBAUTH	0.02 (0.14)	Number	849

a. The numbers in parentheses are standard deviations.

Table 6A-7. *Means of Selected Variables, by Party and Stance, House*[a]

Votes	*PARTY*	*ACA*	*SPENT*	*SPENTZER*	*Number*
Pro	0.45 (0.50)	62.6 (28.4)	8.57 (32.31)	0.68 (0.47)	420
Republican	. . .	81.6 (16.1)	9.62 (35.31)	0.59 (0.49)	187
Democrat	. . .	47.2 (26.8)	7.73 (29.73)	0.76 (0.43)	233
Con	0.37 (0.38)	38.7 (30.8)	3.46 (9.88)	0.62 (0.48)	429
Republican	. . .	69.7 (22.2)	4.36 (12.38)	0.65 (0.48)	160
Democrat	. . .	20.2 (17.5)	2.93 (8.02)	0.61 (0.49)	269

a. The numbers in parentheses are standard deviations.

Table 6A-8. *Means of Selected Variables, by Party, House*[a]

Votes	*ACA*	*COMPET*	*SPENT*	*SPENTZER*	*Percent pro SST*	*Number*
Republican	76.1 (20.0)	10.27 (59.94)	7.19 (27.34)	0.62 (0.49)	53.9 (49.9)	347
Democrat	32.8 (26.0)	4.06 (46.72)	5.16 (21.20)	0.68 (0.47)	46.4 (49.9)	502

a. The numbers in parentheses are standard deviations.

Table 6A-9. *Variable Means by Committee Membership, House*[a]

Committee membership	*Mean*[a]
Authorization subcommittee	
Percent pro SST	50.0
	(51.5)
SPENT	13.87
	(26.29)
Number	18
Authorization, not subcommittee	
Percent pro SST	63.2
	(48.9)
SPENT	5.26
	(14.28)
Number	38
Appropriations subcommittee	
Percent pro SST	57.1
	(51.4)
SPENT	5.31
	(13.14)
Number	14
Appropriations, not subcommittee	
Percent pro SST	53.3
	(50.2)
SPENT	9.51
	(47.40)
Number	92
None of above	
Percent pro SST	48.0
	(50.0)
SPENT	5.37
	(19.33)
Number	687

a. The numbers in parentheses are standard deviations.

Chapter 7

The Applications Technology Satellite Program

LINDA R. COHEN & ROGER G. NOLL

BETWEEN 1963 and 1973 the National Aeronautics and Space Administration conducted the Applications Technology Satellite (ATS) program. This program pioneered advances in satellite technology, laying the groundwork for applications in television transmission, satellite tracking and data relay, communications with ships and aircraft, and direct broadcast satellites. Despite widespread acclaim, the program was canceled in 1973.

The history of the ATS program differs from that of others discussed in this book. The program was canceled despite evident social benefits, rather than clung to in the face of deteriorating economic performance. We argue that the ATS was terminated because of changes in the political calculation of its benefits. The distributive aspects were politically beneficial at the outset but were detrimental by its end. In the beginning, too, the social benefits were immediate, but they became long term and hence politically less relevant. Finally, the political saliency of space research declined. The ATS was a victim of its own success. One cause of the changes in the political climate was the growth of a competitive commercial satellite industry that was in part the result of the program's accomplishments.

History of the Program

The ATS program grew out of earlier federal work on communications satellites. During the 1950s the military and the National Advisory Commission on Aeronautics (NACA), NASA's predecessor, experimented with several satellite concepts.[1] The Soviet launch of Sputnik caused a fervor for space programs that enhanced political support for satellite research. The most important early experimental satellites were the SYNCOM series, built by Hughes Aircraft Company under a 1961 NASA contract and launched in 1963 and 1964.[2] These satellites were the first to be successfully launched into geosynchronous orbit, 22,300 miles above the equator.[3] At that altitude, satellites orbit the earth at the same speed that the earth itself rotates. This has two important implications for satellite system costs. First, only one satellite is needed to cover a given geographic region at all times; otherwise, many satellites would be needed to provide a continuous communications link with any particular location on earth. Second, geostationary satellites allow earth station antennas to be fixed, eliminating the need for expensive tracking equipment and movable antennas for following a satellite as it moves relative to ground stations.

With hindsight the decision to support the Hughes proposal was excellent. At the time, however, it was part of a comprehensive research strategy rather than supported out of recognition that it would provide a commercial breakthrough. The concept of a spin-stabilized spacecraft had been worked out in 1959–60 by Hughes, which proposed the project to, and was turned down by, the Department of Defense.[4] The company then proposed the project to NASA, which was expanding its research on communications satellites in the wake of President Kennedy's communications satellite policy, announced on July 24, 1961, and NASA's substantially increased budget to support the research.[5]

Kennedy's statement is a landmark in U.S. communications satellite policy. The potential of communications satellites inspired a policy debate

1. See Leonard Jaffe, *Communications in Space* (Holt, Rinehart and Winston, 1966); and Delbert D. Smith, *Communication via Satellite: A Vision in Retrospect* (Boston: A. W. Sijthoff-Leyden, 1976).

2. This assessment is voiced by Burton I. Edelson, NASA associate administrator for space science and applications, in a paper that relates these experiments to development of the Intelsat system. See "Communications Satellites: The Experimental Years," *Acta Astronautica*, vol. 11, no. 7–8 (1984), pp. 407–13.

3. SYNCOM I was unsuccessfully launched on February 14, 1963. SYNCOM II was successfully launched on July 26, 1963, into near-geosynchronous orbit. SYNCOM III, launched April 19, 1964, was the first communications satellite in true geosynchronous orbit. Management Support Office, *NASA Pocket Statistics* (Washington: NASA, January 1983), p. B-32.

4. Edelson, "Communications Satellites," p. 411.

5. See "Statement by the President on Communication Satellite Policy," *Public Papers of the President, January 20–December 31, 1961*, pp. 529–31; and Smith, *Communication via Satellite*, pp. 61–87.

over who would own and operate the satellite system (or systems), and how it would fit into terrestrial communication networks and other uses of the radio spectrum. The principles Kennedy set forth became guidelines for U.S. policy. The government was to conduct and encourage R&D, ensure international benefits from the technology, and coordinate the adoption and use of satellites with the United Nations. Although the government would continue to launch spacecraft and regulate the radio frequency spectrum, Kennedy favored private ownership and operation of the U.S. portion of the satellite system—but only if benefits were provided to other countries, procurement was competitive, and any communications carrier could have access to the technology.

The Communications Satellite Act of 1962 established the private Communication Satellite Corporation (Comsat) to be responsible for the U.S. part of an international satellite system.[6] Comsat's operations, pricing policy, ownership, and management structure were regulated by the federal government, with regulatory authority divided among the president, the Federal Communications Commission, and the Department of State to ensure that various domestic and foreign policy objectives would be served.

In 1965 the International Telecommunications Satellite Consortium (Intelsat) was organized to establish a global satellite system. Comsat represents the United States in Intelsat, and during the time period of the ATS program, Comsat developed, launched, and managed the worldwide system for Intelsat. The first Comsat satellite, Intelsat I, also known as Early Bird, which was similar to the SYNCOM satellites, initiated commercial satellite service in 1965. Comsat enjoyed a monopoly in U.S. satellite communications until the mid-1970s.[7]

The 1962 act mandated that NASA advise the FCC and the State Department on technical aspects of communications satellites and provide launch and related services for Comsat on a reimbursable basis. NASA was also directed to "cooperate with [Comsat] in research and development to the extent deemed appropriate by the Administration in the public interest."[8]

The limits of NASA's role in satellite R&D were debated at length in Congress during 1962 and 1963. At issue was whether the agency's program subsidized private industry, particularly Comsat. NASA replied that an

6. 76 Stat. 419.

7. Office of Technology Assessment, "Report on Competition and Telecommunications," draft (June 11, 1984), pp. 6-3, 6-29, 6-31.

8. U.S.C.S. 47 sec. 721 (b)(2).

R&D program was essential to carry out its advisory responsibilities under the 1962 act; the Space Act of 1958 had charged it with ensuring U.S. leadership in space technology, including satellites; and ample precedents existed for federal R&D on technologies considered important to national prestige, foreign policy, and economics.[9] NASA officials testified that financial constraints would limit Comsat's near-term R&D to research on its operating system. In the long term the corporation would perform research applicable only to commercial common carrier communications.

NASA's arguments carried the day. Because Comsat was expected to base its initial system on low-orbit satellites, the agency developed a plan to concentrate on research into geosynchronous satellites, known as Advanced SYNCOM. Hughes Aircraft had been awarded $8 million to write the proposal. Project definition started in fiscal 1962 and continued in fiscal 1963. Between 1963 and 1964, however, NASA's budget for communications satellite R&D was reduced from $48 million to $35.5 million.[10] Mindful of the recent political controversy and Comsat's surprise adoption of geosynchronous technology, NASA redesigned its satellite program to combine communications with other satellite applications.[11] The new Advanced Technology Satellite program included experimental applications for Advanced SYNCOM satellites in meteorology, navigation, and satellite tracking. Hughes was awarded an additional $1.2 million to reorient the proposal. A year later, the program was renamed Applications Technology Satellites to emphasize its broader coverage.[12]

ATS-1 through ATS-5

Because a geosynchronous orbit reduces costs for satellites and ground stations, the ATS program could afford more elaborate spacecraft. A primary

9. *1964 NASA Authorization*, Hearings before the Subcommittee on Applications and Tracking and Data Acquisition of the House Committee on Science and Astronautics, 88 Cong. 1 sess. (Government Printing Office, 1963), pp. 3161–62.

10. *Budget of the United States Government, Appendix, Fiscal Year 1965*, p. 765. Figures show funding for communications R&D and other space applications, including ATS.

11. Memo from assistant administrator for space science and applications to file, January 23, 1964, NASA History Office, ATS Files. The decision to broaden ATS program coverage was made on October 28, 1963.

12. *Independent Offices Appropriations for 1965*, Hearings before a Subcommittee of the House Committee on Appropriations, 88 Cong. 2 sess. (GPO, 1964), p. 1235; and *Independent Offices Appropriations for 1966*, Hearings before a Subcommittee of the House Committee on Appropriations, 89 Cong. 1 sess. (GPO, 1965), p. 1038.

research problem was how to increase the effective power of a satellite.[13] The spin-stabilized SYNCOM satellites rotated and emitted a donut-shaped beam; only a third of the beam hit the earth and at most a third of the solar panels faced the sun. NASA sought methods either to stabilize the spacecraft without spinning or to "despin" the antenna, even though the rest of the spacecraft was spinning. A nonspin stabilization technique, called gravity-gradient stabilization, sought to stabilize a spacecraft by exploiting the decrease in gravitational force at increasing distances from the earth.

Hughes Aircraft developed a proposal for five experimental satellites, and in 1964 NASA awarded the company a contract to build them. The estimated cost of the program was $100 million, of which about $30 million would go to Hughes for the spacecraft.[14] Smaller contracts were awarded to General Dynamics/Convair for launch vehicles and to General Electric Company for the gravity-gradient stabilizers. The program was to test a gravity-gradient stabilized satellite in low orbit, two spin-stabilized geosynchronous satellites, and two geosynchronous gravity-gradient satellites. The specific objectives were

> (1) to provide a capability for performing varied scientific and technological experiments in the synchronous orbit with a single spacecraft; (2) to conduct a carefully instrumented gravity gradient experiment in a 6,500-mile orbit; (3) to develop the technology for an earth-oriented satellite in the synchronous orbit; (4) to investigate in space a directed antenna from a spin-stabilized satellite; and (5) to investigate access to a communication satellite by more than one pair of ground stations at a time.[15]

The Department of Defense considered the medium-altitude, nongeosynchronous gravity-gradient satellite to be of great importance because it was most appropriate for military use. DOD was prepared to provide the launch vehicle and tracking and related services for the satellite.[16]

When the ATS program began, two issues anticipated the problems that led to the program's ultimate cancellation. One concerned the contract awarded to Hughes without competition. The second reflected Comsat's concern about market encroachment.

13. The other main constraint on satellite power was simply how much weight, and what configuration, could be lifted into geosynchronous orbit. This launch research question emerged as a side benefit from the manned space program.

14. NASA release 64-50, March 2, 1964, in NASA History Office, ATS files.

15. *Independent Offices Appropriations for 1965*, Hearings, p. 1235.

16. *Space Daily*, April 24, 1964.

When Hughes finished the ATS proposal, NASA faced the choice of simply awarding the contract to the company or seeking bids. The agency preferred the first choice because it perceived a conflict between competition and efficiency. Holding a competition would set the program back nine months to a year and incur considerable expense. Even more important, Hughes had expertise, a working group, and a good relationship with NASA. Still, the agency's mission included fostering competition in space technology, which caused concern about objections from potential competing firms or from Congress. NASA officials decided against an "arbitrary insistence on competition," and after a thorough discussion, "Mr. Webb, Dr. Dryden, and Dr. Seamans concluded that, despite the serious consideration of exempting this procurement from competition, the government could maximize its chances of getting the best performance, schedule, and cost results on the ATS project by selecting Hughes at this time."[17] In fact, NASA's worries proved unfounded. The sole-source award received only one cursory mention in Congress.[18]

While competition among satellite suppliers was unimportant in 1965, competition with Comsat was more serious. As sole provider of commercial satellite services, Comsat wanted no competition from NASA. Although the corporation stood to be a major beneficiary of the ATS program, experiments with the new technology required that the satellites provide services. Comsat sought assurance that "there would be no operational or commercial traffic carried by the satellites and that the program would remain purely experimental."[19] Its concerns were voiced in a discussion about frequency allocations for the program. NASA planned to use frequencies in the 4–6 GHz range so that R&D frequencies would be in band, that is, they were the frequencies most likely to be selected for commercial use. In late 1964 NASA received a request, instigated by Comsat, to change frequencies to 7–8 GHz, a range reserved for military use. The change was expected to cost more than $15 million and to cause a year's delay in the program.

As an alternative, Comsat stated that NASA's use of 4–6 GHz was acceptable provided

17. "Procurement Approach for Advanced Technological Satellites," memo to record from NASA Executive Officer R. P. Young, March 2, 1964, p. 5, in NASA History Office, ATS files. This discussion summarizes the contents of the memo.

18. *Independent Offices Appropriations for 1965*, Hearings, p. 1237.

19. "Applications Technology Satellite Frequencies," memo to NASA administrator from Homer E. Newell, associate administrator for space science and applications, April 19, 1965, p. 5, in NASA History Office, ATS files.

(1) the authorization for use is limited to the present five flight programs; (2) the program shall remain truly experimental; i.e., the satellite shall not be turned over to another government agency for operational communication use; (3) operation of ATS shall not cause harmful interference to any non-government operation in the 4–6 Gc/s communication satellite bands now licensed or licensed in the future; (4) assignments will be limited to two years, recognizing that renewals would be required; (5) follow-on research can be accomplished at either 4–5 or 7–8; (6) NASA shall not use the satellites for administrative traffic or the communications traffic of any other U.S. agency.[20]

NASA did not oppose these conditions, noting that "the ATS experiments were long lived and would occupy most if not all of the useful life of the satellites."[21]

The initial phase of the ATS program went well. In 1966 the estimated cost of the program rose from $107 million to $138.6 million, which was close to the actual total cost of $149 million.[22] The launchings were also very close to their original schedule: ATS-1 in December 1966, ATS-2 in April 1967, ATS-3 in November 1967, ATS-4 in August 1968, and ATS-5 in August 1969.

The gravity-gradient experiment was a technical failure. ATS-2, the low-orbit gravity-gradient satellite, and ATS-4, a geosynchronous gravity-gradient satellite, had unsuccessful launches. ATS-5, the remaining gravity-gradient satellite, was successfully launched, but a design defect made it impossible to extend the gravity booms and to conduct the gravity-gradient experiment. The two spin-stabilized satellites performed much better than anticipated. The comparative performance of the two stabilization systems greatly influenced the future course of satellite design.[23]

Despite the initial emphasis on the gravity-gradient experiments, their failure did not prevent the program from being considered highly successful.[24] The program laid the bases for weather, aeronautical, and maritime

20. "Applications Technology Satellite Frequencies," p. 3. See also letter to NASA from the Office of the Director of Telecommunications Management, Federal Communications Commission, March 19, 1965, in NASA History Office, ATS files.

21. "Applications Technology Satellite Frequencies," pp. 2–3.

22. *Independent Offices Appropriations for 1965*, Hearings, p. 1237; and *Independent Offices Appropriations for 1966*, Hearings, p. 1040.

23. Edelson, "Communications Satellites," p. 412.

24. See "NASA Project Approval Document," March 10, 1964, NASA History Office, ATS files. "NASA Fact Sheet," November 16, 1965, NASA History Office files, lists the gravity-gradient test as the major goal of the program.

satellite systems and led to the technologies for the first four generations of INTELSAT satellites. ATS-1 demonstrated an electronically despun antenna that improved power utilization over SYNCOM by a factor of ten.[25] ATS-3 demonstrated a mechanically despun antenna, which allowed further refinements and greater efficiency. These satellites tested VHF transmission, transmission in the millimeter frequencies, and simultaneous transmission to multiple earth stations.[26] Greater power and more accurate beaming allowed reception by smaller, cheaper earth stations. The satellites also provided voice and television links over the Pacific and in Australia and were used experimentally to provide service to isolated communities in Alaska. ATS-1 provided the first photographs of earth's cloud cover and was used for meteorological experiments. The satellites (including ATS-5) tested transmission to aircraft and ships, leading to operational aeronautical and maritime satellite systems.

The experiments also influenced domestic and international policy decisions. The results affected the U.S. position at the ICAO meeting in November 1969.[27] They led the Office of Telecommunications policy to favor the use of the ultrahigh frequency spectrum for aeronautical satellites. According to John Naugle, then assistant administrator for Space Science and Applications, NASA's "ability to provide that advice was provided largely by the advanced research in propagation and position location which we had performed with our Applications Technology Satellites."[28]

NASA's description of its successes followed President Kennedy's policy statement in emphasizing cooperation with other countries: television coverage for Italy of the Italian president's trip in 1967 to Australia, the simultaneous broadcast to twenty-six countries of a British Broadcasting Company educational program, and technical experimentation by the Japanese. ATS-1 provided emergency communications during the 1967

25. Burton I. Edelson, Robert D. Briskman, and Robert R. Lovell, "Satellite Communications Systems, Progress and Projections," prepared for the 1982 IAF-COSPAR Forum, Vienna, Austria.

26. The description of experiments on the ATS flights is drawn from Goddard Space Flight Center, *The ATS-F Data Book*, rev. ed. (Greenbelt, Md., May 1974), pp. 1-5 to 1-10.

27. Memo to the associate administrator, Organization and Management, from the associate administrator, Space Science and Applications, November 4, 1968, NASA History Office, ATS files.

28. *1972 NASA Authorization*, Hearings before the Subcommittee on Space Science and Applications of the House Committee on Science and Astronautics, 92 Cong. 1 sess. (GPO, 1971) [no. 2], pt. 3, p. 2.

Alaskan flood. ATS-3 demonstated the first ground-to-satellite-to-aircraft communication over the Atlantic Ocean, took the first color photographs of the earth from space, and provided support for the Apollo program.

Contrary to expectations, both ATS-1 and ATS-3 operated well into the 1980s and were put to numerous uses that justified Comsat's fears of competition. In 1967 the Australian Overseas Telecommunications Commission expressed concern that their charter might have been violated by the use of ATS-1 in emergency support of OGO-IV. NASA felt that its actions were justified, but to maintain goodwill with telecommunications carriers, it discussed the "expeditious phasing out" of such emergency aid.[29] Still, Comsat itself requested the use of ATS-1 and ATS-3 several times in the late 1960s to back up its system.[30] Meteorological experiments on ATS-1 and ATS-3 led the Environmental Science Services Administration (ESSA) to develop the Synchronous Meteorological Satellite (SMS) program. ESSA made "quasioperational" use of the ATS satellites for twelve hours a day from 1969 until the launch of their own satellites in 1974.[31] In 1970 NASA announced that the two satellites were available for further experiments. The approved experiments were for public service applications. For example, after 1971 ATS-1 was used for the Pan-Pacific Educational and Communication Experiments by Satellite (Peacesat), a communications network providing educational, scientific, and cultural services to schools and healthcare facilities in the Pacific. In the 1980s ATS-3 supported emergency medical experiments using mobile communications with the Southern Regional Medical Consortium in Alabama, Louisiana, and Mississippi and on oil rigs in the Gulf of Mexico.[32]

Of the experiments approved in 1971, several provided domestic satellite services that had not yet been approved on an operational basis by the FCC. One major customer was Alaska, which used ATS for broadcasting and communications links with remote areas. Another was the Corporation for Public Broadcasting, which transmitted programming from the East

29. Memo, September, 29, 1967, NASA History Office, ATS files.

30. Memos, 1967 and 1968, NASA History Office, ATS files.

31. *1971 NASA Authorization*, Hearings before the House Committee on Science and Astronautics, 90 Cong. 1 sess. (GPO, 1970), pp. 1087–88, 1091; and Management Support Office, *NASA Pocket Statistics*, p. B-34.

32. *United States Civilian Space Programs*, vol. 2: *Applications Satellites*, Report prepared for the Subcommittee on Space Science and Applications of the House Committee on Science and Technology, 98 Cong. 1st sess. (Congressional Research Service, May 1983), p. 94.

Coast to Los Angeles for three hours a night, Sunday through Thursday. The corporation found that NASA's satellites permitted far cheaper ground terminals than the INTELSAT system.[33]

ATS-F and ATS-G

Planning for ATS-F and ATS-G began in early 1966. This second generation of satellites would demonstrate three-axis stabilization, a non-spin stabilization technique. The satellites would provide an "order of magnitude improvement in spacecraft antenna technology and stabilization and pointing capabilities over the ATS A through E missions" and develop technology for direct and community broadcast.[34] The ATS-F beam was thirty times more powerful than the concurrent INTELSAT beams. In 1974, the year ATS-F was launched, INTELSAT ground stations cost $2.5 million; each ATS-F ground station cost $3,700.[35]

Unlike earlier ATS satellites, ATS-F and ATS-G were planned primarily for communications.[36] According to NASA, "the objectives of this project are to develop the technology for erecting and accurately pointing large antennas in space; to investigate the technology of accurate angle measurement in space using an interferometer; to investigate the technology for multi-beam phased array antennas in space; and conduct other applications technology experiments requiring an accurately stabilized platform in geostationary orbit."[37]

33. *1971 NASA Authorization*, Hearings, p. 1091; and *1972 NASA Authorization*, Hearings [no. 2], pt. 3, p. 106.

34. *Independent Offices Appropriations for 1965*, Hearings, pt. 2, p. 1234; and *Independent Offices and Department of Housing and Urban Development Appropriations for 1968*, Hearings before a Subcommittee of the House Committee on Appropriations, 90 Cong. 1 sess. (GPO, 1967), p. 747. Satellites are known by letter before launch and number thereafter. Thus the ATS-1 through ATS-5 satellites were called ATS-A through ATS-E in 1966 since they had not yet been launched.

"Direct broadcast" refers to broadcasting from the satellite directly to a home or business. "Community broadcast" refers to broadcasting from the satellite to fairly inexpensive receivers, one stage up in cost from those considered reasonable for a home.

35. Jonathan Spivak, "NASA's Planned TV Satellite to Have First Self-Contained Broadcast Ability," *Wall Street Journal*, May 28, 1974, p. 14. The 2,500 ground receivers built in India for the joint U.S.-India experiment conducted using ATS-F (the SITE experiment discussed below) cost $600 each.

36. *1973 NASA Authorization*, Hearings before the Subcommittee on Space Science and Applications of the House Committee on Science and Astronautics, 92 Cong. 2 sess. (GPO, 1972), [no. 15] pt. 3, p. 393. NASA was by then involved in other satellite programs, the ERTS and Nimbus series, dedicated to meteorological and land survey applications.

37. *Independent Offices and HUD Appropriations for 1968*, p. 807.

The development and selection of experiments for ATS-F and ATS-G proceeded into the early 1970s. Major technology experiments on the ATS-F included the position location and aircraft communication experiment (PLACE) in the newly approved L band (1.5 GHz), a millimeter-wave propagation experiment to investigate use of frequencies between 20 GHz and 30 GHz (the Ka band) under various meteorological conditions, and a tracking and data relay experiment to provide communication between earth stations and low-orbiting satellites.[38] The Department of Health, Education, and Welfare and NASA planned the Health, Education, Telecommunications (HET) experiments, the first direct broadcasts. The satellite Instructional Television Experiment (SITE), a joint U.S.-India community broadcast experiment, was the largest ATS-F project. NASA agreed to make the satellite visible for one year to the Indian subcontinent a year after launch. During that time the Indian government would broadcast educational television to 5,000 ground stations. Many of the communities served by SITE had had no television links with the rest of the subcontinent. Some had no communication links at all and had to install electricity to participate. To carry out SITE, India developed a domestic television production industry. In describing the SITE program, NASA stated that the "India/NASA ITV experiment would provide dramatic proof of U.S. interest in applying advanced technology to the problems of less-developed countries."[39] The SITE experiment used the C band at 860 MHz, which was not approved for broadcast use in the United States. The HET experiments used the newly allocated S band (2.5 GHz to 2.69 GHz).

ATS-F and ATS-G were much larger and more expensive than the previous ATS satellites. About two-thirds more was spent on the two spacecraft than on the previous five, even though only ATS-F was actually completed and launched (table 7-1). ATS-F and ATS-G were to be platforms for carrying numerous elaborate experimental packages so that less of the project cost would be attributed to the spacecraft and more to experiments. The cost of experiments on the satellites was to be 60 percent of the cost of the spacecraft; for ATS-1 through ATS-5 the figure was only 33 percent.

ATS-F involved both proof of concept and the development of a user

38. *ATS-F Data Book*, chap. 7. The satellite also carried numerous scientific experiments (see chaps. 8 and 9).

39. Letter from Robert Allnut, assistant administrator for legislative affairs, to Clement Zablocki, April 24, 1969, NASA History Office, ATS-F files, p. 3. The letter continues: "The experiment would offer India an important and useful domestic tool in the interests of national cohesion. . . . Above all, it should provide information and experience of value for future application of educational programs elsewhere in the world."

Table 7-1. *ATS Appropriations, Fiscal Years 1964–75*
Current dollars

Year		*Components of program*			*ATS*[a] *1–5*	*ATS*[a] *F&G*
	Total	*Spacecraft*	*Experiments*	*Launch*		
1964 Actual	15,377	9,832	...	...	...	...
1965 Request	29,900	18,400	...	5,900	...	...
Actual	27,099	20,695	...	4,404	...	...
1966 Request	40,900	24,100	...	14,200	...	...
Estimate	36,240	24,800	...	8,240	...	...
1967 Request	34,603[b]	17,800	...	12,805	...	...
Actual	47,869	15,413	13,400	17,856	...	...
1968 Request	35,500	10,900	13,200	10,301	19,800	15,700
Actual	32,100	12,300	12,200	6,500	...	...
1969 Request	35,400	15,300	13,100	4,200	...	...
Actual	28,296	14,293	9,824	3,596	...	...
1970 Request	50,100	19,100	21,200	5,900	...	...
Actual	45,665	18,450	18,506	6,700	...	...
1971 Request	35,800	15,600	14,200	4,700	...	...
Actual	27,850	1,049	16,985	4,100	3,205	24,645
1972 Request	72,800	26,300	31,000	12,500	...	...
Actual[c]	49,162	26,443	18,543	9,000	1,450	47,612
1973 Request	79,400	25,250	33,250	18,200	1,200	78,200
Actual[c]	53,187	40,189	9,477	5,500	1,333	51,854
1974 Request	16,000	12,500	3,000	...	...	...
Estimate	16,800	12,174	2,619	...	...	...
1975 Request	3,700	...	3,700	...	...	...
Cumulative ATS 1–5[d]	149,000	77,000	25,000[e]	40,000	...	...
Cumulative ATS F&G[d]	227,000	128,000	76,000	25,300	...	...

Source: House Appropriation Committee Hearings, fiscal years 1964–75.
a. Except as noted in these two columns, totals for fiscal years 1964–68 are for ATS-1 through ATS-5 and for fiscal years 1969–75 for ATS-F and ATS-G.
b. Excludes experiment costs.
c. Reflects cessation of ATS-G.
d. Approximate totals.
e. Some "experiment" funds for ATS-1 through ATS-5 may be included in the spacecraft total.

group. Direct broadcast satellites were not expected to compete with over-the-air television in industrialized countries, which had heavy sunk investments in standard broadcasting. In developing areas, the satellites were attractive. Domestically NASA tried to develop new public services rather than introduce the satellites' capabilities as substitutes for existing services.

The user orientation of ATS-F became more pronounced as the program progressed. By the eve of its launch in 1974 a report stated, "The big question ATS-F is seeking to resolve in the U.S. is whether there will be enough demand for TV transmission in remote areas of the nation and for specialized services to schools and hospitals to warrant such a satellite. Government experts figure a market of 4,000 institutions is needed to make such a satellite economical in the U.S."[40]

Uncertain marketability did not dampen the enthusiasm of the companies that sought the contract to build the spacecraft. The awarding of the ATS-F and ATS-G contract was a very different affair from the award to Hughes six years earlier for ATS-1 through ATS-5. On March 1, 1966, NASA announced the selection of three companies to negotiate six-month feasibility studies for the second-generation ATS satellites.[41] The three—Fairchild-Hiller, General Electric Company, and Lockheed Missile and Space Corporation—were chosen from a field of eight that included Hughes Aircraft, Philco-Ford, RCA, TRW, and Westinghouse.[42] In May 1968 NASA announced that the same three companies had been selected for competitive contract negotiations to develop spacecraft designs for ATS-F and ATS-G.[43] Fairchild-Hiller and General Electric proceeded with proposals to build the two spacecraft. NASA awarded the contract to GE on April 8, 1970.

Fairchild-Hiller then filed a lawsuit, claiming unfairness because it had been forced to meet a deadline of February 27, 1970, while GE's deadline was extended to March 6. Fairchild-Hiller also claimed that GE submitted technical data that plagiarized its proposal and alleged that cost information in its proposal could have been leaked to GE before GE's submission. The General Accounting Office investigated the claims, and concluded that Fairchild-Hiller's charges were not supported by the evidence, but that

40. See Spivak, "NASA's Planned TV Satellite to Have First Self-Contained Broadcast Ability."

41. NASA Release 66-45, March 1, 1966, NASA History Office, ATS-F file.

42. *Space Daily,* March 3, 1966, pp. 19–20.

43. NASA Release 68-95, May 22, 1968, NASA History Office, ATS-F files.

NASA had prejudiced the bid by allowing GE to submit its offer one week later. The report concluded that NASA's procedures had been "irregular, deficient and inconsistent," and that "a situation was created where a leak which might have affected the results of the competition was possible." The GAO recommended that NASA reconsider the award of the contract to GE.[44]

NASA reopened the contract, and the NASA administrator appointed an ATS procurement review committee and a selection panel, both made up of senior NASA administrators, to review the award. The committee recommended that the contract be awarded to Fairchild-Hiller. It found that the issues raised by Fairchild-Hiller and GAO did not affect the outcome of the competition. It regarded the two proposals as of similar overall merit and both contractors as capable of undertaking the project. Nevertheless, the committee concluded that the Fairchild-Hiller proposal was superior in "nearly all" aspects of project organization and management and in some important technical matters. The selection panel concurred on Fairchild-Hiller's superiority on technical matters but not on superiority of its project management. Finally, on September 5, 1970, NASA awarded the contract to Fairchild-Hiller.[45]

In 1967 during preliminary planning for ATS-F and ATS-G, the satellites were expected to be launched in 1970 and 1971 for a cost over seven years of about $100 million. Both targets were optimistic. By 1968 the expected launch dates had been moved back two years, and the estimated project cost rose to $174 million. In 1969 the launch dates were revised again, to 1972 and 1974.[46] When ATS-F was finally launched in 1974, total program costs were $227 million, and ATS-G had been canceled.

Superficially, this record appears to describe a failed program. The cost overrun was more than 100 percent for half the planned product, and a four-year delay in the launch doubled the duration of the project. But

44. GAO Report, July 6, 1970.

45. "Applications Technology Satellites F&G Procurement," NASA News Release, September 5, 1970, pp. 2–4.

46. *Independent Offices and Department of Housing and Urban Development Appropriations for 1968*, p. 747; *Independent Offices and Department of Housing and Urban Development Appropriations for 1969*, Hearings before a Subcommittee of the House Committee on Appropriations, 90 Cong. 2 sess. (GPO, 1968), p. 1290; *HUD-Space-Science-Veterans Appropriations for 1973*, Hearings before a Subcommittee of the House Committee on Appropriations, 92 Cong. 2 sess. (GPO, 1972), pt. 1, p. 1282; and *Independent Offices and Department of Housing and Urban Development Appropriations for 1970*, Hearings

ATS-F was not a failure. The project provided substantial technological advances and, most likely, economic benefits that greatly exceeded costs. Moreover, most of the delay had political origins, starting with reconsideration of the contract award and continuing with appropriations lower than the amounts necessary for timely completion.

In 1969 the Nixon administration requested a reduction in project funding of $3.2 million to be accommodated by "reducing and rephasing" the development of experiments.[47] In 1970 the Office of Management and Budget cut NASA's request for space applications from $219 million to $167 million, a reduction made possible in part by a one-year delay in ATS-F and ATS-G.[48] The House Subcommittee on Space Science and Applications restored $5.6 million so that the spacecraft could be launched on schedule, stating that "the ATS project represents one of the most significant research and development efforts undertaken by NASA."[49] The Senate Committee on Aeronautical and Space Sciences trimmed the increase, stating that a timely launch was impossible because the contract dispute was causing delay, and in any event the extra money was inadequate. The House accepted the Senate position, "in view of the fact that the passage of time precluded the possibility of reestablishing the original launch schedule."[50] In 1971 NASA's request for ATS was down $0.8 million because the "delay in award of ATS-F and G contract permits deferring funding to fiscal year 1972."[51] In March 1971 the cost estimate for ATS-F and ATS-G was revised from $180 million to $215 million, plus an additional $50.2 million for launching.[52] This figure held for the next two years until one launch was canceled and the final, lower cost was made firm.[53] Further delays in ATS-F were discussed in early 1972. Technical difficulties with the communications element, according to NASA officials, might cause a delay of a year.[54]

before a Subcommittee of the House Committee on Appropriations, 91 Cong. 1 sess. (GPO, 1969), p. 927.

47. *Independent Offices and Department of Housing and Urban Development Appropriations for 1970*, Hearings, p. 932.

48. *1972 NASA Authorization*, Hearings [no. 2], pt. 1, p. 85.

49. *NASA Authorization for Fiscal Year 1971*, Report of the Committee on Aeronautical and Space Sciences, U.S. Senate on H.R. 16516 (GPO, May 1970), p. 148.

50. *NASA Authorization for FY 71*, H. Rept. 91-1189, 91 Cong. 2 sess. (GPO, 1970), p. 10.

51. *1972 NASA Authorization*, Hearings, p. 46.

52. *1972 NASA Authorization*, Hearings [no. 2], pt. 3, p. 268.

53. *HUD-Space-Science-Veterans Appropriations for 1973*, Hearings, pt. 1, p. 1282.

54. *1973 NASA Authorization*, Hearings [no. 15], pt. 3, pp. 276–77.

During 1972 the communications satellite research program suffered more cuts. A 1969 National Academy of Sciences report had criticized NASA's "leisurely" launch schedule, and concluded that the satellite program was far too small.[55] The Small Applications Technology Satellite (SATS) program was developed to flight-test single experiments on small satellites that would be launched within one year of project approval. The program was designed to fill the launch gap of three or more years in the ATS program. The budget submitted to Congress for fiscal 1973 included no funds for SATS. In addition, NASA's budget included no funds for the follow-on ATS-H and ATS-I satellites. ATS-H and ATS-I, described as the "next logical step," had been planned to develop two-way data and voice communications and multichannel video for small, low-cost receivers. Advanced mission studies for ATS-H and ATS-I were completed in July 1971.[56] Both the SATS and ATS-H and ATS-I projects were deleted by OMB. Although the House Subcommittee on Space Science and Applications voiced its displeasure at these actions, the funds were not restored.

In 1973 NASA responded to a Nixon administration effort to cut government spending by reducing most of its programs, but it cut the communications program the most severely. The agency canceled ATS-G and announced that it would phase out its work on communications satellites.[57] While the House Authorization Committee "strongly" recommended "that NASA reconsider its decision to withdraw from communications research and development," funds were not restored.[58] This left ATS-F, scheduled for launch in 1974, and CTS, a joint U.S.-Canadian communications satellite, as the only remaining projects.[59]

55. National Research Council, *Useful Applications of Earth-Oriented Satellites*, NSR 09-012-909 (Washington: National Academy of Sciences, January 1969), p. 16.

56. Office of Space Science and Applications, *Proposed FY 73 New Effort: Application Technology Satellites H&I*, 1971, NASA History Office files, pp. 1, 2.

57. "NASA Program Reductions," NASA News Release 73-3, January 5, 1973, NASA History Office files. In the same release, NASA announced a slowdown in the space shuttle program, a suspension of the high-energy astronomy observatory, and cessation of the nuclear propulsion and quiet propulsive lift short takeoff and landing (Questol) projects. However, research on nuclear power and Questol technology would continue at a curtailed rate.

58. *1974 NASA Authorization*, Hearings before the House Committee on Science and Astronautics, H. Rept. 4564, 93 Cong. 1 sess. (GPO, 1973), pt. 1, p. 182.

59. CTS, the communication technology satellite, had been approved in 1971. It was built by the Canadians, and NASA supplied the traveling wave tube and launch facilities.

NASA successfully launched ATS-F on May 30, 1974, and in keeping with launch etiquette renamed the spacecraft ATS-6. The planned experiments were undertaken, including SITE and other international scientific experiments. At the conclusion of SITE, the satellite was moved back into orbit above the United States. On the way the Agency for International Development conducted the AIDSAT experiment, which involved demonstrations of satellite applications in twenty-six less developed countries.[60] On July 3, 1975, NASA invited proposals for the third year of ATS-6 in the "societal, communications or technological disciplines—in that order of priorities—beginning in the fall of 1976."[61] The satellite continued to be used for some community and direct broadcast experiments for three years. By 1979 its orbit had degraded, creating a hazard to other satellites, and in August it was moved out of geosynchronous orbit and turned off.[62]

The ATS program pioneered new satellite technology. While no evaluation of the program was ever performed, in a sense it would have been superfluous. A social evaluation should assess whether the government actually had to do the work: if private industry would have done it anyway, the federal effort may have been unnecessary. This question is not resolved here, but from a narrower point of view, the assessment is obvious. The results of work on the first five satellites created several industries and led to innovations on communications satellites worth billions of dollars. Even though only two satellites worked as planned, the experiments obviously justified the investment.

ATS-6 also contributed to development in the satellite industry and can be accorded the same acclaim. However, the pace of the program was slower than for the earlier satellites. ATS-1 through ATS-5 were approved in 1963, and major appropriations came a year later. Launches started in late 1966 and continued through 1969, bringing concurrent commercial and federal policy applications. Planning for ATS-F and ATS-G started in 1966 and intensified in 1968. Major spending began in fiscal 1969, but the single satellite was not launched until 1974.

The SITE and AIDSAT experiments were judged enormously success-

The total cost to NASA of the project was $22 million. *United States Civilian Space Programs*, vol. 2, p. 99. The satellite was an interim step toward ATS-H and ATS-I, testing networking possibilities and investigating properties of transmission in the Ku band, 12 GHz to 14 GHz.

60. *United States Civilian Space Programs*, vol. 2, p. 99.

61. NASA News Release 75-187, July 3, 1975, NASA History Office files.

62. *United States Civilian Space Programs*, vol. 2, p. 108.

ful.[63] Several developing nations proceeded with their own satellite systems. Indsat was launched in 1982; Arabsat and Brazilsat were launched in 1984.[64] These were followed by Indonesian and Mexican national satellites. In addition to benefits to foreign countries and foreign policy, these satellites yield direct benefits for the United States in that they were built in whole or in large part by U.S. companies.[65] Nevertheless the tangible results of the experiments did not materialize for ten to fifteen years after the project began.

The experiments in providing public service in the United States also generated considerable enthusiasm, but they have not led to operational services. In 1975 the Public Service Satellite Consortium was organized and was given a federal grant to aggregate public service satellite users and organize operational services. ATS-6 and CTS were not considered adequate to meet the consortium's needs.[66] In late 1974 Hughes Aircraft suggested replacing twelve channels on a domestic satellite it was building for Western Union with a two-channel S band transponder. The proposal was submitted to HEW with a price of $30 million but was not approved.[67] In the late 1970s PSSC and Hughes considered launching a satellite in the same frequencies as CTS, for which PSSC owned ground equipment. Called SYNCOM IV (or GAPSAT), it was not built because funding problems could not be resolved. Not until 1981 did commercial satellites (ANIK-B and SBS-3) transmitting in the Ku band become available over the United States. PSSC provides some services using these satellites.[68] Some of the S band equipment has been converted to C band to make use of the SATCOM I satellite, launched in 1979.

The latest INTELSAT series (INTELSAT V) uses three-axis stabilization. Space-deployable large antennas have also been incorporated on satellites in the 1980s. With the growth in the use of satellites to distribute television programming and the crowding of the RF spectrum, more interest has arisen in the Ka and Ku bands and the powerful broadcasting technologies developed for ATS-6. In 1981 satellites using the Ku band started commer-

63. *United States Civilian Space Programs*, vol. 2, p. 96.

64. For Indsat see Management Support Office, *NASA Pocket Statistics*, p. B-19.

65. Office of Technology Information, "Satellite Communications," Draft report on Competition and Telecommunications, June 1984, pp. 6-48–6-72.

66. *United States Civilian Space Programs*, vol. 2, p. 105.

67. "Effort for ATS-6 Backup Fails," *Aviation Week and Space Technology*, November 4, 1974, p. 17.

68. *United States Civilian Space Program*, vol. 2.

cial service in Canada and the United States. More recently Japan has launched a satellite using Ku band technology.

ATS and the Technology Pork Barrel

The politics of the ATS program is more difficult to document than it is for the other programs examined in this book. Neither chamber of Congress ever voted on the program, and its budget was never more than 10 percent of the total NASA appropriation. In addition, we have been unable to locate data about the geographic distribution of program expenditures. Nevertheless, sufficient documentation exists to provide a qualitative assessment of the extent to which the program confirms our general view of government commercial R&D programs.

The history of the ATS program differs from that of most other federal technical development projects. While the program was generally regarded as an enormous success, it was canceled when NASA was preparing to pursue the next logical step. This contrasts with other programs, in which efforts appear to have continued in spite of all odds and far past the point of diminishing marginal returns. Our theory of the technology pork barrel claims that the long-run success of a program is only one factor that determines whether it will continue. Other factors are its distributive benefits, ideological appeal, prominence, and present, as opposed to present-value, economic benefits. The communications satellite program is best understood in that context.

The Termination of NASA's Communication Satellite Research

NASA attributed its decision to phase out communications satellite research to two factors: budget pressures and the maturity of the industry. In the early 1970s President Nixon sought cuts in all space programs; however, communications research was accorded lowest priority and the cuts singled it out for complete cancellation. Low priority was rationalized by the second factor: the state of the commercial satellite industry. "NASA has been the catalyst in bringing into being a commercially viable communications satellite business," reads an agency press release on the 1973 program reductions. "Further advances in satellite communications research and development can be accomplished by industry on a commercial basis without Government support."[69] NASA's justification for its fiscal 1974 budget reiterates this position: "Today, intercontinental communications by satellite are sustained through the existence of a privately financed, effective and profitable international communications service which can support its own research and development efforts. Today, private industry is prepared to finance and manage domestic communications satellites to serve the United States. With these conditions established, NASA plans in this calendar and fiscal year to commence to phase out its R&D activities in support of commercial communications satellites."[70]

Certainly both supply and demand in the satellite market had changed since the beginning of the ATS program and the initial planning of ATS-F. However, these changes bore little relation to the original rationale for federal R&D. While the industry might pursue its own R&D, private programs were unlikely to achieve many of NASA's goals. NASA's 1963 arguments for a federal R&D program are only partly connected to the state of the industry. Industry R&D can hardly contribute to NASA's advisory role to the FCC and State Department and to the development of technology for other than commercial, common carrier applications. The other goals are related to the state of the industry: U.S. leadership in space technology. If industry had picked up the R&D effort, as the Nixon administration predicted it would, then this goal might have been achieved. However, in 1973 the NASA associate administrator for applications conceded that industry probably would stick to research with short-term payoffs, so-called evolutionary as opposed to revolutionary research.[71] The House Oversight Committee concluded in its report on the authorizing legislation that

> The Committee has received information that casts grave doubt on the NASA premise that the private sector will take up the slack. While the private sector can be expected to continue its research and development activities at a certain level, it is understood that this work will be directed primarily toward refinement and upgrading of existing technology. There will remain plenty of important research which looks farther into

69. "NASA Program Reductions," NASA News Release 73-3, January 5, 1973, NASA History Office, ATS-F files.

70. *HUD-Space-Science-Veterans Appropriations for 1974*, Hearings before a Subcommittee of the House Committee on Appropriations, 93 Cong. 1 sess. (GPO, 1973), pt. 2, p. 797.

71. *1974 NASA Authorization*, Hearings, pp. 21–22.

> the future and which, in the judgment of the Committee, will require government support if it is to be accomplished. Accordingly, the Committee strongly recommends that NASA reconsider its decision to withdraw from communications research and development.[72]

In fact, the industry did not begin a substitute R&D effort.

For several reasons, such R&D might not be forthcoming from the domestic satellite industry. First, technical innovations are not likely to be appropriable to a satellite developer. New satellite proposals must win approval from the Federal Communications Commission, so that innovating firms must announce their plans to competitors before an experiment is even begun. Moreover, to integrate satellite capabilities with the broader communications system requires revealing technical specifications. Before the experiment is conducted, user groups are not likely to be willing to make substantial financial commitments, and after a successful experiment they have an incentive to encourage competitive bidding. Costs of marketing innovations, in particular, whereby a company undertakes to organize a new user group and to configure a new application especially suited for it, are likely to be especially difficult to recoup.

The telecommunications industry imposes further risks on outside sources of innovation. Much of the market for satellite technology is foreign, and in most of the world, telecommunications and broadcasting are highly protected, nationalized industries. Procurement policies in these countries are driven by political factors, not solely economic and technical considerations. Most industrialized nations promote their own telecommunications equipment industries and limit the extent to which an innovating American firm can capture their domestic market.[73]

U.S. telecommunications policies further distort incentives for innovation in the domestic market. State and federal regulatory authorities determine entry into the industry and allocate types of services among competing technologies, while the FCC allocates the electromagnetic spectrum among competing claimants. These decisions are not made on the basis of economic efficiency and so send inappropriate signals to profit-seeking innovators. For example, a major source of demand for satellite services is to distribute television programming to cable television systems.

72. *1974 NASA Authorization*, Hearings [no. 1], pt. 1, p. 182.

73. See Marcellus S. Snow, ed., *Marketplace for Telecommunications: Regulation and Deregulation in Industrial Democracies* (White Plains, N.Y.: Longmont, 1986), for a comparison of telecommunications policies among OECD countries.

From 1965 until 1972 the FCC prevented the development of cable television in the nation's largest cities out of concern for its economic effects on over-the-air television.[74] From 1972 to 1979 it relaxed its control gradually, until by 1980 both satellite program distribution and cable services were largely deregulated. But the implication of this episode is clear: technical and economic attractiveness are insufficient to produce timely adoption of new technology, especially if it threatens an established part of the telecommunications industry.

Another example of regulation-induced distortion arises from allocation and use of frequencies on the electromagnetic spectrum. The regions of the spectrum that are most suited for inexpensive, high-quality long-distance communications are scarce; however, they are also unpriced, being allocated by an administrative process that for the most part does not permit reallocations among types of users. Hence, the private sector has insufficient incentives to undertake R&D that would use the best parts of the spectrum more efficiently, but has heightened incentives to invent technologies that allow commercial exploitation of the previously unusable frequencies.

Finally, satellite R&D requires investments in flight testing that loom large relative to the sales of a single firm. To test new technology requires launching an experimental satellite. Based on NASA's experience, these experiments are successfully launched about half the time. Total demand for domestic commercial launches is quite small, averaging 3.6 a year in the late 1970s and 4.5 a year in the early 1980s.[75] Consequently, an innovating firm that can retain its technical advantage for a year or two before competitors successfully copy it will capture only a few launches. Thus the risks of a failure are large compared with the possible benefits.

For all of these reasons, it is not surprising that the private sector might underinvest in satellite R&D. In fact, after the cancellation of NASA's programs, doubts about whether private industry would conduct adequate R&D were validated, and a consensus emerged that private R&D was inadequate. Yet NASA's program was not restarted. Following the launch of ATS-6, attempts were made to revive ATS-G, which was nearly complete. An internal NASA memo discussed the value of continuing ATS-6 services after the first year by launching ATS-G, which had been renamed ATS-F

74. Roger G. Noll, Merton J. Peck, and John J. McGowan, *Economic Aspects of Television Regulation* (Brookings, 1973), especially chaps. 6 and 7.

75. Management Support Office, *NASA Pocket Statistics*, p. B-17.

Prime.[76] The memo reported substantial enthusiasm for more health and educational experiments. Users wanted more than one year to assess the usefulness of ATS-6 and to gather a large user group. They could use CTS, but this brought delay and required funds to convert equipment to use CTS frequencies. The technical users also wanted additional experimentation time. Because of the delay in launching ATS-F, the millimeter-wave experiment missed being conducted during the rainy months, which was a major goal of the experiment.

A Hughes Aircraft study supported the experimenters' contention that they were not ready to move to an operational system. Hughes said that it could provide a broadcast capability, but "there is no evidence that any market yet exists or will ever exist to support such a capability."[77] In 1974 the National Academy of Engineering recommended that "ATS-F Prime be launched to provide continuity for current health and educational television experiments."[78] A NASA administrator, James Fletcher, described ATS-F Prime as "at least a quasi-operational satellite, [so] it probably should be funded by the users instead of NASA, which is a research and development agency." The users, however, said that they lacked the funds—launching ATS-F Prime was estimated to cost $45 million. OMB was against HEW's picking up the bill, saying that "the capacity and geographic coverage of the ATS-F prime were too limited to permit expansion of the current services, that the ATS-F prime was too expensive, and that its use 'might discourage the participation of private communications companies in providing the service.' "[79] The Ford administration continued to maintain that the educational uses of ATS-F Prime should be provided by an operational system, even though no one was willing to pay for it.

By 1976 it was clear that industry would not pick up the R&D slack. Industry had launched no flight tests and none were planned. The American Institute of Aeronautics and Astronautics issued its first position paper,

76. Memo to associate administrator from associate administrator for applications, May 28, 1974, NASA History Office files.

77. Memo, May 28, 1974, p. 3, NASA History Office files.

78. William A. Shumann, "NASA, HEW Oppose Backup ATS-6," *Aviation Week and Space Technology*, July 29, 1974, pp. 20–21. See in general *Seventh Applications Technology Satellite*, Hearing before the Senate Committee on Aeronautical and Space Sciences, 93 Cong. 2 sess. (GPO, 1974).

79. "OMB Insists Private Sector Provide Health/Education Satcom," *Defense/Space Daily*, November 27, 1974, p. 147.

arguing for the reestablishment of the NASA flight-test program.[80] Meanwhile, a comprehensive cost-benefit analysis, performed under contract to NASA, estimated large net benefits from a dozen or so possible federal satellite communication technology experiments.[81] Finally, in 1976 a National Academy of Sciences committee (the Davenport Committee) issued a draft of its report on satellite R&D, concluding that current NASA activities were inadequate even to support the agency's advisory role.[82] The committee recommended a greatly expanded satellite R&D effort and that NASA, together with the Office of Telecommunications Policy and the FCC, organize the transition from experimental to operational programs.

The Carter administration initiated a new satellite program at NASA called the Advanced Communications Technology Satellite program, or ACTS. The program had a history reminiscent of ATS-F and ATS-G. By 1978 NASA decided that no private substitutes for the federal program were forthcoming. Moreover, the absence of a federal program was believed to have endangered the dominant U.S. position in satellite technology. The ACTS program would flight-test a satellite that would develop 30/20 GHz technology (Ka band) and would employ on-board switching, a multibeam antenna, and spot-scanning technology. NASA's program plan stated that, "for years, NASA has been encouraged to develop and test such a capability because of the risks involved and the need for actual demonstration and evaluation in orbit. This program will take a bold step in a strongly needed direction, reconfirming the U.S. technological leadership in communications while providing for development worldwide of further public and private services and new industries."[83] As before, the program would help provide public service communications in remote areas and would increase capacity in the geosynchronous arc.

By late 1983 NASA was prepared to award the ACTS contract to RCA.

80. American Institute of Aeronautics and Astronautics, "The Federal Role in Communications Satellite R&D: An Official Position of the American Institute of Aeronautics and Astronautics," February 16, 1976.

81. L. D. Holland and others, *Cost-Benefits of Space Communications Technology*, vol. 2: *Final Report*, NAS 3-19700 (Georgia Institute of Technology, Engineering Experiment Station, August 1976).

82. Committee on Satellite Communications, Space Applications Board, "Federal Research and Development for Satellite Communications, vol. 1: Committee Report," draft report, National Academy of Sciences, 1976.

83. National Research Council, Committee on Satellite Communications, "NASA 5-Year Planning, Fiscal Years 1979 through 1983," draft report, National Academy of Sciences, April 18, 1978, p. E-19. See also "ACTS Advanced Communications Technology Satellite Program," NASA, Office of Space Science and Applications, undated (c. 1980).

In a surprise move, Hughes Communications Galaxy (the successor to Hughes Aircraft) filed with the FCC to launch two satellites in 1988 and 1989. The satellites would have spot beams and on-board signal processing, would use the Ka band, and would be financed entirely by private capital. Hughes then proceeded to argue that NASA's program subsidized RCA to build a similar although somewhat more advanced satellite and so was unfair and would discourage private investment.[84] In the wake of the filing, and despite skepticism as to whether Hughes actually intended to go ahead with the launches, $90 million for the ACTS program was cut from the fiscal 1985 budget. White House science adviser George Keyworth stated that he supported the cancellation. "The more deeply I looked into 30/20 GHz, the more I found substantial development going on in industry and military. . . . I find it extremely difficult to understand why NASA is subsidizing the communications satellite industry."[85] In 1989 the Bush administration reiterated Keyworth's position.[86]

The ACTS program appears victim to the same government ambivalence over funding communication satellite technology that characterized the later years of the ATS program—and probably for the same reasons. The program has been periodically reorganized, subjected to budget cuts, and threatened with outright cancellation. Its technological advances have not been demonstrated elsewhere. As of early 1991 the satellite was under construction and its launch was planned for 1992.

Politics of the ATS Program

Changes in the politics of the ATS program are evident from the history of ATS-F. First, the structure of the satellite industry had changed so that the distributive aspects of the program were a political liability. Second, opposition from firms that owned satellites constrained NASA's programs to narrower applications. Third, there was an evolution toward technologies that had longer R&D lead times before benefits started to accrue. Finally, commercial satellite technology had lost political saliency as a broad national concern.

84. Office of Technology Assessment, "Satellite Communications," pp. 6-77–6-80.

85. "ACTS Battle Continues," *Satellite Week*, vol. 6 (February 6, 1984), p. 2. Although military research on satellites continued unabated during the 1970s, its applicability to commercial applications is disputed because it addresses other questions, in particular, security and communication with a small number of mobile terminals.

86. National Research Council, *NASA Space Communications R&D: Issues, Derived Benefits, and Future Directions* (Washington: National Academy Press, 1989).

Competition among spacecraft manufacturers changed the contracting process for ATS-F from what it had been for ATS-1 through ATS-5. A contract award became a bitter contract dispute. The satellite program was not organized to support the entire industry; rather it provided benefits to the prime contractor. The battle was refought over ACTS. NASA's assistant administrator for applications, Burton Edelson, stated in 1983 (before Hughes filed) that the ACTS contractor would have a "great advantage" over competitors in the communications satellite business.[87] Hughes's contention that it is unfair to have to compete with a federally sponsored company, however self-serving the argument, nonetheless carried political weight. The program's distributive consequences were negative because a competitive industry caused the contracting process to create more political losers than winners.

The satellite program was not organized to minimize distributive liabilities. The alternative strategies are centralization, whereby industry cooperates in the project and shares the technology (as in the case of breeder reactors), or fragmentation, whereby the program is cut up in small pieces to give more firms a piece of the action (as with the space shuttle). The ATS program gave one firm a short-term advantage, even if the technology were made public.

The objections of the satellite common carrier firms to the program are well documented. Comsat voiced concern because a satellite, once launched, becomes a convenient way of providing operational services. Despite radio frequency changes on ATS-F and the effort to restrict experimentation to new users, Comsat and later other satellite-operating firms continued to oppose the program.

ATS-1 and ATS-3 demonstrated that NASA's satellites could be used on a long-term basis for domestic communications. Once domestic satellites were approved by the FCC, continuation of NASA's program threatened private enterprises. By 1973, when the program was canceled, the FCC was approving applications for domestic satellite systems.[88] Comsat and seven other companies proposed domestic satellite services, and they formed a chorus of opposition to NASA's program.[89] In 1978 the Office of Science and Technology Policy surveyed industry and government experts

87. *Defense Daily*, February 23, 1983, p. 286.

88. Federal Communications Commission, "Second Report and Order, in the Matter of Establishment of Domestic Communication Satellite Facilities by Non-Governmental Entities," FCC no. 16495, 35FCC 2d (1972), p. 844.

89. Smith, *Communication via Satellite*, pp. 156–85.

on the federal role in satellite R&D. It reported consensus on the assessment that such R&D was not proceeding rapidly enough and that the United States might well lose its position of technological leadership. The survey concluded that the private sector was reluctant to undertake "many major technology developments that require complex flight tests because of the large cost involved, the long time before financial return, and the high risk." Moreover, the dearth of private R&D investment was due to the difficulty of identifying sufficiently large markets.[90]

NASA's earlier emphasis on demonstrating satellite applications while developing user groups attacked both technical and market uncertainty. Developing new user groups creates a classic free-rider problem for private firms. A single firm could fear that once it developed and marketed a new service to new users, competitors could quickly follow, preventing the innovator from recovering R&D costs. Nevertheless, the satellite carriers unanimously opposed government provision of any user services on a demonstration basis. They argued that government should broaden the application of satellites by supporting applications on existing satellites or, for new public services, by sponsoring applications on a satellite that would be built, operated, and financed by the private sector.[91] The carriers did agree that government had a role in developing new technology (when there was high risk or long-term costs) and even doing experimental flight tests, but not user demonstrations.

The third change for the ATS program was the shift to long-term economic benefits. Because the economic benefits of the first generation of ATS satellites were evident within a very few years after the program began, they could be of consequence in the political calculation of program benefits. In the early 1970s it became clear that economic benefits from ATS-F and subsequent satellites were not coming as soon. While estimates of the discounted net benefits of the program were positive, the payoffs would accrue a decade or more later. Such a delay diminished the political relevance of the program.

Several factors constrained and delayed the benefits of the second generation of ATS satellites. The delay was caused in part by controversy over the use of satellite technology. In 1969 the National Academy of Sciences concluded that the use of communications satellites in the United

90. Office of Science and Technology Policy, "Summary of Survey on the Federal Role in Satellite Communications Research and Development" (Washington, July 1, 1978), p. 8.

91. Office of Science and Technology Policy, "Summary of Survey on the Federal Role."

States was limited by policy rather than technology.[92] The FCC's inquiry into regulations for domestic satellite service lasted from 1966 to 1972.[93] While the commission's 1972 decision established competitive domestic satellite systems, regulation restricted their commercial use. For instance, not until 1977 did the FCC approve small, receive-only earth stations (4.5 meters in diameter). This allowed economical satellite service for small cable television systems with, according to one observer, "dramatic effects upon the entire industry."[94] In 1979 the FCC deregulated receive-only earth stations altogether.

Public service applications were also constrained by politics. In part because of Comsat's initial opposition, the second generation ATS program focused on new users with no previous experience with satellite communications. Moreover, NASA used radio frequencies for domestic experiments that were not used commercially. Consequently the ground equipment developed for the ATS-F experiments could not be used with any other satellite, and the users, however enthusiastic, could not continue operations at the conclusion of the experiment.

The change in the fortunes of the ATS program as the satellite industry grew raises an interesting hypothesis about the relationship of politics to frequently stated objectives of federal R&D programs. The second ATS program produced long-term economic benefits and faced high risks owing to the discrete, all-or-nothing nature of flight tests. Many of the benefits were likely to accrue to new user groups who were as yet not fully aware of the potential of satellite technology, who were too small, diffuse, and disorganized to sponsor their own R&D, and who were unattractive candidates for supplier R&D because, in part, marketing innovations probably were not appropriable. All these circumstances are commonly cited rationales for a federally supported commercial R&D program. Yet these same characteristics made the programs less viable politically. The long-term benefits and high risks proved unappealing to officials as well as private parties, and the constituency of unorganized beneficiaries proved less appealing than the constituency of organized industry opponents. Hence, the sources of a strong economic rationale for the program also constituted causes of political obstacles to its continuation.

92. National Research Council, *Useful Applications of Earth-Oriented Satellites, Point-to-Point Communications* (National Academy of Sciences, 1969), p. 3.

93. See Smith, *Communication via Satellite*, for a discussion of the FCC case.

94. N. J. Nicholas, Jr., "Consumer Demand for Video Alternatives," in Space Applications Board and the Board on Telecommunications-Computer Applications, National

Still another political change hindered the program. In 1961 when President Kennedy outlined his space policy, the nation was still reacting to Sputnik. The manned space program had just begun, and significant accomplishments for the program were still years away. Communications satellites offered important short-term commercial and defense payoffs as well as a means of reclaiming U.S. leadership in a technology having widely recognized economic and political significance. By the 1970s, manned moon flights were a regular event, and the nation was planning the space shuttle, the Viking mission to Mars, and the Voyager grand tour of the outer planets. Moreover, the competition with the Soviet Union for supremacy in space had been won in a rout, diminishing interest in continued spending on space technology. In this milieu, advanced technologies for communications satellites seemed prosaic and unlikely to capture the imagination of the public. Advocates of space had set their sights on far more dramatic challenges. Thus the case for satellites had to be made on the relatively stolid ground of long-term economic payoffs and amidst yawns or even opposition from the leading advocates of the space program. Such is not the stuff of which prominent symbolic political issues are made.

Research Council, *Direct Broadcast Satellite Communications: Proceedings of a Symposium* (National Academy of Sciences, 1980), p. 38.

Chapter 8

The Space Shuttle

JEFFREY S. BANKS

On April 12, 1981, the maiden flight of the space shuttle *Columbia* began, three years past the original launch date, 30 percent over the original estimated R&D costs, and 120 percent over the original estimated average cost per flight for the first three years of operation. On January 28, 1986, a seal failed on the right solid rocket booster of the shuttle *Challenger*, leading to an explosion seventy-three seconds into the launch and the loss of the seven crew members on board.

Once praised as a project combining the desire of the scientists and engineers of the National Aeronautics and Space Administration (NASA) to build a multipurpose, piloted vehicle with the demands of the market for inexpensive access to space, the shuttle had become a costly, noncompetitive system without the performance capabilities initially envisioned. Because of serious doubts about its reliability and no remaining economic justification, continued support for the shuttle could be rationalized only by the necessity of the shuttle for further U.S. endeavors (for example, a space station), and by the public benefits derived from maintaining a manned presence in space. Both aims depended on the future success of shuttle missions. After the *Challenger* disaster, NASA and the U.S. government could not use the shuttle to deliver any benefits, public or private, with a sufficiently high probability.

The theory derived in chapter 4 generates predictions about which potentially commercial projects Congress will undertake and how Congress will react as new information about the costs and benefits of the chosen project comes to light. In the next section I examine the choice of the shuttle as the preferred launch system for the United States, the reaction of Congress to overoptimistic economic estimates about the shuttle, and the sensitivity of Congress to issues of safety and reliability raised during the R&D phase of the program. Subsequently I derive an econometric model of legislative behavior on NASA authorizations during the years 1970–82, treating these as a proxy for congressional preferences regarding the space shuttle program. In particular, I wish to test whether distributive effects and project performance significantly influenced the voting calculus of elected officials in Congress.

Economic History

The Nixon administration in January 1969 inherited an American space program on the threshold of accomplishing the goal set forth at the beginning of the decade, namely, putting an American on the moon. Yet the program lacked a clear understanding of its subsequent plans for the 1970s and beyond. Consequently, President Richard M. Nixon on February 13, 1969, issued a memorandum requesting "direction" for future space programs from a newly formed Space Task Group (STG), headed by Vice-President Spiro Agnew.[1] With public awareness and approval of NASA programs apparently running high, Agnew took to the forefront an "Apollo of the 70s" plan for a manned Mars expedition. Encouraged by such optimistic talk coming from inside the administration, NASA, in August 1969, presented to the STG an ambitious program for a Mars expedition that included earth-orbital and lunar-orbital space stations, a lunar base, and an earth-to-orbit shuttle system.[2] The plan would require annual NASA budgets between $6 billion and $10 billion, whereas previous NASA budgets had peaked at $5.25 billion (fiscal 1965, the height of the Apollo project) and were targeted at $3.715 billion for the new fiscal year (1970).

As NASA's plans for the future became public, reactions by Congress, the press, and the public were swift and decidedly negative. Representative George Miller (Democrat of California), chairman of the House Committee on Science and Astronautics, called the plans premature, as did editorials in the *New York Times*. Senator Clinton Anderson (Democrat of New Mexico), chairman of the Senate Committee on Aeronautical and Space Sciences, commented, "Now is not the time to commit ourselves to the

1. John M. Logsdon, "The Policy Process and Large-Scale Space Efforts," *Space Humanization Series*, vol. 1 (Washington: Institute for the Social Science Study of Space, 1979), p. 67.

2. Logsdon, "Policy Process," p. 71.

goal of a manned mission to Mars."[3] The Gallup Poll reported that 53 percent of those polled disapproved of the project, and only 39 percent were in favor of it.[4]

Taken aback by such vocal and unambiguous dissent, the final STG report, issued in September 1969, referred to a manned Mars expedition only as a "long-range option." NASA initially scaled the program back to include the earth-orbital space station and shuttle system, but in the face of further fiscal restraint and declining near-term budgets, also dropped the space station, leaving the shuttle as the only major NASA program for the 1970s.

Preliminary Considerations

A 1969 joint study by NASA and the Defense Department identified a two-stage (orbiter and booster), fully reusable vehicle as the shuttle configuration with the most promise.[5] Both stages would have a two-person crew and use high-pressure, hydrogen-oxygen engines. The booster stage would burn to forty miles altitude, where it would disengage from the orbiter and be flown back to earth for a horizontal landing. The orbiter would continue to climb to a working altitude of 150 to 300 miles, where it would perform its mission-specific duties (for example, deploying satellites) and then return to earth for a horizontal landing. The shuttle would be designed to replace all existing expendable launch vehicles, except for the (very small) Scout and the (very large) Saturn. Other preliminary characteristics of the shuttle included the following:

—active flight operation up to one week in orbit;

—minimum turnaround time of two weeks on the ground;

—low acceleration during launch and landing, permitting passengers not trained as astronauts;

—a payload compartment on the orbiter measuring 15 feet by 60 feet;

—a 400,000-pound thrust engine;

—a vehicle life of one hundred missions or more;

—a fleet of four to six orbiters;

—a system life of twelve years; and

—first manned orbital flight (FMOF) in 1978, with initial operating capacity (IOC) in 1979.[6]

3. Logsdon, "Policy Process," p. 72.
4. Logsdon, "Policy Process," p. 73.
5. *1971 NASA Authorization*, Hearings before the House Committee on Science and Astronautics, 91 Cong. 2 sess. (Government Printing Office, 1970), vol. 1, p. 57.
6. *1971 NASA Authorization*, Hearings, pp. 188–92.

Another element of the space shuttle system would be a reusable orbit-to-orbit space tug, capable of delivering payloads to geosynchronous orbit and beyond, along with the ability to retrieve payloads from geosynchronous orbit for refurbishment and reuse.[7] As with launch systems, previous technology depended on expendable upper stages to deliver payloads into orbit.

In deciding whether to proceed with the space shuttle, NASA and the administration, for the first time, sought an economic justification of the program. Previous large-scale space programs found justification in the continuation of a manned U.S. presence in space, along with the evolution of accessing space for future use. Those programs did not have to pass a cost-benefit test. The shuttle, however, had its beginnings in an era of increasing skepticism about the usefulness of NASA's manned space program. NASA thus sought a solid economic rationale for the program.

On May 31, 1971, Mathematica, Inc., under contract to NASA's Office of Manned Space Flight, reported an analysis of the economic value of a fully reusable shuttle system, without consideration of which particular configuration would be chosen.[8] Two previous studies, by Lockheed and the Aerospace Corporation, provided data on optimal payload configurations using the shuttle and tug and identified the payload and launch costs and benefits of using the shuttle system as opposed to an expendable launch system.[9] The decision criterion employed by Mathematica would be the comparison programs' total life-cycle costs. These costs include launch system life-cycle costs (research and development, investment, twelve-year operating costs for the shuttle and tug) and payload system life-cycle costs (research and development and investment costs for the payload).

Using a baseline mission model of 736 flights (over a twelve-year period) and a social discount rate of 10 percent, Mathematica determined that a fully reusable shuttle was economically preferable if the present value of

7. President's Science Advisory Committee, "The Next Decade In Space: A Report of the Space Science and Technology Panel of the President's Advisory Committee," March 1970.

8. Mathematica, Inc., *Economic Analysis of the Space Shuttle System: Executive Summary*, prepared for the National Aeronautics and Space Administration (Washington, January 31, 1972).

9. Lockheed Missiles and Space Company, "Payload Effects Analysis," June 30, 1971; and Aerospace Company, "Integrated Operations/Payload/Fleet Analysis: Final Report," January 1972.

the nonrecurring design, development, test, and evaluation (DDTE) and investment costs of the shuttle and tug were less than $17.0 billion. At the time, these costs were estimated at $12.8 billion; hence the decision to proceed with development of the shuttle and tug was economically justified. Further, the study indicated that "the direct costs (payload and transportation) of space activity by a space shuttle system are expected to be about one-half of the direct costs of the current expendable transportation system."[10] As the Aerospace study noted, 49 percent of the net economic benefits of the shuttle system compared with an expendable system were due to the payload retrieval capability of the tug. Both the Aerospace and Mathematica studies assumed that a fully reusable tug would be developed concurrently with the shuttle.

In mid-1971 NASA was dealt a severe blow when the Office of Management and Budget (OMB) announced that NASA's annual budget would be held fixed at its fiscal 1971 level of $3.2 billion (adjusted for inflation) for the foreseeable future.[11] With a peak annual spending of $1.8 billion, the two-stage, fully reusable shuttle left meager funds for other (nonshuttle) projects. Consequently NASA began to reexamine the shuttle design for one more suited to the imposed budgetary constraints. The main alternative configurations studied were manned, flyback booster stages with expendable external fuel tanks, and unmanned booster stages with expendable external tanks using either pressure-fed boosters or solid rocket motors, aligned either in series or in parallel (twin).[12] The twin pressure-fed and solid rocket designs together were labeled thrust-assisted orbiter system (TAOS). Moving from reusable to expendable fuel tanks and manned to unmanned booster stages lowered the nonrecurring costs while raising per flight costs, as depicted in figure 8-1.

Subsequently, Mathematica was recontracted to address the question of the efficient shuttle design among those deemed feasible when compared with an expendable launch system.[13] On January 31, 1972, Mathematica issued an updated study of the shuttle system, which showed that a TAOS shuttle system was economically justified compared with an expendable

10. Mathematica, *Economic Analysis*, p. 5.

11. John M. Logsdon, "The Space Shuttle Decision: Technology and Political Choice," *Journal of Contemporary Business*, vol. 7 (1979), p. 15.

12. *Space Shuttle–Skylab: Manned Space Flights in the 1970s*, Status report for the Subcommittee on NASA Oversight of the House Committee on Science and Astronautics, 92 Cong. 2 sess. (GPO, 1972), pp. 262–66.

13. Mathematica, *Economic Analysis*, pp. 14, 26.

Figure 8-1. *Trade-offs between Nonrecurring and Recurring Costs, Median Values*

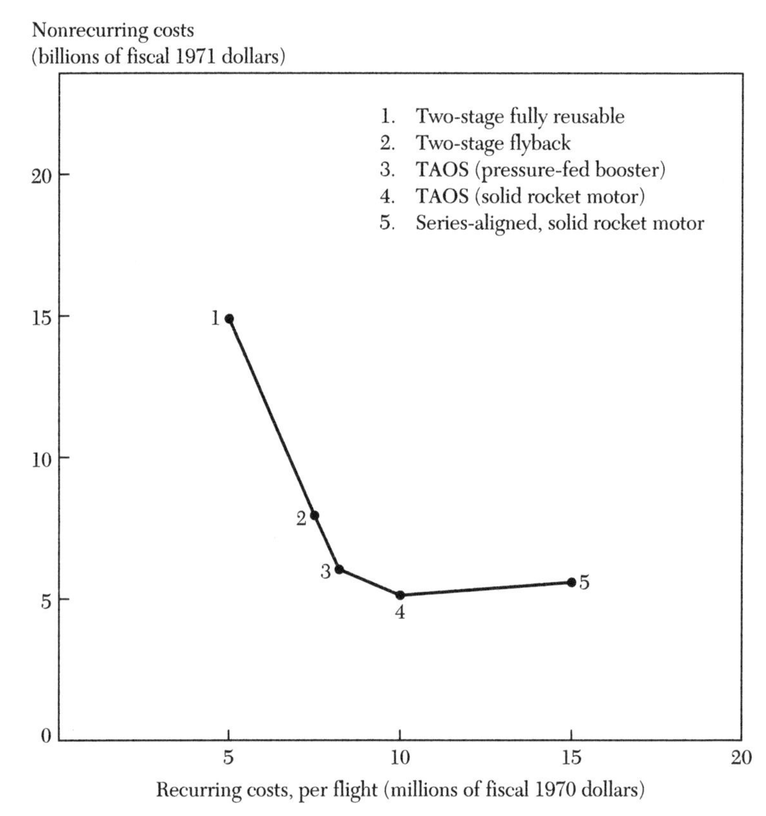

Source: Mathematica, Inc., *Economic Analysis of the Space Shuttle System: Executive Summary*, prepared for the National Aeronautics and Space Administration (Washington, January 31, 1972).

system (using a 10 percent discount rate), and a TAOS shuttle system was the most economic configuration among those studied.[14]

The Early Phase

On March 15, 1972, after President Nixon's go-ahead for the shuttle in January, NASA announced that the twin solid rocket motor configuration would be the chosen shuttle design. On March 17, 1972, NASA requested proposals for the DDTE phase of the project, which was scheduled to last six years and would include production of the first two orbiters. Along with

14. Mathematica, *Economic Analysis*, p. 27.

the choice of design, NASA issued a Space Shuttle fact sheet, listing the following cost estimates for the shuttle (fiscal 1971):[15]

DDTE	$5.15 billion
Facilities	$300.0 million
Each additional orbiter	$250.0 million
Cost per flight	$10.5 million
Peak annual spending	$1.2 billion

These cost estimates were derived from a shuttle system that differed in two ways from the design studied by Mathematica. First, the space tug would begin operations in 1985 rather than 1979; second, the system would have 581 flights rather than 736.

The economic justification of the shuttle was obviously sensitive to the point estimates used in the cost-benefit analysis of Mathematica. A report issued by the General Accounting Office in June 1972 concluded that, with the above-noted changes in the system, an increase of only 20 percent in DDTE, production, and operation costs would eliminate the shuttle's economic superiority over an expendable launch system.[16] That did not leave much room for error on the part of NASA and its contractors.

By November 1973 all the main shuttle contracts were secured. North American Rockwell of Downey, California, was awarded the prime shuttle contract, which included development of the orbiter as well as overall system integration. Rocketdyne in Canoga Park, California, a division of Rockwell, would provide the space shuttle main engines; Martin Marietta of Denver, Colorado, would provide the external fuel tanks; and Thiokol Chemical of Brigham City, Utah, would provide the solid rocket motors. The subsystem breakdown of the DDTE expenditures is found in table 8-1.

As noted, the space tug was an integral factor in the economic justification for the space shuttle. Not surprisingly, the tug was subject to the same trade-offs between DDTE and cost per flight as was the shuttle. The tug was also subject to greater technical uncertainty, because retrieving payloads would require novel propulsion and guidance systems. For example, a tug that was itself recoverable but that did not possess the ability to recover payloads would have a DDTE cost of $200 million and a per flight cost of $2 million, and would be able to deploy 5,000 pounds into a

15. *Second Supplemental Appropriations for FY 72,* Hearings before the Senate Committee on Appropriations, 92 Cong. 2 sess. (GPO, 1972), pp. 159–65.

16. General Accounting Office, *Cost-Benefit Analysis Used in Support of the Space Shuttle Program,* June 1972.

Table 8-1. *Design, Development, Test, and Evaluation Estimates for the Space Shuttle Subsystems, Fiscal Year 1972*
Millions of fiscal 1971 dollars

Subsystem	*Estimate*
Orbiter	3,513
Space shuttle main engines	580
External tanks	331
Solid rocket boosters	390
Launch and landing	336
Total	5,150

Source: *1981 NASA Authorization*, Hearings before the Subcommittee on Space Science and Applications of the House Committee on Science and Technology, 96 Cong. 2 sess. (Government Printing Office, 1980), vol. 4, p. 1304.

geosynchronous orbit. A full capability tug would be able to deploy 8,000 pounds and retrieve 4,000 pounds from geosynchronous orbit, and it would have a DDTE cost of $700 million and a per flight cost of $1.5 million.[17]

Delaying the expected completion date of the space tug to 1985 opened a six-year gap from the planned first launch in which the shuttle would be incapable of launching many of its payloads. To fill this gap, NASA and the Defense Department agreed to a plan, in November 1973, whereby the department would develop an initial upper stage (IUS), without payload retrieval capability, which would be a modification of an existing expendable stage.[18] The reusability of the IUS would remain under study. The IUS would be available for both defense and civilian payloads, and NASA would establish the shuttle and ground system interfaces with it. In addition, NASA would continue the study of, and planning for, the fully reusable space tug. A primary factor in the decision to phase in the tug to full reusability status was the timing of the DDTE funds. As NASA noted, "Development of a (fully reusable) tug based on current technology and available for use on initial shuttle flights required resources considerably beyond those expected to be available to NASA during the interval prior to the shuttle operation capability date."[19]

In 1976 NASA began formulating its pricing policy for users of the space shuttle. The objective of the policy would be to recover all operation costs

17. *Space Tug—1973*, Hearings before the Subcommittee on Manned Space Flight of the House Committee on Science and Astronautics, 93 Cong. 1 sess. (GPO, September 1973), p. 20.

18. *1975 NASA Authorization*, pts. 2 and 3, Hearings before the Subcommittee on Manned Space Flight of the House Committee on Science and Astronautics, 93 Cong. 2 sess. (GPO, 1974), pp. 186–87.

19. *1975 NASA Authorization*, Hearings, pp. 186–87.

of the shuttle over its lifetime, as well as the pro rata share of depreciation for the use of facilities and equipment and the amortization of the space shuttle investment from commercial and nongovernment users.[20] At the time it was thought that the operation costs of the shuttle would decrease significantly over time; hence, to maintain incentives for early use of the shuttle, NASA would set the price for the first three years at the estimated average operation cost over the twelve-year life of the shuttle, plus the capital charge. In 1977 NASA determined this price to be $18 million (fiscal 1975 dollars).[21]

The Test Flight Phase

On February 18, 1977, the test flight phase of the shuttle program began as the (unmanned) orbiter *Enterprise*, mounted on a Boeing 747, flew about Edwards Air Force Base in California. In July the piloted flights began, and on August 12 the first orbiter free flight was successfully completed. In October the fifth and final orbiter free flight was completed; at the time, the first piloted orbital flight was set for March 1979, with an initial operating capacity in August 1980.

The original plan for DDTE spending was for a peak in fiscal 1977 and fiscal 1978 followed by a fairly sharp reduction. In addition, fiscal 1977 would be the first year of "production" funds aimed at the manufacture of the third, fourth, and fifth orbiters. For fiscal 1977, $1.288 billion (then-year dollars) was authorized by Congress for shuttle DDTE. An increase of $55 million was provided from the following sources: Economic Stimulus Appropriations Act of 1977, $25 million; Residual Apollo-Soyuz Test Project Fund, $27 million; and various space flight operations funds, $3 million.[22]

For fiscal 1978, another $1.2 billion was authorized for DDTE, to which was added $100 million from "shuttle production."[23] When NASA's fiscal 1978 budget was submitted to Congress in early 1977, the DDTE runout estimates were still within 1.4 percent of the original estimate of $5.15

20. *Operational Cost Estimates, Space Shuttle*, Report prepared by the Subcommittee on Space Science and Applications of the House Committee on Science and Technology, 94 Cong. 2 sess. (GPO, 1976), pp. 2–3.

21. *Space Transportation System*, Hearings before the Subcommittee on Space Science and Applications of the House Committee on Science and Technology, 95 Cong. 1 sess. (GPO, 1977), pp. 151–54.

22. *Oversight: Space Shuttle Cost, Performance, and Schedule Review*, Hearing before the Subcommittee on Space Science and Applications of the House Committee on Science and Technology, 96 Cong. 1 sess. (GPO, 1979).

23. *Oversight*, Hearing, p. 142.

Table 8-2. *Supplemental Appropriations for the Space Shuttle Subsystems, Fiscal Year 1979*
Millions of dollars

Subsystem	*Total funds*	*DDTE*	*Production*
Orbiter	61.5	118.4	−56.9
Space shuttle main engine	48.0	−15.3	63.3
External tanks	27.1	27.1	...
Solid rocket boosters	36.7	36.7	...
Launch and landing	19.5	18.1	1.4
Spares and equipment	−7.8	...	−7.8
Total	185.0	185.0	...

Source: *1980 NASA Authorization*, Hearings before the House Committee on Science and Technology, 96 Cong. 1 sess. (GPO, 1979), p. 687.

billion (fiscal 1971 dollars). In January 1978, NASA submitted its fiscal 1979 budget to Congress, maintaining a March 1979 FMOF. By June, delays in production schedules led to rescheduling the FMOF for June 1979. In addition, by May 1978, subcontractors began reporting impending cost overruns to Rockwell. On September 29, 1978, Rockwell officials notified the Johnson Space Center that these overruns would cause budget problems in fiscal 1979 and 1980. In January 1979, after a destructive failure of a main engine on December 27, 1978, NASA again revised the FMOF to November 1979 and requested fiscal 1979 supplemental appropriations of $185 million, which Congress subsequently approved on March 28, 1979.[24] The subsystem breakdown of the supplemental is contained in table 8-2. In addition, shuttle DDTE for fiscal 1979 received another $70 million to $90 million from shuttle production.[25]

In April 1979, NASA issued the following revisions in the shuttle program: a flight operating plan to provide for four orbital test flights, instead of six; a production slip of undetermined duration; a reduction in the number of flights; and a consideration of transferring $200 million in fiscal 1980 funds from production to DDTE.[26]

Instead of transferring more production funds, NASA on May 14, 1979, submitted a $220.0 million fiscal 1980 budget amendment to Congress, which was subsequently approved, raising the fiscal 1980 DDTE funds to

24. *1981 NASA Authorization (Program Review)*, Hearings before the Subcommittee on Space Science and Applications of the House Committee on Science and Technology, 96 Cong. 1 sess. (GPO, 1979), vol. 1, p. 456.

25. *Oversight*, Hearing, p. 142.

26. *1981 NASA Authorization*, Hearings.

$830.5 million. The budget amendment also eliminated $27.0 million for the fifth orbiter previously approved by thc full House. In subsequent years there would be no funds allocated to a fifth orbiter. At the time the estimated DDTE cost was at $5.986 billion (fiscal 1971 dollars), an increase of 16 percent over the initial estimates, whereas in January 1979 the numbers were $5.654 billion and 9.8 percent, respectively.[27]

In response to these mounting problems, NASA, in May 1979, commissioned a group of consultants to undertake a factfinding study to provide information about the shuttle management processes. Their major findings were as follows:

—The original cost commitment for shuttle development established an austere fiscal environment at the beginning of the program. This environment become more constraining under the annual budgets established in subsequent years.

—In the effort to live with budget limitations while still progressing acceptably toward completion, shuttle management had generally set up work schedules that demanded more performance than could be delivered.

—There had been a lack of adequate long-range planning and timely status reporting. Emphasis was on the current fiscal year, with only secondary attention to succeeding years and estimates of completion.[28]

These findings dealt only with the continuing difficulty of NASA and its contractors to meet cost limitations and production deadlines. They did not examine the implications for the economic viability of the shuttle program as originally conceived.

In November 1979, as a result of continuing shuttle problems, the program began seriously to erode NASA's civilian space program, when the OMB rejected three of the four new starts requested by NASA in its fiscal 1981 budget. These were as follows:

—the solar electric propulsion system, with a proposed fiscal 1981 budget of $20 million and three-year R&D runout of $262 million (this effectively killed the U.S. involvement in the proposed U.S.-European Halley's Comet/Tempel 2 probe, which carried an additional NASA price tag of $400 million over three years);

—NASA involvement in a NASA–Defense Department–Department of Commerce national oceanic satellite system, with proposed fiscal 1981 funds of $6 million and six-year runout of $179 million; and

27. *Oversight*, Hearing, p. 146.
28. *1981 NASA Authorization*, pp. 409–10.

Table 8-3. *Supplemental Appropriations for the Space Shuttle Subsystems, Fiscal Year 1980*
Millions of dollars

Subsystem	*Funding*
Orbiter	140.1
Space shuttle main engine	0
External tank	11.0
Solid rocket booster	3.7
Launch and landing	45.2
System upgrading	100.0
Total	300.0

Source: *1981 NASA Authorization*, Hearings, vol. 4, pp. 1283–84, 1287.

—the shuttle power extension package, a device to extend shuttle orbiter stays from seven to twenty days, with $17 million proposed for fiscal 1981 and a three-year runout of $159 million.[29]

In November 1979 there was a second malfunction of a main engine during testing, delaying the FMOF to August 1980.[30]

In February 1980 NASA was forced to request another $300 million in fiscal 1980 DDTE supplemental appropriations, with the subsystem breakdown as shown in table 8-3. The estimated DDTE was now $6.185 billion (fiscal 1971 dollars), an increase of 20 percent, with the subsystem increases as shown in table 8-4. In addition, the estimated number of flights for the twelve-year operational life of the shuttle stood at 487.[31]

The development of the IUS by the Defense Department was encountering problems similar to those of the shuttle. In September 1976, Boeing was awarded the initial validation phase contract. Also in 1976 McDonnell Douglas began development of a spinning solid upper stage (SSUS). With a smaller payload capability than that of the IUS, the SSUS was aimed more specifically at the commercial communications satellite market. The IUS would be able to put 5,000 pounds into a geosynchronous orbit, while the SSUS would come in two versions: the SSUS-D, capable of putting 2,200 pounds into geosynchronous orbit, and the SSUS-A, able to put 4,400

29. *Aviation Week and Space Technology*, November 26, 1979, p. 20; and *1981 NASA Authorization*, Hearings before the Subcommittee on Space Science and Applications of the House Committee on Science and Technology, 96 Cong. 2 sess. (GPO, 1980), vol. 4, pp. 1307–08.

30. *Aviation Week and Space Technology*, November 12, 1979, p. 20.

31. *1981 NASA Authorization*, Hearings, p. 1325.

Table 8-4. *Changes in Estimated Costs of the Space Shuttle Subsystems, Fiscal Years 1973 and 1981*
Millions of fiscal 1971 dollars

Subsystem	*1973 estimate*	*1981 estimate*	*Percent change*
Orbiter	3,513	4,097	17
Space shuttle main engine	580	903	56
External tank	331	381	15
Solid rocket booster	390	343	−12
Launch and landing	336	461	37
Total	5,150	6,185	20

Source: *1981 NASA Authorization*, Hearings, vol. 4, p. 1304.

pounds into geosynchronous orbit.[32] All Defense Department missions requiring an upper stage were to use the IUS, as would some NASA missions.

The full-scale development of the IUS by Boeing began in March 1978, with an estimated completion date of July 1980, and a first shuttle usage in May 1980 for the tracking and data relay satellite (TDRS).[33] By late 1979 Boeing began reporting problems with the solid rocket booster motor and the software, necessitating a revision of the schedule that allowed for a first shuttle interface in July 1981. Fortunately, the problems with the shuttle and its subsequent rescheduling gave a September 1981 date for the TDRS launch. However, the technical problems on the IUS also led to cost overruns of over 45 percent in the development phase. Initially, it was estimated that the DDTE runout for the IUS would be $270 million (fiscal 1978 dollars). By early 1980 the Defense Department was reporting a cost overrun of $144 million, or about $125 million in fiscal 1978 dollars.[34] The department was forced to reprogram $39 million in fiscal 1980 funds from various other projects, raising fiscal 1980 IUS spending from $66 million to $105 million, and placed the balance of the overrun in their fiscal 1981 and fiscal 1982 budget requests.

In March 1981 the Reagan administration announced cuts totaling $603.5

32. *1978 NASA Authorization*, Hearings before the Subcommittee on Space Science and Applications of the House Committee on Science and Technology, 94 Cong. 2 sess. (GPO, 1976), pp. 346–47.

33. *Hearings on Military Posture and H.R. 6974*, Hearings before the House Committee on Armed Services, 96 Cong. 2 sess. (GPO, 1980), pt. 4, bk. 1, pp. 1833–41.

34. *Hearings on Military Posture*, p. 1835.

million in former President Jimmy Carter's original fiscal 1982 budget, cuts that included start-up funds (again) for the National Oceanic Satellite System and the service-module electric power system, and delayed development funds for the gamma ray observatory (the fiscal 1981 start that made it past the OMB) and the Venus orbiting imaging radar mission. The cuts also freed $60 million as insurance against further shuttle delays.[35]

The Post-Launch Phase

On April 12, 1981, the first launch of the shuttle *Columbia* was successfully accomplished. *Columbia* landed two days later at Edwards Air Force Base after a nearly perfect mission. In June Congress overwhelmingly approved NASA's fiscal 1982 budget, which included $740 million in shuttle DDTE funds and $1.19 billion for production. Fiscal 1982 would be the last year in which DDTE appeared as a line item in the budget. In 1982 NASA declared the shuttle operational, and starting in fiscal 1983 all NASA funds specifically for the shuttle (that is, excluding "space flight operations") were designated as for either production or "systems upgrading." To the $740 million in DDTE funds for fiscal 1982 was added another $158 million as the year progressed. The actual DDTE expenditures would thus turn out to be $6.748 billion (fiscal 1971 dollars), an increase of 31 percent over the original estimate of $5.15 billion.

Table 8-5 gives the time series of the nonrecurring cost estimates for the shuttle, tug, and related facilities. Beginning in fiscal 1979, the line item "shuttle investment" included two production orbiters where previously it had included three, and after fiscal 1980 NASA was not even reporting a line item "space tug DDTE and investment," for by this time the tug had been abandoned as infeasible during the foreseeable horizon of the shuttle program.

Figure 8-2 shows the actual time series of DDTE expenditures, along with the estimated time series from fiscal 1973. Figure 8-3 gives the changes over time of the estimated FMOF and IOC; figure 8-4 gives the changes in the number of missions estimated for the twelve-year operational life of the shuttle system.

For the recurring costs of the shuttle, a General Accounting Office report released in February 1982, and based on data from September 1980, showed that the average cost per flight of the shuttle had increased 73

35. *Aviation Week and Space Technology*, March 16, 1981, p. 24.

Figure 8-2. *Actual and Estimated Expenditures for Design, Development, Test, and Evaluation, Fiscal Years 1971–82*

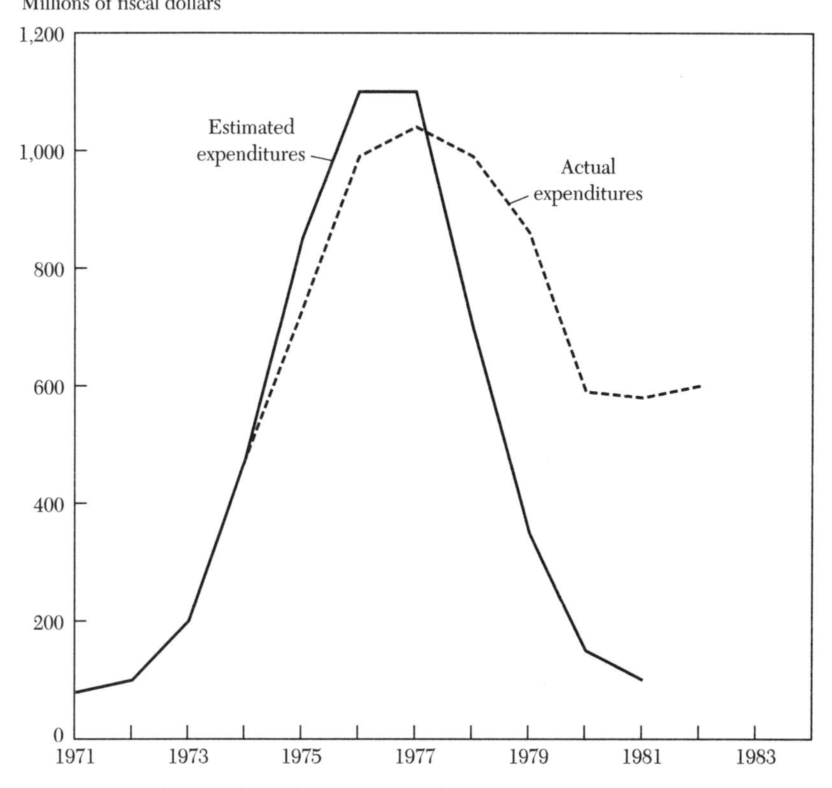

Source: *NASA Authorization for Fiscal 1974*, Hearings before the Senate Committee on Aeronautical and Space Sciences, 93 Cong. 1 sess. (GPO, 1973), p. 150; and *Congressional Quarterly* data.

percent, from $16.1 million to $27.9 million (fiscal 1975 dollars).[36] Table 8-6 gives the breakdown of the cost increases. In addition, the average cost per flight for the first three years of the shuttle's operational life (now fiscal 1983–85) had more than doubled, from $26.2 million in June 1976, to $57.5 million in September 1980.[37] With the price for a shuttle launch fixed for non-NASA users during the period, NASA would end up subsidizing Defense Department and other users a whopping $1.18 billion from fiscal 1983–85 (fiscal 1975 dollars).

36. General Accounting Office, *NASA Must Reconsider Operation Pricing Policy to Compensate for Cost Growth on the Space Transportation System: Report to the Congress* (February 1982), p. 7.

37. GAO, *NASA Must Reconsider*, p. 14.

Table 8-5. *Nonrecurring Cost Estimates for the Space Shuttle, Tug, and Related Facilities, Fiscal Years 1973–82*
Billions of fiscal 1971 dollars

Year	*Shuttle DDTE*	*Shuttle investment*[a] *Total*	*Shuttle investment*[a] *Per orbiter*[b]	*Facilities*	*Tug DDTE and investment*	*ELVs in transition*[c]	*WTR*[d]	*Total*
1973	5.15	1.0	0.25	0.3	0.809	0.291	0.5	8.050
1974	5.15	1.0	0.25	0.3	0.809	0.291	0.5	8.050
1975	5.2	1.0	0.25	0.3	0.809	0.291	0.5	8.110
1976	5.2	1.0	0.25	0.3	0.809	0.291	0.5	8.100
1977	5.22	1.2	0.3	0.3	0.809	0.163	0.5	8.192
1978	5.22	1.347	0.337	0.3	0.809	0.179	0.427	8.282
1979	5.43	1.273	0.424	0.3	0.52	0.226	0.490	8.239
1980	5.986	1.321	0.442	0.3	0.52	0.234	0.565	8.926
1981	6.185	1.658	0.552	0.3	. . .	0.334	0.610	9.087
1982	6.654	1.712	0.570	0.3	. . .	0.449	0.910	10.025

Source: For 1974–80, *Space Shuttle 1980*, Status report for the House Committee on Science and Technology, 96 Cong. 2 sess. (GPO, 1980, pp. 40–43); *1981 NASA Authorization*, Hearings, vol. 5, p. 3155; and *1982 NASA Authorization*, Hearings before the Subcommittee on Space Science and Applications of the House Committee on Science and Technology, 97 Cong. 2 sess. (GPO, 1981), note 42, p. 24.

a. For 1973–78, includes refurbishment of two orbiters from DDTE and three production orbiters; for 1979, includes two production orbiters.

b. In the March 1972 shuttle fact sheet, each production orbiter would cost $0.250 billion; thus $0.250 billion would be used to refurbish the first two orbiters.

c. Expendable launch vehicles.

d. NASA-specific western test range facilities.

Figure 8-3. *Estimated First Manned Orbital Flight and Initial Operating Capacity, Fiscal Years 1971–82*

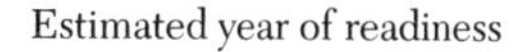

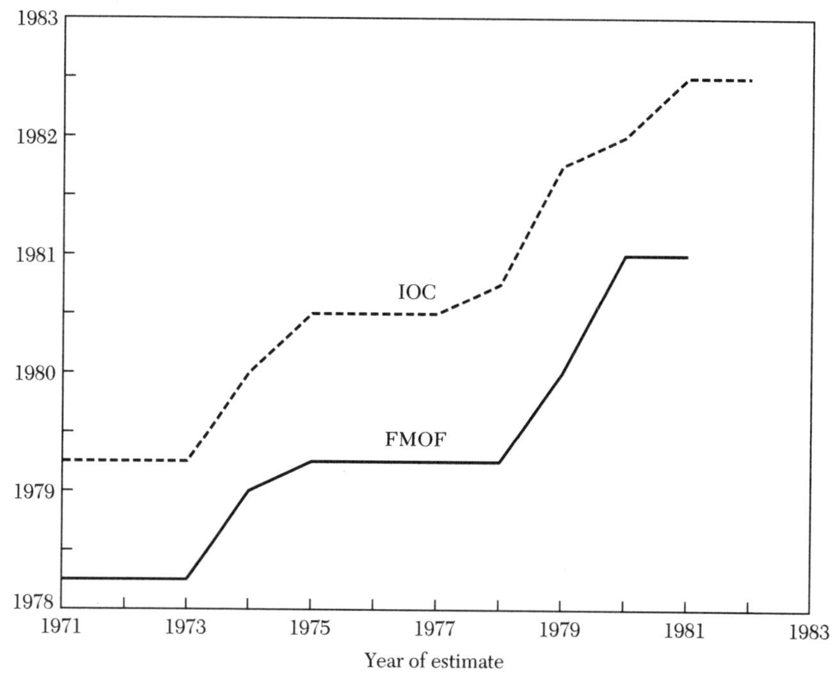

Sources: *Oversight: Space Shuttle Cost, Performance, and Schedule Review*, Hearings before the Subcommittee on Space Science and Applications of the House Committee on Science and Technology, 96 Cong. 1 sess. (GPO, 1979), p. 146; and *1982 NASA Authorization (Program Review)*, Hearings before the Subcommittee on Space Science and Applications of the House Committee on Science and Technology, 96 Cong. 2 sess. (GPO, 1980), p. 174.

In June 1982 NASA announced its pricing policy for the second three years of operation, from fiscal 1986 through fiscal 1988. The main characteristic of the new policy was that NASA would seek to recover shuttle "out of pocket" costs on an incremental basis, as opposed to recovering the full operational costs over the twelve-year period.[38] Thus the $1.18 billion subsidy to non-NASA users just mentioned would not be recovered. NASA's charge for a full payload during this period would be $38 million (fiscal 1975 dollars) or $71 million (fiscal 1982 dollars), as opposed to $18 million and $35 million, respectively, for the fiscal 1982–85 period. In addition, NASA announced a greatly reduced number of shuttle flights; including a total of seventy-two flights from 1981 to 1987 and a 24 flight-

38. *Aviation Week and Space Technology*, June 21, 1982, pp. 16–18.

Figure 8-4. *Estimated Number of Flights of the Space Shuttle for Its Twelve-Year Operational Life, Fiscal Years 1971–83*

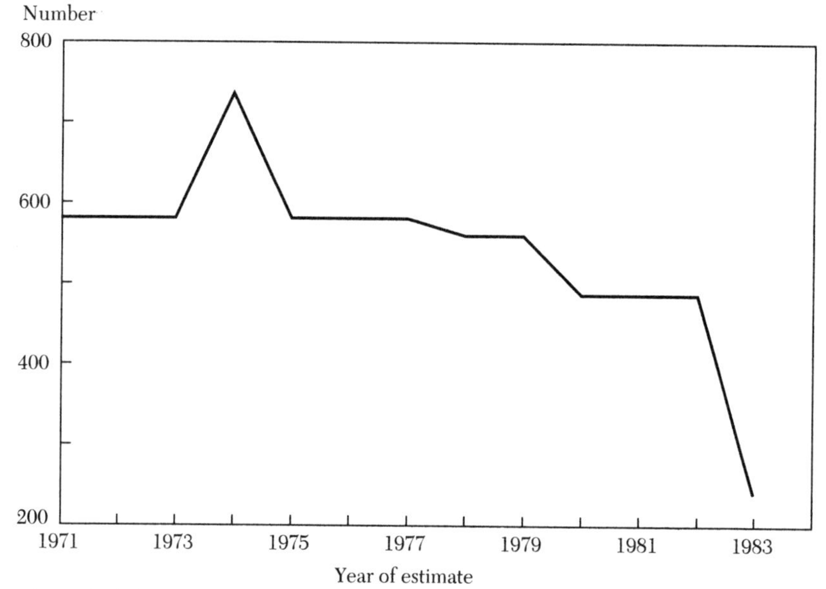

Source: See text.

per-year rate for 1988 to 1994, the total stood at only 240 flights, about one-third of the initial plan.[39]

By 1985 the shuttle program was sufficiently developed to compare the projected cost and performance parameters with their actual values. Table 8-7 contains the most relevant data. It is obvious from these data that the shuttle program as a means of accessing space fell far short of the lofty expectations generated by NASA in the early 1970s. In addition, the program experienced serious failures in two of the main components of the program, one with tragic consequences. The sixth flight of the shuttle *Columbia* was launched in April 1983, carrying with it the first of the tracking and data relay satellites, along with the first operation of the IUS. After leaving the cargo bay, a breakdown of the thermal protection system on the IUS led to a collapse of the oil-filled seals in the steering mechanism, throwing the satellite into the wrong orbit.[40] That led to delays in further

39. *1983 NASA Authorization*, Hearings before the Subcommittee on Space Science and Applications of the House Committee on Science and Technology, 97 Cong. 2 sess. (GPO, 1982), pp. 31, 744.

40. *Aviation Week and Space Technology*, April 11, 1983, p. 20.

Table 8-6. *Average Cost Estimates for the Space Shuttle Program, Fiscal Years 1976 and 1980*
Millions of fiscal 1975 dollars

Item	*June 1976*[a]	*September 1980*[b]
Consumables		
External tank	3.04	6.22
Solid rocket booster	3.55	6.98
Space shuttle main engine	0.31	0.48
Crew equipment	0.26	0.18
Propellants	0.76	0.66
Spares (orbiter and ground support equipment)	1.08	1.18
Contract administration	0.14	0.18
Subtotal	9.14	15.88
Additional operation costs		
Flight operations	2.82	5.01
Launch operations	4.23	6.94
Subtotal	7.05	11.95
Total	16.19	27.83

Source: General Accounting Office, *NASA Must Reconsider Operation Pricing Policy to Compensate for Cost Growth on the Space Transportation System: Report to the Congress* (February 1982), p. 8.
a. 572 flights, 399 from Kennedy Space Center and 173 from Vandenberg Air Force Base.
b. 487 flights, 362 from Kennedy Space Center and 125 from Vandenberg Air Force Base.

operations of the IUS, forcing NASA to consider using a modification of the McDonnell Douglas SSUS-D for subsequent TDRS launches.[41] Until the suspension of flights after the *Challenger* disaster, the IUS had not been used on any other shuttle missions, because of its uncertain performance.

The loss of the *Challenger* led to the conclusion that the probable cause of the accident was the failure of a seal in the right solid rocket booster, which previous studies had already identified as a problem. Thus the booster rocket had to be redesigned before shuttle flights could resume. The next launch did not take place until 1989.

From the economic history of the shuttle program it is clear that during the years 1978–81, as new information about the economics of the shuttle came to light, there was indeed an unwillingness to change the program. Unquestionably, much of the technical work had been completed on the first two orbiters and was progressing on the subsequent two. Nevertheless, no effort was made to reevaluate the program from an economic, cost-benefit perspective despite the meteoric rise in operation costs, the fall of benefits from the lack of a tug and a greatly reduced mission set, the

41. *Aviation Week and Space Technology*, July 11, 1983, p. 18.

Table 8-7. *Performance and Cost Characteristics*

Item	*Estimated*	*Actual*
Maximum payload capacity (lbs.)	65,000	53,000
Turnaround time (manhours)	160	1,240
Number of flights in 1984	60	22
Average cost of a 1984 launch (millions of fiscal 1975 dollars)	16	140
Cost per payload lb. in 1984 (fiscal 1975 dollars)	246	2,641

Sources: Stuart Diamond, "NASA Wasted Billions, Federal Audits Disclosed," *New York Times*, April 23, 1986, p. A1; and David H. Moore, *Pricing Options for the Space Shuttle* (Congressional Budget Office, March 1985), p. 19.

decrease in the reliability of the shuttle due to safety cutbacks, and the ensuing competition in the market from the development of the Ariane launch vehicle by Europe. One feature of the shuttle program that makes it amenable to such a reevaluation at any time during the program is the self-prescribed incremental approach of NASA to the development and production of the full five-shuttle fleet. Shuttles 1 and 2 would be built in the DDTE phase of the program, and shuttles 3, 4, and 5 would be built starting in the late 1970s, each with different start and completion dates. One implication is that, in the late 1970s, while most nonrecurring costs of shuttles 1 and 2 were already appropriated, such costs for shuttles 3 through 5 were in the future.

A Cost-Benefit Analysis

In analyzing the costs and benefits of the program, the approximate homogeneity of the shuttles enables one to distribute the payloads across the shuttles according to the economic benefits of using the shuttle to launch them. Thus the following analysis assumes that shuttles 1 and 2 would handle all payloads incompatible with an expendable launch vehicle (ELV) (for example, spacelab missions), whereas shuttles 3 through 5 would deploy only Delta-class payloads into geosynchronous orbit (this includes almost all commercial satellites). The question is whether the decision to proceed with the production of shuttles 3 and 4, and not to proceed with shuttle 5, was correct if one considers the new cost estimates of the late 1970s and ignores the production funds previously appropriated (that is, ignores sunk costs).

I focus initially on the fall of 1979 (the beginning of fiscal 1980) and consider the shuttle in comparison only to the current Delta ELV (as with earlier cost-benefit studies) and not with an upgraded ELV or the Ariane,

Figure 8-5. *Assumed Distribution of Remaining Nonrecurring Costs for Shuttles 3 and 4, circa 1979*

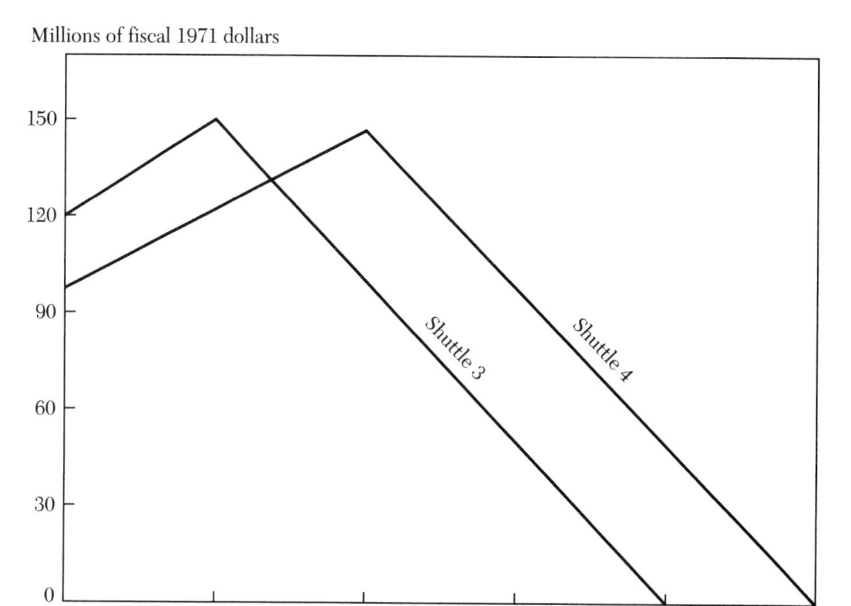

whose technical future and cost future were then in doubt. At the time shuttle 3's arrival was four years hence and shuttle 4's was five.[42] Each had a nonrecurring cost estimated at $552 million, and $287 million had been previously spent on the pair[43] (all dollar figures are in fiscal 1971 dollars). It is assumed that these funds were divided two-thirds to shuttle 3 and one-third to shuttle 4, and that the remaining nonrecurring costs were to be distributed as in figure 8-5. A Delta rocket would cost $11.6 million,[44] and a shuttle flight, which can accommodate three Delta-class payloads, would cost $15.2 million.[45] To the shuttle cost is added $6 million for three SSUS-Ds and $1.5 million in optional services.[46] Thus the shuttle would

42. *Space Shuttle 1980,* Status report for the House Committee on Science and Technology, 96 Cong. 2 sess. (GPO, January, 1980), pp. 3–5.

43. *Space Shuttle 1980,* p. 66.

44. *Aviation Week and Space Technology,* October 29, 1979, p. 14.

45. *Space Shuttle 1980,* p. 66.

46. Barbara Stone, "Development, Implementation, and Implications of Space Shuttle Pricing Policies," 1983, p. 41.

Table 8-8. *Cost-Benefit Analysis of Space Shuttles 3 and 4, 1979 and 1980*
Millions of fiscal 1971 dollars

Item	*September 1979*		*September 1980*	
	Shuttle 3	*Shuttle 4*	*Shuttle 3*	*Shuttle 4*
Benefit of shuttle − benefit of Delta ELV	380	376	65	57
Nonrecurring cost of shuttle	351	385	42	220
Net benefit of shuttle	29	−9	23	−163

Source: Author's calculations.

save $4 million per Delta payload, according to the 1979 estimates. The flight rate would increase after a three-year buildup to ten flights a year and remain at that rate for seven years, an estimate of shuttle performance that would prove optimistic but that was commonly used in this period. Finally, a 10 percent discount rate is used in all calculations.

The first two columns of table 8-8 give the (discounted) nonrecurring costs and net benefits of shuttles 3 and 4 under this scenario. Completing production of shuttle 3 is justified as opposed to launch on a Delta rocket, but shuttle 4 production is not. This implies that, with benefits further into the future and no funds spent on production, shuttle 5 could not be justified as well. Of course, the decision was made to continue production of both shuttles 3 and 4.

In similar fashion, one can update the cost estimates and ask whether production should have been continued as of fall of 1980, given the funds invested in fiscal 1980 and the revised cost estimates. The last two columns in table 8-8 provide the answer. The data are derived from assumptions similar to those just described. The dramatic fall in benefits is due to the upward revision in the operation cost estimates (see table 8-6). Thus, although NASA had spent more than $500 million in the production of shuttles 3 and 4 in fiscal 1980 and their arrival dates remained unchanged, the rise in operation costs argued for the same decision as the year before: continue production of shuttle 3 but suspend production of shuttle 4.

This conclusion about shuttle 4 is robust to the assumption that shuttles 1 and 2 could handle all the ELV-incompatible payloads, for even if shuttle 3 was needed to support these missions, shuttle 4 would still not stand the test of cost-benefit analysis. Furthermore, a negative conclusion for either of the shuttles in the analysis can be regarded as only a sufficient condition for cancellation, since I have examined only one alternative launch system.

Thus, although it is clear that shuttle 4 was not economically justifiable, it is not conclusive that shuttle 3 was economically justifiable. In summary, then, NASA and Congress had ample opportunity for altering the shuttle program when the "bad news" about costs came to light and a compelling justification for doing so as well. NASA and Congress made a correct decision in not pursuing a fifth shuttle but did not make the best decision, which would have at least entailed cancellation of the fourth shuttle in mid-production.

This argument has relevance to later shuttle decisions. After the *Challenger* disaster, President Ronald Reagan, in August 1986, gave the go-ahead for production of a replacement shuttle. Furthermore, the administration decided that the shuttle system would no longer serve the commercial satellite market but would launch only NASA, Defense Department, and other government payloads. The updated cost-benefit analysis described above shows that, on the basis of the shuttle cost estimates from 1980, a replacement shuttle could not be justified on economic grounds if it were in fact to serve the commercial market. The same obviously holds true after further upward revisions in the cost estimates.[47] However, the arguments for a replacement shuttle revolved instead around implementing NASA's Space Station and the Defense Department's Star Wars defense system, both of which presumably require the greater launch capacity and the manned presence of the shuttle. Thus, as the demands on the shuttle changed over the years the justification of the shuttle in relation to current ELV technology becomes more compelling in that the revised demands attempt to exploit the unique capabilities of the shuttle. In evaluating the shuttle program, then, it becomes necessary to consider how predictable these changes in demand were in the early 1970s, or equivalently, how decisive NASA was in leading the debate away from these issues.

Influence of NASA

In 1969 NASA revealed its preferences when it presented its comprehensive plan for the exploration and development of the moon and Mars to the Space Task Group. Given sufficient leeway in determining America's space program for the 1970s, NASA would choose the initial element from the plan, namely, the shuttle. In "selling" the shuttle program to Congress, NASA found it necessary to downplay this feature of the shuttle and instead

47. Congressional Budget Office, "Pricing Options for the Space Shuttle," March 1985.

argued its merits as a way to access space for commercial and scientific missions. These services could have been supplied by several different launch systems; however, options such as a new generation of ELVs or even a mixed system of ELVs and shuttles, where the shuttles could be configured to handle only person-intensive missions, were never fully considered in the early economic analyses. The Mathematica report of 1971–72 attempted to show the economic feasibility of a shuttle program and the economic optimality of the final design, yet any notion of optimality is only as pervasive as the set of alternatives is all-encompassing. The TAOS system was economically superior to the alternative designs and launch demands that were studied, but not to all possible launch systems, all potential cost or demand scenarios or, for that matter, all alternative uses for the funds. Thus, although Congress and the OMB could and did enforce changes in the shuttle program, these changes were incremental and did not address the issue of the space system NASA envisioned.

NASA's relative informational advantage in space technology enabled the agency to control the political agenda in two ways: first, by defining which program would be considered in relation to a specified "reversion level"; and second, by determining that the reversion level was the existing state of expendable launch vehicles, as opposed to a system that might have been developed in place of the shuttle. This approach stacked the deck of the economic evaluation in favor of the alternative preferred by NASA on other grounds. It also allowed NASA to choose a demand scenario compatible with congressional pessimism about space programs, that is, a heavy orientation toward commercial satellites and no space station. Moreover, the leverage gained from the use of economic analysis in support of the shuttle allowed NASA to direct governmental decisions away from whether there should be a shuttle to the specifics of shuttle system and subsystem design, an area in which NASA could proclaim expertise. However, by the late 1970s, if NASA had continued to stress the economics of the shuttle in the commercial satellite market, it would have been burned by the fire it fanned to initiate the program. Economic analysis such as that just discussed would have suggested programmatic changes inconsistent with NASA's goals for the shuttle. Justification for the shuttle in the later years had to come from another source, namely, the new Defense Department and NASA programs that had been designed to depend on the shuttle's unique attributes—programs that had played no role in the initial justification.

Political Support for the Space Shuttle

In the following analysis I deal with full House authorization votes on NASA's annual budget. Actions on authorizations by the executive and congressional branches of the government are shown in table 8-9. In no year did the final bill contain less for shuttle expenses than the OMB asked for, although the OMB request to Congress was frequently less than NASA's initial request to the OMB.

The choice of votes is dictated by available roll call data. Though shuttle appropriations came in for some heated debate in the early 1970s, amendments directly relating to the shuttle were subjected to only three roll call votes. Two instances were in the Senate (amendments deleting funds from the appropriations bills were offered by Senator Walter Mondale in 1971, when it was defeated 22 to 64, and in 1972, when it was defeated 21 to 61). One standing vote took place in the House (an amendment deleting funds from the authorization bill was offered in 1972 by Les Aspin (Democrat of Wisconsin) and rejected 11 to 103).[48] By contrast, NASA authorizations were voted on annually. Because the shuttle consumed a large share of NASA's entire budget over the 1970–82 period (see figure 8-6), and because it was the flagship of NASA's program in those years, voting by House members on these bills is a reasonable indicator of political attitudes toward the shuttle program.

Unlike the other statistical analyses in this book, the votes analyzed are the final action on a bill. Like most final actions, the votes are overwhelmingly in favor, as shown in figures 8-7 and 8-8. In contrast to amendment voting, which offers members of Congress a relatively easy way to express preference on a detail of a program, a no vote on a final bill sets in motion the process to report out a new bill, with modifications to all parts possible. Consequently, opposing votes arise only when a legislator has serious difficulties with the bill overall, whereas supporting votes may indicate satisfaction with compromises, or with the entire package represented by the bill. One would therefore expect that the explanatory power of regression results on final bills will be less, owing to the difficulty of distinguishing strong supporters from the rest of the group. In any event, these votes and regression results are not fully comparable to the amendment votes that are examined in other chapters.

48. *Congressional Quarterly Weekly Report*, July 2, 1971, p. 1455; May 20, 1972, p. 1160; April 29, 1972, p. 974.

Table 8-9. *Shuttle Design, Development, Test, and Evaluation and Production Authorization, Fiscal Years 1971–81*
Millions of current dollars

Year	*NASA request to OMB*	*NASA request to Congress*	*House committee*	*Full House*	*Senate committee*	*Full Senate*	*Final*
1971	78.5	78.5	78.5	78.5	78.5	78.5	78.5
1972	190[a]	100	125	125	100	100	115
1973	200	200	200	200	200	200	200
1974	560	475	500	500	475	475	475
1975	889	800	820	820	800	800	805
1976	1,251	1,206	1,206	1,206	1,206	1,206	1,206
1977	1,388	1,288	1,288	1,288	1,288	1,288	1,288
1978	1,354	1,349[b]	1,349	1,349	1,354	1,354	1,354
1979a	1,443	1,439[c]	1,443	1,443	1,443	1,443	1,443
1979b[d]	185	185	185	185	185	185	185
1980a	1,366	1,366	1,393	1,393	1,393	. . .	. . .
1980b[e]	. . .	. . .	. . .	. . .	1,586	1,586	1,586
1980c[f]	300	300	300	300	300	300	300
1981	1,893	1,873	1,878	1,878	1,873	1,873	1,873

Source: *Congressional Quarterly Weekly Report*, 1970–81.
a. Includes $90 million in space station DDTE funds.
b. Excludes $5 million in start-up funds for fifth orbiter.
c. Excludes $4 million in start-up funds for fifth orbiter.
d. Supplemental funding for fiscal 1979.
e. Amended 1980 budget excludes $27 million for fifth orbiter.
f. Supplemental funding for fiscal 1980.

Figure 8-6. *Shuttle DDTE and Production Funds as a Percentage of the NASA Budget, Fiscal Years 1971–83*[a]

Source: *Congressional Quarterly* data.
a. The 1979 percentage includes the fiscal 1979 supplement; the 1980 percentage includes the fiscal 1980 amendment and supplement.

Precise definitions of variables are contained in figure 8-9. Included are variables for party, *ACA* score, and membership on the authorizing committee. Before 1974 NASA budgets were authorized by the House Committee on Science and Astronautics; since 1974 the Science and Technology Committee has been responsible for oversight. In regressions not reported here, subcommittee variables were included in the analysis, but these were significant in none of the regressions.

The variable *FMOF* measures years remaining until the estimated date of the first manned orbital flight, an event that occurred just before the fiscal 1982 vote, in the spring of 1981. Because this sequence ends in fiscal 1982, a dummy variable for 1983 is included in the pooled regressions, and *FMOF* in the fiscal 1982 and 1983 votes is coded as zero. Additional longitudinal variables measuring efficiency aspects of the program are also discussed.

Distributive effects of the space shuttle program are captured by the *PRIME* dummy variable, which codes one for districts containing one of

Figure 8-7. *Percentage of House Votes in Favor of NASA Authorization, Fiscal Years 1971–83*[a]

Source: *Congressional Quarterly* data.
a. The 1979 percentage does not include the fiscal 1979 supplemental; the 1980 percentage includes the amended budget but does not include the fiscal 1980 supplemental. "Pairs" are counted as voting for or against the bill. (In House voting, two members can "pair" up, with one voting one way on a bill and one the other way, registering this pairing before the vote and then not having to show up for the actual vote.)

the primary contractors on the program and zero otherwise. These are listed in table 8-10. More detailed expenditure data could not be located.

Separate regressions, excluding longitudinal variables, were estimated for each vote, and then groups of consecutive votes were tested for pooling. No loss in explanatory power results from pooling the votes into four groups: fiscal 1971–73, 1974–78, 1979–82, and 1983.[49] In contrast, tests reject pooling across groups.[50] Tests for pooling specific variables across vote

49. The hypothesis that votes pool cannot be rejected at the 5 percent level. For the fiscal 1971–73 group the relevant statistic is $\chi^2(8) = 7.88$; for fiscal 1974–78, $\chi^2(13) = 4.42$; and for fiscal 1979–82, $\chi^2(13) = 9.71$. The discrepancy in degrees of freedom in these tests arises because in some years certain variables perfectly explained votes—for example, in fiscal 1973–77 all committee members voted for the bill, and in fiscal 1977–78 and fiscal 1980 all primary group members voted for the bill. These variables had to be excluded from the annual regressions in the identified years.

50. For fiscal 1971–78, $\chi^2(6) = 17.0$; for fiscal 1974–80, $\chi^2(6) = 33.67$; and for fiscal 1979–83, $\chi^2(4) = 112.5$.

Figure 8-8. *Percentage of House Votes in Favor of NASA Authorization by Party, Fiscal Years 1971–83*

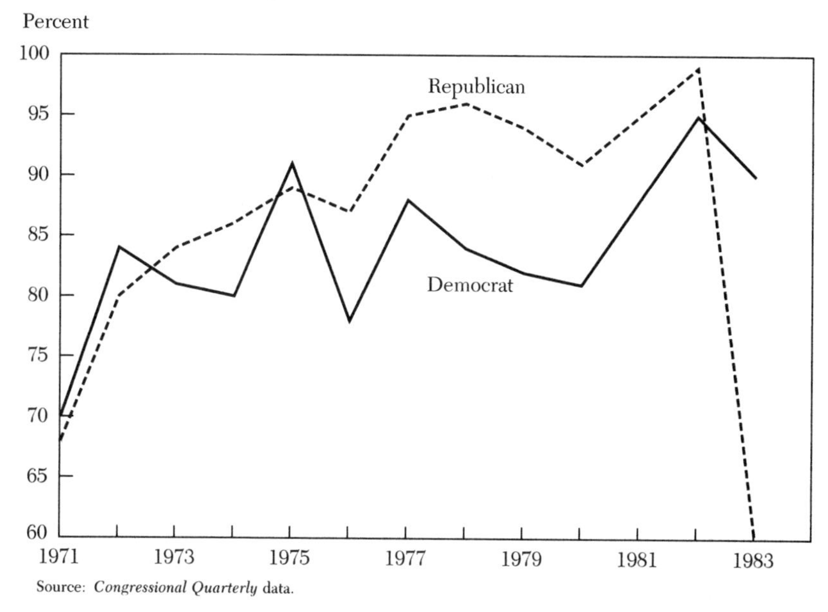

Source: *Congressional Quarterly* data.

groups are also discussed. Sample statistics for each year are shown in table 8-11, and the results of the pooled regressions are shown in table 8-12.

Party and Ideology

The *PARTY* variable is not a strong indicator of support or opposition to the program in these regressions. However, because *ACA* and *PARTY* are highly correlated, the regressions indicate that Republicans, overall, were more likely to support the program than Democrats through 1981, and that

Figure 8-9. *Variable Definitions*

PARTY: 1 if Democrat or Independent; 0 if Republican.
COMM: 1 if member of committee overseeing NASA authorization.
PRIME: 1 if congressman's district contains a primary shuttle contract.
ACA: Americans for Constitutional Action score.
FMOF: Years until estimated first manned orbital flight.
83 Dummy: 1 for fiscal 1983 vote; 0 otherwise.
DDTE: Estimated cost overruns for design, development, test, and evaluation.

Table 8-10. *Location of Principal Contractors for Space Shuttle Program*

Firm or center	*City and state*	*District*[a]
Marshall Space Flight Center	Huntsville, Ala.	[8],5
Martin Marietta	Huntsville, Ala.	[8],5
Sperry Rand	Phoenix, Ariz.	[1],4
Aerojet–General Corp.	Sacramento, Calif.	3
Ames Research Center	Mountain View, Calif.	[10],(9),12
Lockheed Missiles and Space	Sunnyvale, Calif.	[17](9),12
United Technologies	Sunnyvale, Calif.	[17],(17),12
North American Rockwell	Downey, Calif.	[23],(23),33
	Palmdale, Calif.	[27],(24),18
Rocketdyne	Canoga Park, Calif.	[28],(27),20
Marquardt	Van Nuys, Calif.	[28],(27),21
Menasco	Burbank, Calif.	[28],(26),22
TRW	Redondo Beach, Calif.	[28],(31),27
McDonnell-Douglas	Huntington Beach, Calif.	[34],(32),34
Vandenburg AFB	Lompoc, Calif.	[13],(36),19
General Dynamics	San Diego Calif.	[37],(40),41
Rohr	Chula Vista, Calif.	[37],(41),42
Martin Marietta	Denver, Colo.	1
Beech Aircraft	Boulder, Colo.	2
Pratt and Whitney	West Hartford, Conn.	1
Hamilton Standard	Windsor Locks, Conn.	6
Martin Marietta	Orlando, Fla.	[11],5
Honeywell	Tampa, Fla.	[12],7
Kennedy Space Center	KSC, Fla.	[11],9
IBM	KSC, Fla.	[11],4
Sundstrand	Rockford, Ill.	16
Michoud Assembly Facility	New Orleans, La.	1
Martin Marietta	New Orleans, La.	1
Honeywell	Minneapolis, Minn.	5
National Technological Science Laboratory	Kiln, Miss.	5
McDonnell-Douglas	St. Louis, Mo.	2
Bendix	Teterboro, N.J.	9
Fairchild Industries, Republic Aviation Division	Farmingdale, N.Y.	3
Grumman Aerospace	Bethpage, N.Y.	1
IBM	Oswego, N.Y.	[33],27
American Airlines	Tulsa, Okla.	1
Ling-Temco, Vought	Dallas, Tex.	5
Manned Spacecraft Center	Houston, Tex.	22
Thiokol	Brigham City, Utah	1
Boeing	Seattle, Wash.	7
Ladish	Cudahy, Wis.	4

Sources: *NASA Authorization for FY 74*, Hearings before the Senate Committee on Aeronautical and Space Services, 93 Cong. 1 sess. (GPO, 1973); *Space Shuttle 1976*, Status report for the House Committee on Science and Technology, 94 Cong. 1 sess. (GPO, 1975), p. 139; *Science Shuttle, Space Tug, Apollo-Soyuz Project—1974*, Status report for the House Committee on Science and Astronautics, 93 Cong. 2 sess. (GPO, 1974), pp. 426, 575; *1975 NASA Authorization*, Hearings before the House Committee on Science and Astronautics, 93 Cong. 2 sess. (GPO, 1974), pp. 240, 781; *1976 NASA Authorization*, Hearings before the Subcommittee on Space Science and Applications of the House Committee on Science and Technology, 94 Cong. 1 sess. (GPO, 1975), pp. 1869, 2331; *Space Shuttle 1977*, Status report for the House Committee on Science and Technology, 94 Cong. 2 sess. (GPO, 1976), pp. 506–07; Congressional Quarterly, *Congressional Districts in the 1970s* (Washington, 1973); and Congressional Quarterly, *Congressional Districts in the 1970s*, 2d ed. (Washington, 1974).

a. The numbers in brackets are the relevant districts before the 1972 reapportionment; the numbers in parentheses for California are the relevant districts between the 1972 and 1974 reapportionments.

Table 8-11. *Sample Statistics of the Regression Variables, 1971–83*[a]

Year	*VOTE*	*PARTY*	*COMM*	*PRIME*	*ACA*	*FMOF*
1971	0.672	0.529	0.081	0.070	49.347	8.000
	0.469	0.499	0.273	0.255	28.612	0.000
1972	0.824	0.574	0.063	0.069	50.629	7.000
	0.381	0.494	0.243	0.253	32.292	0.000
1973	0.806	0.572	0.073	0.082	49.383	6.000
	0.396	0.495	0.261	0.274	33.027	0.000
1974	0.812	0.556	0.071	0.081	46.939	6.000
	0.391	0.497	0.257	0.273	31.026	0.000
1975	0.900	0.557	0.069	0.090	42.554	5.000
	0.300	0.497	0.253	0.286	31.663	0.000
1976	0.815	0.664	0.082	0.082	44.923	4.000
	0.388	0.472	0.274	0.274	32.091	0.000
1977	0.899	0.668	0.076	0.076	40.899	3.000
	0.301	0.471	0.265	0.265	33.086	0.000
1978	0.885	0.654	0.089	0.084	43.089	2.000
	0.319	0.476	0.285	0.277	31.348	0.000
1979	0.860	0.651	0.082	0.080	48.282	2.000
	0.347	0.477	0.275	0.271	32.044	0.000
1980	0.853	0.632	0.100	0.074	45.326	2.000
	0.354	0.482	0.300	0.261	35.418	0.000
1981	0.905	0.636	0.111	0.079	46.035	1.000
	0.293	0.481	0.314	0.270	27.744	0.000
1982	0.969	0.549	0.089	0.084	50.273	0.000
	0.174	0.498	0.284	0.277	30.306	0.000
1983	0.765	0.533	0.099	0.077	48.710	...
	0.424	0.499	0.299	0.267	31.666	...

Source: Author's calculations.
a. The top number for each year is the mean; the bottom number is the standard deviation.

Democrats were more favorable in 1982 (the fiscal 1983 vote). The NASA program accorded well with conservative platforms through fiscal 1982 but switched in 1983. Tests for pooling these variables across vote groups fail, marginally for *PARTY* and spectacularly for *ACA*.[51]

The results for committee support show a similar pattern. Committee

51. The procedure for testing pooling of specific variables is as follows. A logit estimation was performed for the entire sample with a constant term; *PARTY*, *COMM*, *PRIME*, and *ACA* interacted with four dummy variables for the four different groups; *FMOF* interacted with three dummy variables for the first three time periods, and dummy variables for the fiscal 1974–78, fiscal 1979–82, and fiscal 1983 groups. This yields an equation with twenty-three independent variables, 4,844 degrees of freedom, and coefficients identical to those reported in the first four columns of table 8-12. The regression was then run with the interacted variables in question replaced by a single variable for the entire time series, for

Table 8-12. *Logit Estimation of the Effects on Vote of Political and Economic Variables, Fiscal Years 1971–83*[a]

Variable	*1971–73*	*1974–78*	*1979–82*	*1983*	*All years*	*All years*
Constant	3.18	2.11	2.59	2.72	2.09	1.99
	(0.70)	(0.31)	(0.41)	(0.54)	(0.16)	(0.20)
PARTY	0.35	−0.29	−0.27	0.44	0.043	0.044
	0.19	(0.20)	(0.29)	(0.38)	(0.113)	(0.113)
COMM	1.77	2.02	0.71	−0.02	1.14	1.14
	(0.52)	(0.59)	(0.38)	(0.49)	(0.23)	(0.23)
PRIME	1.75	1.62	1.06	1.49	1.46	1.46
	(0.52)	(0.46)	(0.52)	(0.66)	(0.26)	(0.26)
ACA	0.008	0.009	0.017	−0.031	0.007	0.007
	(0.003)	(0.003)	(0.004)	(0.007)	(0.002)	(0.002)
FMOF	−0.39	−0.14	−0.71	. . .	−0.19	−0.18
	(0.09)	(0.05)	(0.13)		(0.02)	(0.03)
83 Dummy	. . .	. . .	. . .	. . .	−1.40	−1.51
					(0.16)	(0.20)
DDTE	. . .	. . .	. . .	. . .	. . .	(0.006)
						(0.008)
−2L	1,102.37	1,456.96	899.18	325.05	3,952.80	3,952.14
Degrees of freedom	1,070	1,906	1,508	357	4,857	4,856

Source: Author's calculations.
a. The numbers in parentheses are standard deviations.

members voted as a block in favor of the program in all years up to fiscal 1983. In many of those years all committee members voted for the program. It therefore enjoyed strong institutional support within Congress, as is usually true for final actions on committee bills. In fiscal 1983 the committee coalition fell apart; the coefficient becomes negative and insignificant. The committee variable fails a pooling test across all four groups at the 1 percent level.[52]

The fiscal 1983 results, when taken in isolation, appear anomalous. Indeed, one interpretation is that the shift in support is a matter of timing

example, *ACA71–73*, *ACA74–78*, *ACA79–82*, and *ACA83* replaced by a single *ACA* variable. The difference in −2 times the log-likelihood functions for the two regressions is a χ^2 statistic with degrees of freedom equal to the difference in degrees of freedom of the two equations, and tests for an increase in explanatory power from adding the interacted variables. The relevant statistic for *PARTY* is $\chi^2(3) = 7.483$, and for *ACA* is $\chi^2(3) = 43.096$.

52. $\chi^2(3) = 10.78$.

and luck: the vote was taken immediately after a Republican-proposed amendment to bar authorization of funds that would violate a proposed law (PL95-435) requiring a balanced budget.[53] But this result is consistent with the findings in other chapters of the book (see especially chapters 7 and 10, on the satellite and synthetic fuel programs). Although the conservative wing of the Republican party supports R&D programs, it tends to oppose them once commercialization is imminent. In this case, the shift coincides as well with the change in administration, so that the antigovernment precept may have received a boost from either the election of President Reagan or the apparent growth among the population in acceptance of his theories of the role of government, as shown by his election.

The space shuttle achieved manned orbital flight a year before the fiscal 1983 vote. By 1982 NASA was reclassifying the program as operational rather than R&D. Even though subsequent events—most notably the *Challenger* disaster—have shown this assessment to be incorrect, as of 1982 commercialization was under way. Consequently, as with the satellite and synfuel programs, conservatives' support came into conflict with their belief that government should not be involved in commercial activities. Indeed, in January 1982 the Space Transportation Company was proposing a down payment of $200 million to $300 million to buy a fifth shuttle to service the commercial and foreign markets, with an eye toward eventually purchasing other shuttles as well. Administration officials noted that such a policy "dovetails nicely" with Reagan's stated economic and political goals for space development.[54]

Distributive Benefits

Pork barrel support for the space shuttle program shows up clearly in the regression results. *PRIME* is positive and significant in all time periods. Pooling across all groups is not rejected for the *PRIME* variable, despite the growth and subsequent decline of actual contract dollars flowing into the specified districts for the shuttle program.[55]

Two explanations are consistent with the apparent absence of retrospective support. First, the result may reflect a data problem in dealing with an agency authorization rather than an amendment for funds on a specific

53. The balanced budget bill failed 121 to 248: Republicans 99 to 70; Democrats 22 to 178. *Congressional Quarterly Weekly Report*, May 22, 1982, p. 1222.

54. *Aviation Week and Space Technology*, January 18, 1982, p. 20.

55. $\chi^2(3) = 0.933$.

program. Although the shuttle was a large fraction of NASA's authorization, other programs were included as well, especially in the early years of the time series, before shuttle spending was large. The primary contractors on the shuttle had big Apollo program contracts as well as other NASA-related contracts. Support by these districts in the years preceding heavy shuttle expenditures may derive from the contract dollars flowing at that time to other projects.

Second, the pooling result may reflect the particularly superb way that the pork got carved up on the shuttle program. Every major space technology firm got a juicy contract, and furthermore, NASA's reputation for big contracts had been well established in the Apollo program. Thus the primary contractors formed a permanent support group (a space-industrial complex) in favor of big space projects. Hence the identity of the agency's beneficiaries did not change as the composition of the agency's programs shifted. As a result, the core group favoring the shuttle may have voted on the basis of recent expenditures on other contracts, and with an expectation that funds would be carried over to the next program. If this explanation is correct, the next big space program, the space station, ought to exhibit pork barrel support before significant funds are expended.

Longitudinal Variables

The voting data span a period when the first manned shuttle launch varied between eight and zero years into the future. The data thus allow a test of the hypothesis that near-term payoffs are weighed heavily (that is, in excess of market discounting) by members of Congress. The hypothesis is borne out in the statistical analysis: *FMOF* is negative and significant for all groups. Furthermore, the coefficient does not pool across groups.[56] The coefficient is significantly bigger when the shuttle launch date gets down to the wire, which shows the difference in support between visible benefits two years in the future (that is, after the next congressional election) and concurrent benefits.

The *FMOF* variable measures the arrival of benefits, from either the economic value of further launches or the symbolic value of astronauts and American technology in space. An additional variable related to efficiency is *DDTE*, which measures the expected nonrecurring design, development, test, and evaluation cost overruns (calculated from table 8-5). This measure

56. $\chi^2(2) = 21.716$.

of economic performance is highly correlated with other performance indicators and enjoyed prominence and regularity in the congressional hearings.

Including the *DDTE* variable adds no explanatory power to the pooled regression; indeed, the coefficient is the wrong sign (bigger cost overruns lead to greater support) and insignificant. When *FMOF* is deleted from the regression, the coefficient on *DDTE* remains positive and becomes significant, owing to its correlation with *FMOF*. Thus at least over the life of the program, deterioration of program efficiency did not diminish congressional support.

Testing *DDTE* on subgroups of years is problematic owing to lack of variation with the inclusion of more than one longitudinal variable on a set of data encompassing three to five years. For both the fiscal 1974–78 group and the fiscal 1979–82 group regressions (not reported here) *DDTE* had a positive coefficient. The conclusion remains that efficiency considerations were not particularly relevant.

Additional support is provided by the fact that the dummy variable for 1983 is negative in the pooled regression. The peak in support was in fiscal 1982. After the demonstration of the feat of manned shuttle flight, support declined, although commercial operation began and dollars started flowing to NASA for launch services. Thus, throughout the period spanned by the votes, shuttle efficiency never became important.

Conclusion

The preceding history of the space shuttle program illuminates the interaction and influence of economic and political variables on government-sponsored R&D programs. The political environment surrounding the initial decision to proceed with the shuttle program was such that further manned adventures into space would not have the widespread support of previous programs such as Apollo. Hence NASA was required to justify the need for the shuttle program on economic grounds; in addition, this justification could not rely on the shuttle's prospective role in future manned space endeavors. Political considerations therefore framed the decision to proceed with the shuttle in a straight stand-alone, cost-benefit analysis. Such an analysis revealed that the best NASA could hope for was not the fully reusable spacecraft they sought, but rather a partially reusable system. Even this conclusion required a rosy scenario from a technological standpoint (for example, sixty flights a year at a $10 million cost per flight).

Once the shuttle program was under way and the contractors were selected, the program picked up political steam as the relevant constituencies became enfranchised and the contract benefits of the program began to accrue. These political considerations then were sufficient to keep the shuttle program going when, in the late 1970s, the economic rationale initially offered for the shuttle was in jeopardy. In particular, simple cost-benefit analysis implies that the program should have been scaled back and a more efficient mix of shuttles and expendable launch vehicles employed. At the time, however, the core political constituency, which by then also included those with a stake in maintaining NASA's "prestige" as an organization, was sufficiently influential to overcome any arguments about the economic shortcomings of the program. In other words, even if NASA had wished to scale back the program (and there is no evidence that it did), political considerations alone generated constraints on the ability of decisionmakers to implement the economically correct course of actions. As seen in the regression analysis, economic considerations had a negligible effect on the decision calculus of Congress.

Another political effect on the history of the program can be deduced from the rush in the late 1970s to get the first shuttle flight airborne. At that time NASA was frantically amending budgets and transferring funds from production (for shuttles 3 and 4) to DDTE (for shuttles 1 and 2) in an effort to launch the first shuttle as quickly as possible. Further, the regression analysis highlights the fact that such behavior is consistent with the revealed preferences of Congress, in that the time to the first launch was a significant variable in congressional voting decisions. Hence political pressure implied that if the program as initially conceived was to continue, the timing of the first launch must take precedence in the decisionmaking regardless of the future consequences of such actions.

One result of a project schedule dictated by political considerations was that in the early 1980s the United States found itself with a launch technology that was clearly not efficient. A relevant question then concerns the continuing justification of the shuttle program subsequent to the early launches and in the face of severe technological shortcomings (for example, the *Challenger* disaster). What should not be too surprising, given NASA's attempts in the early 1970s to proceed with a number of large-scale projects, is that one of the principal justifications for the shuttle's current existence is its role in developing the next major NASA project, namely, the earth-orbiting space station. The space station is predicated on the existence of the shuttle, because (it is argued) the only means of developing it is with a shuttle-based technology. Thus, as the economic shortcomings of the shuttle became apparent, the justification for the program shifted from the efficacy of the shuttle in launching satellites toward its role in the larger space development program envisioned by NASA.

Chapter 9

The Clinch River Breeder Reactor

LINDA R. COHEN & ROGER G. NOLL

THE FEDERAL government has persistently taken the lead in developing commercial uses of nuclear power. In the 1950s research and development (R&D) focused on the commercialization of light-water reactors, the nuclear power technology originally designed for submarines. In the early 1960s attention shifted to the breeder reactor program, which dominated federal energy R&D until the cancellation of the Clinch River Breeder Reactor Demonstration in 1983.

The legitimacy of federal support for R&D in nuclear power has always been taken for granted. A case can be made for market failure on the basis of regulation. First, because radiation is a significant hazard should nuclear power facilities be improperly constructed or operated, the federal government has always asserted far greater control over commercial uses of nuclear energy than over other methods of generating electric power. Second, the users of nuclear generation facilities are largely regulated public utilities, and economic regulation distorts choices about capital investments and new technologies. Extensive regulation of nuclear power is a source of uncertainty and costs that may retard utility R&D efforts to develop advances in commercial nuclear technology.

The most compelling political reasons for federal support rest on the link between nuclear power plants and nuclear weapons. In the 1950s the government effort was justified by, first, the advantages of exploiting federal expertise and classified information from the atomic defense programs and, second, the international prestige and recognition derived from developing valuable peaceful applications of atomic energy. A decade later, a third reason emerged: the goal of nonproliferation. Nuclear power plants (especially breeder reactors) can provide know-how and materials needed to build nuclear weapons. The ability of the United States to influence international nuclear policies, and especially to control the nuclear fuel cycle so as to thwart the proliferation of nuclear weapons, depends on its continuing to be a primary supplier of commercial nuclear technology.

Between 1969 and 1983 the federal government spent more than $5 billion on research and development for the liquid metal fast breeder reactor. Although this amount was about twice the initial estimates of the expenditures that would be necessary to bring the technology into commercial use, the goal of commercialization proved elusive. Indeed, commercialization was a more distant goal when the breeder demonstration program ended than when it had begun. In 1969 the projected date for initiating a commercial breeder reactor industry was 1984, or fifteen years into the future. By 1983 the estimate was 2025 at the earliest, or forty-two years into the future.

The failure of the breeder reactor to obtain anything approaching commercial status in the 1980s has little to do with the nature and quality of federal development efforts. Estimates of the demand for electricity in 2000 and beyond plummeted while the program was in progress, delaying the optimal date of introducing all new sources of electricity. Moreover, the demand for light-water reactors fell even more sharply than total demand projections. Because the demand for liquid metal fast breeder reactors is related to that for light-water reactors, the fall in light-water demand diminished the commercial desirability of the breeder.

Even so, political oversight of the breeder reactor program was a failure. The first section of this chapter discusses the three main changes in the economic underpinning of the program: the change in demand for breeders as reflected in official benefit-cost studies, the changes in licensing requirements, and the (related) increases in cost estimates of components of the federal program, particularly for the Clinch River Demonstration Plant. Managerial decisions in the program are shown to diverge from efficient responses to these changes. The second section presents a statistical analysis of a time series of votes, in the House and in the Senate, on the Clinch River Demonstration that tests for sources of political support for the program. Program choices are shown to be systematically distorted to correspond to political incentives, consistent with the theory of political oversight presented in chapter 4. The consequence was a program that was more costly than it had to be, that managed to slow the pace of development of breeder technology, and that delayed the likely date at which it might prove commercially attractive. This result was at odds with both economic efficiency and the attainment of other goals that motivated the initiation of the program.

Politics, Economics, and the Clinch River Breeder Reactor

We first develop the initial logic and goals of the Liquid Metal Fast Breeder Reactor (LMFBR) Program and the role of the demonstration plant within the program. We then turn to the official benefit-cost analyses and the changes in the environment that affected the commercial desirability of LMFBRs as well as the other original program goals. We conclude the section with an assessment in light of the efficiency considerations discussed in chapters 2 and 3.

Initiation of the LMFBR Program

The U.S. nuclear power research program began in the military, specifically with the development of the nuclear-powered submarine. From this time on nuclear technologists have been optimistic about the commercial value of nuclear energy. As a result, cognizant government agencies have persistently overestimated the demand and underestimated the cost of nuclear power technologies. The breeder was no exception, despite a strong technical argument in its favor: its ability to produce more fuel than it uses.

Since first investigated during World War II, breeder reactor technology has been a critical element in the U.S. nuclear energy development strategy. Because of constraints on uranium supply, scientists believed long-term use of nuclear power should rely on breeder technology. All nuclear reactors obtain energy from reactions of fissionable isotopes of uranium, U-233 and U-235, and from plutonium, Pu-239. Besides releasing energy, the fission reactions result in the conversion of so-called fertile material (U-238 or thorium-232) to plutonium or U-233.[1] In breeder reactors the conversion ratio—the ratio of new fission isotopes formed per fission reaction—exceeds one, so that breeders create more fuel than they consume. Since natural uranium contains over 99 percent U-238, which is fertile but cannot fission, and only 0.7 percent fissionable U-235, current

1. See, for example, U.S. Atomic Energy Commission (AEC), *Civilian Nuclear Power: A Report to the President, 1962* (November 1962), pp. 35–39. (Hereafter *1962 Report to the President.*)

Table 9-1. *U.S. Fast Breeder Reactors*

Milestone plants	*Capacity*	*Year*
EBR-1	150 kWe	1951–64
EBR-2	16 MWe	1963[a]
Fermi	61 MWe	1963–73
Sefor	(20 MWt)[b]	1968
Fast Flux Test Facility	(400 MWt)[b]	1979
Clinch River Demonstration	350 MWe	. . .

Source: *Review of the Liquid Metal Fast Breeder Reactor Program*, Hearings before the Ad Hoc Subcommittee to Review the Liquid Metal Fast Breeder Reactor Program of the House Committee on Atomic Energy, 94 Cong. 1 sess. (GPO, 1975), vol. 1, p. 16.
a. Operating, 1975.
b. No electricity generated.

light-water reactors (LWRs), which have low conversion ratios, use less than 1 percent of the energy in uranium. In theory, breeders can do a hundred times better. To put this in perspective, on the basis of 1977 estimates LWR technology using domestic uranium reserves could provide one-tenth the electricity that could be generated from domestic coal reserves; with breeder reactor technology, domestic uranium reserves could provide ten times the electricity that could be generated from domestic coal.[2] Breeder technology is thus the key to an essentially infinite energy supply based on nuclear power. The current generation of light-water reactors, by contrast, offers a relatively minor supplement to national energy supplies.

Initially breeder technology was considered the best nuclear technology for generating electricity. In the late 1940s and early 1950s, known uranium reserves were largely committed to military applications.[3] The first experimental power reactor built by the Atomic Energy Commission (AEC) was a breeder reactor (table 9-1). However, additional uranium was discovered. And after the experimental breeder reactor suffered a meltdown accident, it became clear that there was still substantial research to be done. The federal government then took advantage of a shorter-term developmental opportunity in light-water reactors based on Admiral Hyman Rickover's successful nuclear submarine propulsion program.[4]

By 1967 the AEC's Light-Water Reactor Development Program was

2. Spurgeon M. Keeny, Jr., and others, *Nuclear Power Issues and Choices*, Report of the Nuclear Energy Policy Study Group (Ballinger, 1977), p. 32.
3. George T. Mazuzan and J. Samuel Walker, *Controlling the Atom: The Beginnings of Nuclear Regulation, 1946–1962* (University of California Press, 1984), p. 16.
4. Mazuzan and Walker, *Controlling the Atom*, p. 17.

largely finished and was hailed as a success.[5] LWR orders by utilities had far exceeded the expectations of even five years before, and the AEC was predicting that 23 to 30 percent of U.S. generating capacity in 1980 would be nuclear, rising to 50 percent by 2000. The latter prediction could be realized only with the discovery of additional low-cost uranium resources, technological progress in mining, and the successful commercial development of advanced convertors and breeder reactors.[6]

The assumptions about energy growth, uranium availability, and nuclear costs and acceptability justified a wide range of developmental strategies in the late 1960s, including different high- and low-gain breeders (breeders with different conversion ratios), nuclear power plants that used a thorium–U-233 fuel cycle, and more mundane, but still advanced, LWRs with higher conversion ratios than then available. The AEC maintained small programs in other exotic technologies at this time (see table 9-5) but dropped out of LWR research and development. It established the LMFBR as the highest-priority development program, marshaling impressive economic evidence to support its position. In a cost-benefit study issued in 1968, the commission estimated R&D costs for the LMFBR at $3.8 billion and benefits at $13.9 billion, for a benefit-cost ratio of 3.6 to 1.[7]

The choice of LMFBR technology rested on several additional considerations.[8] The LMFBR produces fuel more rapidly than the other breeder technologies that were considered and also has excellent thermal efficiency. This second characteristic meant that the sodium-cooled breeder would generate less waste heat and thermal pollution than LWRs and other alternatives. An independent economic analysis performed at the time concluded that another high-gain breeder—the Gas-Cooled Fast Breeder Reactor—had a higher expected benefit-cost ratio. However, since less was known about the GCFBR, it also had a higher variance on outcome and development time.[9] The LMFBR design was in accord with AEC

5. *1962 Report to the President*, pp. 1–15; and Joint Committee on Atomic Energy, *Nuclear Power Economics, 1962 through 1967*, Committee Print, 90 Cong. 2 sess. (Government Printing Office, 1968), pp. 95–96.

6. AEC, *Civilian Nuclear Power: A Report to the President, 1962; The 1967 Supplement* (1967), p. 6. (Hereafter *1967 Supplement.*)

7. AEC, Division of Reactor Development and Technology, *Cost-Benefit Analysis of the U.S. Breeder Reactor Program*, WASH-1126 (April 1969).

8. *1967 Supplement*, p. 32.

9. Paul W. MacAvoy, *The Economic Strategy for Developing Nuclear Breeder Reactors* (MIT Press, 1969).

traditions of technological sophistication and uranium conservation and satisfied the criterion of keeping the United States at the forefront of nuclear power technology.

The LMFBR was also the design that most interested U.S. manufacturers. The AEC explicitly rejected the idea of investigating subjects that were not interesting to manufacturers.[10] Both manufacturers and the AEC preferred the LMFBR over its alternatives on three counts. First, it most easily tied into the light-water reactor industry. LWRs create relatively small amounts of plutonium that can be used for the initial fuel loading for LMFBRs. Subsequently, the plutonium generated in breeders could be used instead of U-235 to power LWRs. Second, experience had already been gained with the LMFBR. As is shown in table 9-1, the United States had by then operated three experimental LMFBR facilities, including Fermi-1, a utility-sponsored project built under the AEC's cooperative power reactor development program. While their experiences may not seem very positive—both EBR-I and Fermi-1 had meltdown accidents, and Fermi-1 occasioned enormous licensing battles[11]—the liquid sodium design was the closest to commercialization of the advanced breeder designs. The chairman of the AEC, Glenn Seaborg, cited experience as the main reason for choosing the sodium-cooled concept for development.[12] Finally, the same design was under investigation in Europe, the Soviet Union, and Japan. Thus the LMFBR program gave U.S. manufacturers a crack at applying experience with the U.S. program to foreign markets.

The U.S. program and European breeder programs were quite different. The United States emphasized developing the technology base, testing components in special research facilities, and constructing large, complete demonstration plants. The European and Japanese programs planned sequential construction of complete prototypes with midsize breeders at the core of their R&D programs. Although the U.S. demonstrations were

10. As Milton Shaw, the director of the (AEC) Division of Reactor Development and Technology, stated: "If those organizations [such as manufacturers and utilities] feel that our ideas and programs are not feasible or practical from an engineering standpoint or if concepts are so sophisticated or costly they cannot build them or use them in their environment, then we feel we should adjust our reactor development program accordingly." *Public Works for Water, Pollution Control, and Power Development and Atomic Energy Commission Appropriation Bill, 1971,* Hearings before a Subcommittee of the House Committee on Appropriations, 91 Cong. 2 sess. (GPO, 1970), pt. 4, p. 472.

11. Mazuzan and Walker, *Controlling the Atom.*

12. Glenn T. Seaborg, "Nuclear Power: Status and Outlook," *Atomic Energy Law Journal,* vol. 12 (Spring 1970), pp. 36–58.

important to the commercialization effort, they were initially not expected to be the primary central component. The first demonstration, for example, was projected to cost about 2 percent of the breeder budget (see table 9-4). The United States planned three demonstrations specifically to transfer technology to private industry. The demonstration plants were supposed to emphasize "maximum use of existing technology to reduce technical risks and assure safe, reliable operation," and their main objective was to "demonstrate the technical performance, reliability, maintainability, safety, environmental acceptability and economic feasibility in a utility environment of an LMFBR central electric power station."[13]

The Cost-Benefit Studies

Over the course of the breeder program the economic underpinnings of the program changed. The extent of these changes can be traced in the benefit-cost analyses issued periodically by the Atomic Energy Commission, the Energy Research and Development Administration (ERDA), and the Department of Energy (DOE). These studies show when changes in underlying information became official, in the sense that the agency was assuming them for planning purposes.

The result of each study is determined by the assumptions made about the relative capital and fuel costs of different types of generating capacity and the growth in electricity demand. Breeder reactors become competitive with LWRs when the price of uranium becomes high enough to offset the breeder's higher capital costs. Table 9-2 gives estimates of breeder-LWR capital cost ratios from 1968 to 1982. Because nominal capital costs for LWRs increased much more rapidly than inflation over this period, they became a significantly larger component of total costs. Thus an identical ratio in later years implies a much greater cost differential in constant dollars, and consequently the price of uranium had to be corrresspondingly higher for breeders to become competitive with LWRs. The capital cost differential in both the 1968 and 1970 studies is about $68 per kilowatt. The range in the 1975 study is from $0 to $179, and in a 1978 DOE study the differential was estimated to be from $200 to $400, all in 1978 dollars.[14] The price of uranium depends on estimates of high- and low-

13. *Modifications in the Proposed Arrangements for the Clinch River Breeder Reactor Demonstration Project*, Hearings before the Joint Committee on Atomic Energy, 94 Cong. 2 sess. (GPO, 1976), p. 71. (Hereafter *Modifications*, Hearings.)

14. Brian G. Chow, "Comparative Economics of the Breeder and Light Water Reactor," *Energy Policy*, vol. 8 (December 1980), pp. 293–307.

Table 9-2. *Primary Factors Affecting Economic Assessment of Breeder Reactors, Cost-Benefit Studies, 1968–82*

Expected date of commercial introduction	*Year and LMFBR-LWR capital cost per kw ratio*[a]	*Uranium projection: millions of tons available at forward cost (in 1978 dollars)*	*Electrical energy demand (billions of kwh a year)*	*Year and nuclear capacity projections (GWte)*[b]
1968 study: 1984	1990: 1.23[c] 2000: 1.20[c]	1.3 (30.00)[d] 1.9 (90.00)[d]	1,980–2,000[e]	1990: LWR, 185; total, 370[f]
			2,000–8,000[e]	2000: LWR, 185; total, 823[f]
			2,020–18,500[e]	2020: LWR, 239; total, 3,000[f]
1970 study: 1986	1990: 1.10[g] 2000: 1.09[g]	1.8 (30.00)[h] 4.0 (70.00)[h]	1,980–3,070[i]	1990: LWR, 385; total, 534[j]
			2,000–10,000[i]	2000: LWR, 385; total, 1,384[j]
			2,020–23,000[i]	2020: LWR, 1,008; total, 4,000[j]
1975 study: 1987	1990: 1.16[k] 2000: 1.00[k]	3.0 (35.00)[l] 4.0 (72.00)[l]	1,980–3,000[m]	1990: LWR, 445; total, 500[n]
			2,000–12,500[m]	2000: LWR, 700; total, 1,200[n]
			2,020–27,600[m]	2020: LWR, 950; total, 3,300[n]
1977 ERDA update: 2020	First plants: 1.23[o] Mature plants: 1.08[o]	2.5–4.0 (80.00)[p]	. . . 2,000–6,100[q] 2,020–13,800[q]	1990: 190[r] 2000: 380[r] 2020: 650[r]
1980 DOE: after 2020	First plants: 1.4–1.7[s] Mature plants: 1.25–1.5[s]	4.4 (50.00)[t]	. . . . . .	1990: 131[u] 2000: 180[u]
1982 DOE: after 2020	First plants: 1.38[v] Mature plants: 1.2[v]	4.1 (50.00)[w]	. . . . . .	2000: 165[x] 2020: 285[x]

a. LMFBR stands for the Liquid Metal Fast Breeder Reactor Program, LWR for the Light-Water Reactor Program.

b. Gigawatts of electricity.

c. AEC, Division of Reactor Development and Technology, *Updated (1970) Cost-Benefit Analysis of the U.S. Breeder Reactor Program,* WASH-1184 (January 1972), p. 37.

d. AEC, Division of Reactor Development and Technology, *Cost-Benefit Analysis of the U.S. Breeder Reactor Program,* WASH-1126 (April 1969), p. 71.

e. AEC, *Updated (1970) Cost-Benefit Analysis,* p. 72.

f. AEC, *Updated (1970) Cost-Benefit Analysis,* p. 49.

g. AEC, *Cost-Benefit Analysis,* p. 37.

h. AEC, *Cost-Benefit Analysis,* p. 42.

i. AEC, *Cost-Benefit Analysis,* p. 45.

j. AEC, *Cost-Benefit Analysis,* p. 34.

k. AEC, *Final Environmental Impact Statement for LMFBR Program,* vol. 4, sec. 2: *Cost Benefit Analysis,* WASH-1535 (December 1975), p. 11.2-79.

l. AEC, *Final Environmental Impact Statement,* pp. 11.2-105, 11.2-109.

m. AEC, *Final Environmental Impact Statement,* pp. 11.2-5, 11.2-9.

n. AEC, *Final Environmental Impact Statement,* p. 11.2-115.

o. *ERDA Updated Cost-Benefit Analysis of the LMFBR* (June 1977); reprinted in *1978 ERDA Authorization,* Hearings before the Subcommittee on Fossil and Nuclear Energy Research Development and Demonstration of the House Committee on Science and Technology, 95 Cong. 1 sess. (GPO, 1977), vol. 4, p. 500.

p. *1978 ERDA Authorization,* House Hearings, p. 492.

q. *1978 ERDA Authorization,* House Hearings, p. 499.

r. Testimony of James R. Schlesinger (Assistant to the President), *1978 ERDA Authorization,* House Hearings, p. 13.

s. Robert Civiak, "The Economic Competitiveness of Breeder Reactors Compared to Light Water Reactors," Congressional Research Service, Library of Congress, September 13, 1982, p. CRS-5.

t. Civiak, "Economic Competitiveness," p. CRS-19.

u. *Energy and Water Development Appropriations for 1982,* Hearings before the Subcommittee on Energy and Water Development of the House Committee on Appropriations, 97 Cong. 1 sess. (GPO, 1981), pt. 8, p. 1262.

v. Civiak, "Economic Competitiveness," p. CRS-5.

w. Civiak, "Economic Competitiveness," p. CRS-19.

x. Civiak, "Economic Competitiveness," pp. CRS-17, CRS-18.

Table 9-3. *Cost-Benefit Results of Early Case Studies of Breeder Reactor*
Billions of 1975 dollars

Item	*1968*	*1970*	*1975*
Energy cost	313.3	572.3	354.4
Gross benefits	13.9	29.6	53.9
R&D cost	3.8	3.3	5.3
Benefit-cost ratio	3.6	9.0	10.2

Source: For 1968: AEC, *Cost-Benefit Analysis;* for 1970: AEC, *Updated (1970) Cost-Benefit Analysis;* and for 1975: AEC, *Final Environmental Impact Statement.*

grade uranium ore reserves and on uranium demand, which essentially depends on projections for the expansion of LWR capacity. These projections are determined by the relative economics of LWRs and coal plants and by projections of electricity growth.

The early studies find that breeder reactors are competitive with LWRs as soon as they are commercially available (which was presumed to be in the mid-1980s).[15] The studies then examine optimal capacity expansion with and without the breeder option over the next forty to fifty years and the difference in the cost of providing electricity for the two cases. Appropriately discounted, this figure becomes the program benefits, which are compared to the discounted R&D costs incurred on the LMFBR program.

Each study considers a range of values for the main parameters in order to do sensitivity analyses. Table 9-2 contains the values for the base cases, that is, the values considered to be most likely. The results of the benefit-cost studies for the base cases in 1968, 1970, and 1975 are shown in table 9-3. The 1970 study concludes that the breeder program's benefits are more than twice the 1968 estimates and that the benefit-cost ratio is two and a half times as large. The increase in benefits over 1968 is due primarily to the large increase in projected demand for LWRs, which more than offsets a projected increase in low-cost uranium reserves. About half the change in LWR demand is attributed to higher energy demand and half to higher costs of fossil fuel. The 1975 study assumes more optimistic capital

15. AEC, Division of Reactor Development and Technology, *Updated (1970) Cost-Benefit Analysis of the U.S. Breeder Reactor Program,* WASH-1184 (January 1972); and Energy Research and Development Administration, *Final Environmental Statement, LMFBR Program,* ERDA-1535 (December 1975).

costs for breeders and higher projections of electricity growth. The difference between the results of the 1970 and 1975 studies, however, is due almost entirely to increases in the expected real cost of electricity (from all generating sources).

The first sign of no commercial need for breeder reactors emerged in 1975. That was the first year in the history of nuclear power in which the number of new orders for nuclear generating facilities fell below the number of plants that were canceled. Indeed, in no year after 1974 were more plants ordered than were canceled.[16] Of course, the meaning of the 1975 data was hotly contested. One plausible explanation was that it was a temporary setback, reflecting the more general disruptions following the Arab oil embargo in 1973. But each subsequent year told the same story—never again during the life of the CRBR would the annual number of orders for new nuclear power plants exceed four. This was devastating news for the breeder program, since the point of breeder technology was to satisfy the demand for fuel of a rapidly expanding nuclear industry.

The reality of the market for breeders soon became apparent. ERDA issued its first critique of the LMFBR program in 1977 in an abbreviated update of the 1975 study. Besides using the revised parameters shown in table 9-2, it claimed that the current LWR and LMFBR programs suffered from construction slippages, so that the earliest date at which breeder reactors could be introduced commercially would be 1995. The update concluded that the LMFBR program had no net benefits. Approximately two-thirds of the reduction in benefis from the 1975 estimate was due to reduced electricity demand and one-third to the delayed potential introduction of the LMFBR. Moreover, the study noted that a further decline in electricity use was anticipated and that other energy alternatives would further reduce demand so that net benefits to the LMFBR program were likely to be negative.

In testimony before Congress in 1977, James R. Schlesinger, then the "energy czar," argued that, on the basis of LWR projections and probable uranium supplies, the breeder reactor certainly would not be needed before 2020.[17] Under questioning, he conceded that if pessimistic uranium

16. See AEC, *Annual Report to Congress* (1968 through 1974), and the August issues of *Nuclear News* (after 1974).

17. *1978 ERDA Authorization* (the Clinch River Breeder Reactor Program), Hearings before the Subcommittee on Fossil and Nuclear Energy Research, Development and Demonstration of the House Committee on Science and Technology, 95 Cong. 1 sess. (GPO, 1977), vol. 4, pp. 4–13.

supply projections were assumed, the breeder could be important twenty years sooner, around the turn of the century. He argued, however, that the cost of such a mistake would be relatively low because a commercially oriented breeder R&D program could be restarted in the mid-1980s without great expense. His more likely scenario implied no need for a commercial program for many years; thus the Carter administration's proposal to cancel the Clinch River project would save large sums of R&D money as well as further the administration's nonproliferation goal.

The 1977 study was a rough extrapolation from the 1975 study and subject to easy criticisms from proponents of the breeder reactor as well as from the General Accounting Office. ERDA and the DOE followed up the study with analyses aimed at better integration of the data to determine when breeders might be commercially attractive. A National Academy study, drafts of which were available in early 1979, concluded that breeder reactors would not be competitive until 2020.[18] Similarly, DOE studies issued in 1980 and 1982 found successively lower LWR projections and concluded that the LMFBR would not be commercially attractive at any time during the planning period considered, that is, not until sometime after 2020. The Congressional Research Service summarized much of the post-1979 data in late 1982 and found consistent evidence for the 2020 date as a minimum.[19] Indeed, CRS found it was very unlikely that breeders would be competitive by 2020, but a fair chance that they would be competitive by 2040 or 2050.

During the fourteen-year period that the energy agencies were analyzing the program, a lively industry developed among academics and utility industry analysts for the same purpose.[20] These studies were consistent among themselves and with the government studies. Before 1975 the studies supported the economic rationality of the breeder program. By 1977 almost all the studies concluded that it would be well into the next century before breeders would be commercially attractive. The private studies recognized the implications of low demand for energy growth earlier than the government: between 1975 and 1977 their results were mixed, but projections for late commercialization predominated.

18. National Research Council, Committee on Nuclear and Alternative Energy Systems (CONAES), *Energy in Transition 1985–2010* (Washington: National Academy of Sciences, 1979).

19. Robert L. Civiak, "The Economic Competitiveness of Breeder Reactors Compared to Light Water Reactors," Congressional Research Service, September 13, 1982.

20. See references in Chow, "Comparative Economics of the Breeder," and Civiak, "Economic Competitiveness of Breeder Reactors."

The demise of the economic rationale for breeders arose from external conditions. Results from R&D played almost no role in the overall assessment of the program, in part because the program proceeded slowly, but mainly because the external shifts were far more dramatic in their implications than the technical progress of the program. Had the program attained its original technical goals precisely, it would have been a failure as a commercial R&D program.

Program Performance

Of course, the program did not proceed as planned. In 1969 the AEC initiated the project definition stage of the demonstration. By 1970 three manufacturers and seventy-seven utilities were involved, and authorization had been obtained for one demonstration. The first demonstration was to have cost $200 million, of which the government would put up $80 million and industry $120 million. Cost overruns were to be abosrbed by the reactor manufacturers. By 1971 the plans for a demonstration were completely stalled, with vendors and utilities unwilling to accept the kind of risk they had assumed for the LWR demonstration program a decade earlier.

The vendors' reluctance to accept financial risk does not necessarily imply that they were skeptical about the breeder concept. Manufacturers lost heavily on the LWR development program—perhaps several billion dollars—and Westinghouse, one of the largest manufacturers, underwent a corporate reorganization as a result of its participation.[21] Moreover, the reactor industry was strategically advantaged in negotiating with the government because of the latter's perceived need to have the U.S. industry a major player in the world market so as to contain nuclear proliferation. Thus industry could expect to strike a better deal in breeder development than it did with LWRs. Meanwhile the estimated total cost of the demonstration plant had escalated from $200 million to at least $400 million (see table 9-4).

On June 4, 1971, President Richard M. Nixon emphasized his administration's commitment to nuclear power, saying that the LMFBR was the nation's "best hope" to meet growing electricity demand and calling for a successful demonstration of LMFBR technology by 1980. Schlesinger, then chairman of the AEC, improved on this, saying that "what we want to

21. See H. Stuart Burness, W. David Montgomery, and James P. Quirk, "The Turnkey Era in Nuclear Power," *Land Economics*, vol. 56 (May 1980), pp. 188–202.

accomplish by 1980 is a breeder that has the possibility of commercial acceptability so that it is an economic breeder rather than merely a demonstration of technology."[22] The utility impasse was quickly resolved when the AEC agreed to pay up to 50 percent of the cost of the demonstration, excluding end-capital items,[23] and to indemnify the other participants from various financial risks, including project termination. Legislation would be needed to accomplish these agreements, but the AEC's promise to seek such legislation was enough to make the utilities decide to continue with the project. In November 1971, five proposals were submitted to the AEC.

The second and third demonstrations had a short life. Although the AEC conducted design activities on the second demonstration in 1971, industry had said it was willing to support only a single demonstration because of the financial risks. Over the objections of the AEC, the Office of Management and Budget cut funds for these design activities in 1972.[24]

Progress on the demonstration continued in 1972. In January the Tennessee Valley Authority–Commonwealth Edison (CE) plan was selected by the AEC as a basis for discussion. In March the Breeder Reactor Corporation and the Project Management Corporation were formed to manage and coordinate activities. Chief management authority rested with TVA and CE, but the AEC was to have representation on the corporations as well. Legislative approval would be necessary for AEC participation. The Clinch River site was selected in August. In November Westinghouse was chosen to be the lead contractor, and in December Burns and Roe was selected to manage construction activities. By the end of 1972 the utilities had pledged some $250 million to the project, representing half the total estimated cost.

Progress was slower on the government side. Three efforts were under way: a safety review, an environmental review, and legislative approval for the contracts and various indemnification promises. Licensing was a crucial part of the demonstration project because of its commercial orientation. The licensing requirement also paralleled the demonstrations that were part of the LWR development program. With the significant exception of

22. *Public Works for Water and Power Development and Atomic Energy Commission Appropriation Bill, 1973,* Hearings before the Subcommittee on Public Works of the House Committee on Appropriations, 92 Cong. 2 sess. (GPO, 1972), pt. 4, p. 21.

23. See *Modifications,* Hearings, p. 78.

24. *Public Works for Water and Power Development and Atomic Energy Commission Appropriation Bill, 1974,* Hearings before the Subcommittee on Public Works of the House Committee on Appropriations, 93 Cong. 1 sess. (GPO, 1973), pt. 4, p. 402.

the one breeder reactor included in the Cooperative Power Reactor Development Program,[25] licensing in the early cases had been straightforward. However, because of the growing environmental movement, licensing at the AEC had changed dramatically between the beginning of the LMFBR program in 1968 and the beginning of licensing activities in 1971. In 1970 environmental assessments were added to the licensing process. Moreover, since these assessments were new, their standards and processes were being developed through AEC activities and federal court reviews.[26] A second change was the extent of controversy in individual licensing cases. Two-thirds of the LWR licensing cases initiated in 1970 had active intervenors who were opposed to the proposed license; less than a quarter of the applications filed between 1966 and 1968 faced similar opposition. Licensing for construction permits took about three years to be concluded for cases filed in 1970 and 1971, up from an average of thirteen months in 1967, and approximately two years in 1968 and 1969.[27] Licensing for Clinch River proved to be a major hurdle.

In 1972 the Scientists' Institute for Public Information (SIPI) challenged the Clinch River environmental impact statement, claiming that it should include LMFBR programmatic issues, such as the environmental impact of plutonium recycling. This view was supported by the Atomic Safety and Licensing Board, overturned on review by the commission, and taken to court. In 1973 the District of Columbia Circuit Court supported SIPI, requiring the AEC to include programmatic issues in the impact statement.[28] The new draft statement was issued in March 1974, but the Environmental Protection Agency objected to it. The AEC and the EPA met in August 1974, resulting in EPA approval. The EPA's position was that there was inadequate experience or knowledge to assess the environmental impact of a commercial LMFBR industry. However, the impact of the Clinch River project was acceptable, and the other issues could be reevaluated before the commercialization of the industry. Essentially, the EPA approved the original parts of the environmental impact

25. Mazuzan and Walker, *Controlling the Atom.*

26. See, for example, Steven Ebbin and Raphael Kasper, *Citizen Groups and the Nuclear Power Controversy: Uses of Scientific and Technological Information* (MIT Press, 1974); and Elizabeth S. Rolph, *Nuclear Power and the Public Safety: A Study in Regulation* (Lexington Books, 1979).

27. Linda R. Cohen, "Innovation and Atomic Energy: Nuclear Power Regulation, 1966–Present," *Law and Contemporary Problems*, Winter–Spring 1979, pp. 84, 72.

28. *Scientists' Institute for Public Information, Inc.* v. *AEC*, 481 F.2d 1979 (D.C. Circuit 1973).

statement, and those parts required by the 1973 court decision, though too speculative to warrant approval, were not judged to be vital to the CRBR project.[29] The proposed final environmental impact statement was issued in January 1975, but its approval was delayed another two years because of problems with the safety analysis report.

According to the 1973 schedule, by 1975 the safety review would be finished, and construction would begin.[30] A GAO report issued in 1974 alluded to regulatory problems, saying that the AEC regulatory staff was not getting enough safety information and that "there exists a difference of opinion between AEC's regulatory organization and the CRBR project participants concerning the timeliness and sufficiency of AEC's current efforts to resolve a CRBR safety issue."[31] In 1975 a finished safety review was not necessary for initial construction, but an environmental review was. The environmental analysis was tied up because the AEC belatedly had decided to include analysis of a (parallel) alternative safety system in the safety review, and it would have to be considered in the environmental impact statement as well. According to a frustrated congressman, Mike McCormack, "We are being delayed for a year by regulatory matters, by the need to go out and write an additional environmental impact statement because we are afraid somebody might bring a lawsuit because we are designing the thing with or without a core catcher at the moment."[32] McCormack was unduly optimistic: the final environmental statement was issued in January 1977, clearing the way for a limited work authorization.

Costs and Management

During 1974 the AEC reviewed its entire breeder program and came up with startling new cost estimates.[33] The estimated cost for the Fast Flux Test Facility, a critical component of the base program, rose from $187.8

29. *Review of the Liquid Metal Fast Breeder Reactor Program,* Hearings before the Ad Hoc Subcommittee to Review the Liquid Metal Fast Breeder Reactor Program of the House Joint Committee on Atomic Energy, 94 Cong. 1 sess. (GPO, 1975), vol. 1, pp. 171–78. (Hereafter *Review of LMFBR*, 1975 Hearings.)

30. *Public Works for Water and Power Development, 1974,* Hearings, pt. 4, p. 50.

31. General Accounting Office (GAO), *Problem Areas Which Could Affect the Development Schedule for the Clinch River Breeder Reactor* (December 1974); quoted in *Public Works for Water and Power Development and Energy Research Appropriation Bill, 1977,* Hearings before a Subcommittee on Public Works of the House Committee on Appropriations, 94 Cong. 2 sess. (GPO, 1976), pt. 6, p. 335.

32. *Review of LMFBR*, 1975 Hearings, vol. 2, p. 993, and appendixes.

33. *Review of LMFBR*, 1975 Hearings.

Table 9-4. *Clinch River Breeder Reactor Estimated Costs, Selected Years, 1969–83*

Year	*Estimated costs, government share (millions of dollars)*[a]	*Estimated year of commercial breeder industry*[b]	*Estimated government share of CRBR (cost as percent of LMFBR R&D)*[c]	*Cost of LMFBR (billions of dollars)*[b]
1969	80	1984	2.4	3.3
1972	250	1986	6.7	3.7
1974a	450	1987	6.8	6.6
1974b	1,450	1987	19.3	8.8
1978	2,200	?2000–2025	18.5	11.9
1981	3,200	2025	30.0	10.0
1983	4,000	?2025	?	?

a. For 1969: *Public Works for Water and Power Development and the Atomic Energy Commission Appropriation Bill, 1970*, Hearings before the Subcommittee on Public Works of the House Committee on Appropriations, 91 Cong. 1 sess. (GPO, 1969), pt. 4, p. 358. For 1972: *Public Works . . . Appropriation Bill, 1973*, Hearings, pt. 4, p. 316. For 1974a: *Public Works . . . Appropriation Bill, 1975*, Hearings, pt. 4, p. 152. For 1974b: *Public Works for Water and Power Development and Energy Research Appropriation Bill, 1976*, Hearings before a Subcommittee of the House Committee on Appropriations, 94 Cong. 1 sess. (GPO, 1974), pt. 5, p. 101. For 1978: *Public Works . . . Appropriation Bill, 1979*, Hearings, pt. 5, p. 1295. For 1981: *Energy and Water Development Appropriations for 1982*, Hearings before the Subcommittee on Energy and Water Development of the House Committee on Appropriations, 97 Cong. 1 sess. (1981), pt. 8, p. 1237. For 1983: Robert Civiak, "DOE Estimates of the Cost and Time to Complete CRBR," CRS memo, Library of Congress, Congressional Research Service, September 13, 1983, table 1.

b. For 1969–74b: *Public Works . . . Appropriation Bill, 1976*, Hearings, pt. 5, p. 323. For 1978: *Public Works . . . Appropriation Bill, 1979*, Hearings, pt. 5, pp. 1295, 1442. For 1981: *Energy and Water Development Appropriations for 1982*, Hearings, pt. 8, pp. 1237, 1274.

c. CRBR stands for the Clinch River Breeder Reactor Demonstration; LMFBR for the Liquid Metal Fast Breeder Reactor Program.

million to $420 million,[34] and the estimated cost for Clinch River went from $700 million to $1.7 billion (see table 9-4). Of this increase, $553 million was attributed to better project definition and $600 million to real escalation in construction costs and slippages in the project's schedule.[35] The estimated cost of the entire LMFBR program escalated as well, from $3.7 billion in 1972 to $6.6 billion in March 1974, and up to $8.8 billion in December 1974 (table 9-4).

The industry was firm in committing no more than the previously agreed $250 million to the Clinch River project. The AEC was still seeking authorization to assume financial risks for the program and was now also seeking authorization of an extra billion dollars for the demonstration. The fiscal 1976 authorizing legislation provided the financing conditions that

34. *Public Works for Water and Power Development and Atomic Energy Commission Appropriation Bill, 1975*, Hearings before the Subcommittee on Public Works of the House Committee on Appropriations, 93 Cong. 2 sess. (GPO, 1974), pt. 4, p. 137.

35. *Public Works for Water and Power Development and Atomic Energy Commission Appropriation Bill, 1976*, Hearings before the Subcommittee on Public Works of the House Committee on Appropriations, 94 Cong. 1 sess. (GPO, 1975), pt. 5, pp. 315, 334.

the agency sought and indemnified industry and utility participants from cost overruns. If the project were terminated, their liability would be only for funds previously due and expended. The legislation also changed the project's management structure. ERDA assumed lead management control of the project, and the Division of Reactor Development and Demonstration, concerned solely with the LMFBR program, was established.

The 1974 review led ERDA to decide to revamp the development strategy. Commercialization was still desired by 1987, but cost overruns and delays cast doubt on the original plan for a series of demonstrations. The base program was delegated the additional job of developing and testing large-scale sodium components, which previously had been assigned to subsequent demonstrations. ERDA now planned to move directly from Clinch River to a plant of commercial size.[36]

Despite the apparent deemphasis of demonstration plants in the new strategy, a sharply increasing share of the LMFBR budget (and the entire civilian reactor development budget) was going to Clinch River. As CRBR entered the component purchasing stage, it absorbed a larger and larger fraction of a fairly stable budget allocation for LMFBR research and development (tables 9-4, 9-5, and 9-6). In reality, the new plan put a much greater emphasis on the Clinch River demonstration. Since the government was preparing to pay for at least 85 percent of the cost of Clinch River and was assuming management control, industrial participation was no longer the overriding justification for the project. Rather, the demonstration was having to stand on its own merits to resolve uncertainty about the practical future economic value, reliability, environmental effects, and safety characteristics of breeder reactors.[37]

All these attributes were the subject of controversy. Among the issues discussed were whether the Clinch River design was state of the art, the extent to which it went technologically beyond the fast flux test facility, and its value in testing steam sodium loops compared with a test plant designed for that purpose.[38]

The EPA was an important source of early criticism. Its qualified approval for Clinch River, issued in 1975, stated there was great uncertainty about

36. *Public Works for Water and Power Development, 1976,* Hearings, pp. 315, 334.

37. *Review of LMFBR,* 1975 Hearings, vol. 1, p. 144; see also the remarks of Chairman Pastore: "We have to make sure about the environment, I grant you that. We have to make sure about the cost-benefits aspects of the program. We have to look into that. How will we ever know those questions until we get a demonstration plant?" Ibid., pp. 171–72.

38. *Review of LMFBR,* 1975 Hearings, vol. 2, p. 1071; vol. 1, p. 397.

Table 9-5. *Breeder Program Appropriations, Fiscal Years 1968–85*
Millions of current dollars

Year	*Advanced convertors and low-gain breeders*	*Gas-Cooled Fast Breeder Reactor (GCFBR)*	*Molten-Salt Breeder Reactor (MSBR)*	*Liquid Metal Fast Breeder Reactor (LMFBR)*
1968	51.6	1.3	4.6	68.4
1969	28.7	1.9	5.2	80.1
1970	22.1	1.3	5.0	87.8
1971	32.0	0.6	5.0	88.9
1972	28.8	1.0	4.8	123.2
1973	30.5	1.0	4.6	144.1
1974	12.6[a]	1.8	1.6	206.6
1975	31.2[b]		4.8	308.9
1976	52.9	8.2	3.3	351.6
Transition quarter	14.8	3.4	0.2	101.1
1977	53.0	12.8	0.0	564.5
1978	69.6	14.7	0.0	412.8
1979	90.4	21.0	0.0	456.0
1980	57.9	14.6	0.0	476.3
1981	95.0	0.0	0.0	459.5
1982	84.0	0.0	0.0	500.4
1983[c]	81.8	0.0	0.0	451.2
1984[c]	70.5	0.0	0.0	303.6
1985[d]	64.4	0.0	0.0	236.9

Sources: For 1968: *Public Works Appropriations for 1970 for Water and Power Resources Development and the Atomic Energy Commission*, Hearings before a Subcommittee of the House Committee on Appropriations, 91 Cong. 1 sess. (GPO, 1969), pt. 4, pp. 223, 236; and *AEC Authorizing Legislation, Fiscal Year 1970*, Hearings before the Joint Committee on Atomic Energy, 91 Cong. 1 sess. (GPO, 1969), pt. 2, p. 1421. For 1969: *Public Works for Water, Pollution Control, and Power Development and Atomic Energy Appropriation Bill, 1971*, Hearings before a Subcommittee of the House Committee on Appropriations, 91 Cong. 2 sess. (GPO, 1970), pt. 4, pp. 579, 590; and *Public Works for Water, Pollution Control, and Power Development and Atomic Energy Commission Appropriations for Fiscal Year 1971*, Hearings before a Subcommittee of the Senate Committee on Appropriations, 91 Cong. 2 sess. (GPO, 1970), pt. 3, vol. 2, p. 975. For 1970: *Public Works for Water and Power Development* and *Atomic Energy Commission Appropriation Bill, 1972*, Hearings before a Subcommittee of the House Committee on Appropriations, 92 Cong. 1 sess. (GPO, 1971), pt. 4, pp. 435, 452. For 1971: *Public Works . . . Appropriation Bill, 1973*, Hearings, pt. 4, pp. 218, 222, 227. For 1972: *Public Works . . . Appropriation Bill, 1972*, Hearings, pt. 4, p. 330; and *Public Works . . . Appropriation Bill, 1974*, Hearings, pt. 4, pp. 317, 325, 327. For 1973: *AEC Authorizing Legislation, Fiscal Year 1975*, Hearings before the Joint Committee on Atomic Energy, 93 Cong. 2 sess. (GPO, 1974), pt. 1, pp. 386, 392, 394, 397. For 1974: *Public Works for Water and Power Development and Energy Research Appropriation Bill, 1976*, Hearings before a Subcommittee of the House Committee on Appropriations, 94 Cong. 1 sess. (GPO, 1975), pt. 5, pp. 175, 186. For 1975: *Public Works . . . Appropriation Bill, 1977*, Hearings, pt. 6, p. 211. For 1976 and transition quarter: *Public Works . . . Appropriation Bill, 1978*, Hearings, pt. 5, pp. 156, 217. For 1977: *Public Works . . . Appropriation Bill, 1979*, Hearings, pt. 5, pp. 794, 1327. For 1978: *Public Works for Water and Power Development and Energy Research Appropriation Bill, 1979*, H. Rept. 95-1247, 95 Cong. 2 sess. (GPO, 1978), p. 15. For 1979: *Energy and Water Development Appropriations for 1981*, Hearings before a Subcommittee of the House Committee on Appropriations, 96 Cong. 2 sess. (GPO, 1980), pt. 5, p. 497; and *Energy and Water Appropriation Bill, 1980*, S. Rept. 96-242, 96 Cong. 1 sess. (GPO, 1979), p. 9. For 1980: *Energy and Water Development Appropriations for 1982*, Hearings, pt. 5, p. 530. For 1981: *Energy and Water Development Appropriations Bill, 1982*, S. Rept. 97-256, 97 Cong. 1 sess. (GPO, 1981), p. 87. For 1982: *Energy and Water Development Appropriation Bill, 1983*, S. Rept. 97-673, 97 Cong. 2 sess. (GPO, 1982), pp. 24, 31, 33, 73. For 1983: *Energy and Water Development Appropriation Bill, 1984*, H. Rept. 98-217, 98 Cong. 1 sess. (GPO, 1983), p. 84; and *Energy and Water Development Appropriation Bill, 1984*, S. Rept. 98-153, 98 Cong. 1 sess. (GPO, 1983), pp. 113, 114. For 1984 and 1985: *Energy and Water Development Appropriation Bill, 1985*, S. Rept. 98-502, 98 Cong. 2 sess. (GPO, 1984), pp. 115, 116; 121, 122.

a. High-temperature gas reactor.
b. Light-water breeder.
c. Estimated.
d. Requested.

Table 9-6. *Budget for Clinch River Breeder Reactor, Fiscal Years 1974–84*
Millions of current dollars

Year	*Administration request*[a]	*Appropriation (budget authority)*[b]	*Expenditures (public and private)*[c]
1974	20.0	21.9	29.9
1975	14.0	56.2	70.1
1976	57.0	107.0	140.2
Transition quarter	15.1	24.1	. . .
1977	171.0[d]	237.6	184.4
1978	33.0	80.0	166.2
1979	13.4	172.4	175.1
1980	0	172.4	189.4
1981	0	172.0	192.9
1982	254.0	193.9	194.1
1983	252.5	187.9	275.7[e]
1984	270.0	40.0	. . .

a. For 1974: *Public Works for Water and Power Development and Atomic Energy Commission Appropriation Bill, 1974*, Hearings before a Subcommittee of the House Committee on Appropriations, 93 Cong. 1 sess. (GPO, 1973), pt. 4, p. 330. For 1975: *Public Works . . . Appropriation Bill, 1975*, Hearings, pt. 4, p. 98. For 1976 and transition quarter: *Public Works for Water and Power Development and Energy Research Appropriation Bill, 1976*, S. Rept. 94-505, 94 Cong. 1 sess. (GPO, 1975), p. 23. For 1977: *Public Works and Power Development and Energy Research Appropriation Bill, 1977*, Hearings before a Subcommittee of the House Committee on Appropriations, 94 Cong. 2 sess. (GPO, 1976), pt. 6, p. 215. For 1978: *Public Works for Water and Power Development and Energy Research Appropriation Bill, 1978*, S. Rept. 95-301, 95 Cong. 1 sess. (GPO, 1977), p. 19. For 1979: *Public Works . . . Appropriation Bill, 1979*, Hearings, pt. 5, p. 1327. For 1980: *Energy and Water Development Appropriation Bill, 1980*, S. Rept. 96-242, 96 Cong. 1 sess. (GPO, 1979), p. 10. For 1981: *Energy and Water Development Appropriations for 1981*, Hearings before a Subcommittee of the House Committee on Appropriations, 96 Cong. 2 sess. (GPO, 1980), pt. 5, p. 499. For 1982: *Making Appropriations for Energy and Water Development for the Fiscal Year Ending September 30, 1982, and for Other Purposes: Conference Report*, H. Rept. 97-345, 97 Cong. 1 sess. (GPO, 1981), p. 73. For 1983: *Energy and Water Development Appropriation Bill, 1983*, S. Rept. 97-673, 97 Cong. 2 sess. (GPO, 1982), p. 73. For 1984: *Energy . . . Appropriation Bill, 1984*, H. Rept. 98-217, 98 Cong. 1 sess. (GPO, 1983), p. 84.

b. For 1974, 1975: *Public Works . . . Appropriation Bill, 1976*, Hearings, pt. 5, pp. 176, 186. For 1976 and transition quarter: *Public Works . . . Appropriation Bill, 1978*, Hearings, p. 156. For 1977: ibid., p. 1327. For 1978: *Public Works . . . Appropriation Bill, 1979*, Hearings, p. 40. For 1979: *Energy and Water Development Appropriations for 1981*, Hearings pt. 5, p. 499. For 1980: *Energy and Water Development Appropriation Bill, 1981*, S. Rept. 96-927, 96 Cong. 2 sess. (GPO, 1980), p. 10. For 1981: *Energy and Water Development Appropriation Bill, 1982*, S. Rept., p. 87. For 1982: *Energy . . . Appropriation Bill, 1983*, S. Rept., p. 73. For 1983: *Energy . . . Appropriation Bill, 1984*, H. Rept., p. 84. For 1984: *Energy . . . Appropriation Bill, 1985*, S. Rept., p. 122.

c. For all years: Robert Civiak, "DOE Estimates of the Cost and Time to Complete CRBR," CRS memo, Library of Congress, Congressional Research Service, September 13, 1983, table 3.

d. Budget obligation.

e. Projected.

the environmental, safety, and economic impact of an LMFBR industry, and even about the CRBR. Because the Project Independence estimates of electricity use (also issued in 1975) were lower than previous forecasts, the EPA suggested that delaying commercialization of LMFBRs by four to twelve years would be both feasible and desirable.[39]

Nevertheless, the new authorization for Clinch River was approved, and the purchasing of components began. Actual construction at the Clinch

39. *Review of LMFBR*, 1975 Hearings, vol. 2, p. 1025.

River site was still held up by the Nuclear Regulatory Commission (NRC), but in 1976 and 1977 ERDA signed contracts with many companies for construction of components.

The Carter Opposition

Jimmy Carter, despite his background as a nuclear engineer, was the only president to oppose Clinch River. But his opposition should be placed in perspective. Throughout the life of Clinch River before Carter became president, the various administrations' budget proposals were always substantially lower than what Congress ultimately appropriated (table 9-6). Throughout the 1970s presidents persistently sought a more balanced program—more for R&D, less for demonstration—than Congress approved.

In 1977 the Carter administration started its campaign to cancel the Clinch River project. Licensing at the NRC was suspended for the duration of his administration, but Carter was not successful at halting the program. Each year he requested no appropriation for the plant; each year Congress approved several hundred million dollars. In 1977 the president vetoed the energy appropriation bill because it included funds for Clinch River. Nevertheless, the funds continued to flow, albeit through supplementals and continuing resolutions. Funds for Clinch River did not appear in regular appropriation bills for the duration of the project.[40]

Carter wanted to put greater emphasis on the base program (although his budget request for this category dropped as well) and the design of a large breeder. The large breeder would be developmental, would be built on a government reservation, and would not go through the commercial licensing reviews.

Carter's opposition to the Clinch River project rested on the conclusions from the cost-benefit studies that the breeder would not be competitive for decades and that it was not needed for energy independence, on questions about the safety and technical obsolescence of Clinch River, and on a concern that plutonium-based fuels could lead to an international proliferation of nuclear weapons.[41] All these claims were denied by propo-

40. For an excellent discussion on the confusing subject of energy appropriation bills in the late 1970s, see Eric M. Uslaner, *Shale Barrel Politics: Energy and Legislative Leadership* (Stanford University Press, 1989), pp. 24–49.

41. GAO, "U.S. FBR Program Needs Direction," Report to the Congress by the Comptroller General of the United States, EMD-80-81 (September 1980), p. 1. The most influential of the nongovernmental critiques with the Carter administration was Keeny and others, *Nuclear Power*.

nents of the breeder.[42] They insisted that the benefit-cost study results were subject to uncertainty and therefore that Clinch River should be built as an insurance policy in case commercialization turned out to be desirable twenty years earlier than projected.

Two other rationales for keeping Clinch River appeared at this time: the "sunk cost" argument and the public confidence issue. Representative Michael Myers elaborated on the former at the fiscal 1979 hearing: "If we were back at the point now of first making a decision as to whether or not we should enter into fast breeder reactor research, then we would be more inclined toward [CRBR cancellation]. But there has been so much invested, more than ten years of research in the project that we are now talking about, that we are way past that point of making that decision."[43]

The public confidence issue was raised in connection with Carter's proposal to build a large, unlicensed facility. This proposal raised great concern among utilities, which considered licensing to be the biggest hurdle facing nuclear technology.[44] The utilities worried that the "retreat from public scrutiny" of the Carter plan would undermine public confidence, which, in the wake of the Three-Mile Island accident, had degenerated.

The 1980s mark the end of the Clinch River project, although President Ronald Reagan supported it. The project by then was justified by rationales that retreated even further from commercial demonstration. Secretary of Energy James B. Edwards argued that Clinch River fit into the administration's energy policy because it was a "long-term high-risk R&D venture," it enhanced national security, and it contributed to nonproliferation goals by allowing the United States to dominate the field internationally. Finally, he claimed it was necessary to support Clinch River to show utilities that the government still had a commitment to nuclear power.[45] The administration recommended completing NRC licensing activities. A limited work authorization was finally issued in 1983, and site preparation began at Clinch River.

The 1980s brought more bad news about the projected cost of the Clinch

42. See, for example, GAO, *The Clinch River Breeder Reactor: Should the Congress Continue to Fund It*, Report to the Congress by the Comptroller General of the United States, EMD-79-62 (May 1979).

43. *Public Works for Water and Power Development and Energy Research Appropriation Bill, 1979*, Hearings before the Subcommittee on Public Works of the House Committee on Appropriations, 95 Cong. 2 sess. (GPO, 1978), pt. 5, p. 1294.

44. See GAO, *The Clinch River Breeder Reactor*, p. 16.

45. *1981 Economic Report to the President*, Hearings before the Joint Economic Committee, 97 Cong. 1 sess. (1981), pt. 3, p. 242.

River project (table 9-4). All the increases in the cost estimates for Clinch River after 1975 were attributed to delays, which were said to add $17 million a month to the cost of the project. In contrast, cost estimate increases in the other R&D programs discussed in this book are mainly attributed to initial low cost estimates arising from technological optimism. This difference suggests three possibilities about Clinch River. First, the plant may really have held to its initial goals of incorporating known equipment; that is, it did not include a great deal of unproved R&D components. Second, program delays may have prevented the project from progressing to the point where new sources of cost overruns would appear. Third, the difference may show that technological optimism is not necessarily true for R&D projects. The bulk of evidence, pertaining to other programs analyzed in this book, other nuclear programs, and the breeder history before 1975, strongly suggests that the second possibility is the most likely.

By 1981 a coalition of fiscal conservatives and environmentalists began lobbying Congress to end the project.[46] Finally, in 1983 the House of Representatives required the Department of Energy to devise a plan for private parties to assume at least half the cost. The Breeder Reactor Corporation drew up a plan to sell bonds and equity in the plant. The utilities would contribute no further resources. The government would bear all the risks, in that a return on equity would be guaranteed regardless of whether the project was completed and successful. Congressional Budget Office and GAO analyses of the proposal concluded that this plan would cost the government more money than outright financing, because investors would be able to take tax deductions and because the government itself could borrow money at a cheaper rate than the return they would have to pay to private investors for the sale of bonds.[47] The plan was not approved.

Appropriations for Clinch River were cut off in fiscal 1984. In December 1983 the NRC terminated licensing and vacated the limited work authorization that had been issued the previous year.[48] In January the licensing

46. See, for example, William J. Lanouette, "Clinch River Breeder Project Draws Opposition of Strange Bedfellows," *National Journal*, vol. 14 (October 2, 1982), p. 1678; and Alan Murray, "Congress Moving to Scuttle Clinch River Breeder Reactor," *Congressional Quarterly Weekly Report*, Special Report, May 21, 1983, p. 995.

47. Congressional Budget Office, "Comparative Analysis of Alternative Financing Plans for the Clinch River Breeder Reactor Project," Staff working paper, September 20, 1983; and GAO, "Analysis of Studies on Alternative Financing for the Clinch River Breeder Reactor," GAO/RCED-83-151 (May 12, 1983).

48. Nuclear Regulatory Commission, "In the Matter of U.S. Department of Energy Project Management Corporation Tennessee Valley Authority (Clinch River Breeder Reactor Program)," 18 NRC 1337 (1983), ALAB-755 (December 15, 1983).

board on the case issued a special decision that would have authorized a construction permit had the case still been open.[49]

Assessment of the Program

The history of the breeder reactor program spans three changes in the environment for energy policy that should have affected the course of the program, particularly of the demonstration plant at Clinch River. The first fits in the context of the benefit-cost analyses: changes in electricity demand, in light-water reactor demand, and in uranium supply estimates, all of which influenced the commercial immediacy of breeder technology. The change in uranium supply estimates after 1970 suggests that continuing to emphasize LMFBR technology, whose logic rested on efficient uranium use, might have been fruitfully reconsidered. The second factor affecting Clinch River was the change in regulatory requirements that made commercial licensing more difficult for facilities having uncertain, or at least unproved, environmental and safety characteristics. The third was the performance of the breeder program itself, specifically the enormous escalation in estimated costs between 1970 and 1975 and the unwillingness of the utilities and manufacturers to assume any additional financial burden or risk.

These three changes suggest that the program's commercial orientation should have been abandoned no later than 1977, and probably a year or more earlier. The post-1975 benefit-cost studies showed that commercialization was likely to be far in the future, although some controversy surrounded the actual estimates of benefits, costs, and date of commercial need. But also important was the fact that by 1975 Clinch River had ceased to be a private project. The government had been forced to take over management of the project because of a loss of private interest. If its position on cost sharing had been purely strategic, industry would have been reluctant to give up hands-on experience with a coming dominant technology. Hence industry's loss of interest in management should have led to a serious reassessment of the commercial orientation of Clinch River.

Maintaining the commercial goal of the program proved extraordinarily expensive, a result that could have been predicted from the changes in the licensing environment by the mid-1970s. If commercialization was not imminent, the main headache of obtaining licenses could be dispensed

49. Nuclear Regulatory Commission, "In the Matter of U.S. Department of Energy Project Management Corporation Tennessee Valley Authority (Clinch River Breeder Reactor Program)," 19 NRC 288 (1984), LBP-84-4 (January 20, 1984).

with. The plant itself was a government facility rather than an industry project and therefore did not require a commercial license. But more important, the changes in licensing requirements and processes meant that an experimental facility would face great difficulty in gaining approval. In effect, operating experience had become necessary before licenses could be issued. A sensible strategy was to reorient Clinch River as a test facility for generating data that could be used to obtain commercial licenses for subsequent plants. But to continue a commercial orientation dictated a catch-22: licenses were required before operation, but operation was necessary for licensing.

As Clinch River became increasingly expensive, the rest of the breeder R&D program shrunk. Recall that the U.S. breeder strategy included a broad, research-oriented base program in which commercial demonstration was to play a fairly small role. The first casualties of the cost overruns for Clinch River were the other demonstration projects. Eventually the base program was also curtailed. Federal expenditures on Clinch River were initially estimated at 2.4 percent of the total that would be spent on the LMFBR (table 9-4). By 1972 the figure was 6.7 percent. After the first major reestimation of Clinch River costs, the figure rose to 19.3 percent in 1974 and to 40 percent by 1983.

As performance deteriorated, the justification for Clinch River became increasingly vague and grandiose. From commercial demonstration and industry participation in 1969, the justification for the project changed to an emphasis on resolving uncertainties. In 1975 proponents claimed it to be a demonstration of environmental and safety characteristics; in 1981 they described it as a "long-term, high-risk R&D project." They also gave noneconomic justifications for the project, such as the magnitude of the sunk costs and the value of expressing public confidence in and support for the nuclear industry. In 1983 a report prepared for the Project Management Corporation examined the consequences of canceling the Clinch River project. The report discussed the effects of cancellation on employment and the taxes paid by companies and employees working on the project. The consequences of delaying technology development were not discussed at all.[50]

The LMFBR program received a push for early commercialization from technologists in the bureaucracy and put emphasis on industrial

50. "Clinch River Breeder Reactor Plant Project: An Economic Impact Assessment," Report prepared for the Project Management Corporation by JWK International Corporation, June 1983.

Table 9-7. *House Votes on Funds for Clinch River, Fiscal Years 1975–83*[a]

	All votes			*Republicans*			*Democrats*		
Year	*Support*	*Oppose*	*Abstain*	*Support*	*Oppose*	*Abstain*	*Support*	*Oppose*	*Abstain*
1975	237	146	52	92	26	27	145	120	35
1976	218	182	33	95	37	13	123	145	20
1977a	215	205	15	100	41	5	115	164	10
1977b	249	165	21	114	24	8	135	141	13
1978	200	155	78	93	21	31	107	134	47
1979	241	186	8	128	28	2	113	158	6
1981	211	191	32	107	70	13	104	121	19
1982	196	217	21	99	80	10	97	137	11
1983	1	388	44	1	147	18	0	241	26

Sources: *Congressional Quarterly Almanac* (Washington, 1975–83).

a. The votes are as follows. 1975: HR3474, vote to prohibit use of funds in fiscal 1976 on Clinch River; 1976: HR13350, vote to limit the federal share of cost overruns over $2.5 billion to 50 percent; 1977a: HR6796, vote to limit federal share of cost overruns over $2.5 billion to 50 percent; 1977b: HR6796, vote to authorize $33 million, instead of $150 million, for Clinch River; 1978: HR12163, vote to reduce Clinch River funding by $159 million; 1979: HR3000, vote to allow termination of Clinch River; 1981: HR4144, vote to delete $228 million for Clinch River; 1982: HJRes631, vote to bar use of funds for Clinch River in continuing resolution; 1983: HR2587, vote to deauthorize fiscal 1984 funding for Clinch River. Note that in each case a "nay" vote was pro–Clinch River.

participation. When the initial plans proved to be overoptimistic, President Carter turned against the program, though Congress continued to support it. Reflecting greater congressional concern for current demonstration over long-term R&D, the result was to transfer R&D funds to a commercially useless demonstration.

President Reagan's support in the 1980s is not consistent with the view that the president should be more inclined to favor long-term R&D, and be less concerned about distributive politics, than Congress. We suspect that the political polarization that took place in the late 1970s over nuclear power may provide an explanation for Reagan's behavior.

Political Support for Clinch River

The preceding history of the breeder program shows that Congress caused the commercial demonstration program to be continued after 1977. This section analyzes roll call votes in the House of Representatives and the Senate to determine the sources of congressional support for the program.

House Votes

The voting record on Clinch River is exceptionally rich. Roll call votes were taken in the House of Representatives nearly every year between 1975 and the cancellation of the project in 1983. The distribution of votes is shown in table 9-7. Although the precise content of each vote varies, the practical effect of all the amendments would have been the cancellation of the project at that time (see the footnotes to table 9-7). Because only one representative voted for Clinch River in 1983, we include only votes between 1975 and 1982 in the statistical analysis.

A logistic function is estimated to predict support for the Clinch River project. The independent variables are designed to measure programmatic distributive benefits (that is, contracts awarded in each congressional district), distributive benefits from increased nuclear power options, social benefits, and institutional and ideological support in Congress. Precise definitions are found in figure 9-1, and table 9-8 shows sample statistics for these variables.[51]

Four sets of regressions are estimated for this data set. Table 9-9 contains annual regressions; table 9-10 a pooled regression over all votes, and separate equations for three groups of votes that fail a between-group pooling test but pool within years, as described in chapter 5. The three groups correspond to the different presidential administrations and also to the distinct project phases discussed below. Table 9-11 contains separate regressions for each party, in which the independent variables are interacted with dummy variables for the three project phases.

Distributive Politics

CONTRACTS. Pork barrel politics played a significant role in the Clinch River program. One indication of its importance is that good data are available: in 1978 the Department of Energy provided information to Congress about contract expenditures in each district. Consequently, the test for pork barrel support is much cleaner for this program than for the other cases studied in this book.

Representatives from districts with large contracts are more likely to support the project; however, the pattern of support shows a distinct threshold effect. The coefficients on the *PORK* variables are consistently of

51. As table 9-7 shows, abstentions are not constant over the period, but rather shift substantially over time. Abstentions may complicate an analysis if they are correlated with the independent variables and vary over time. For instance, if abstentions occurred during the Carter administration among Southern Democrats who would have voted for the project but are influenced by the president to abstain, then this analysis would understate the importance of the president. A two-stage model that incorporates a model of voting, as well as voting for or against a proposal, is presented in Linda R. Cohen and Roger G. Noll, "How to Vote, Whether to Vote: Strategies for Voting and Abstaining on Congressional Roll Calls," *Political Behavior*, vol. 13, no. 2 (1991), pp. 97–127. The results are similar for the variables analyzed here.

Figure 9-1. *Variables for the Clinch River House Roll Call Analysis*

Xit: 1 if member *i* votes for Clinch River in year *t*, and 0 if member *i* votes against, where being paired but not present is regarded as voting.

PARTY: 1 if Republican, 0 if Democrat or Independent.

ACA: the legislator's score on the Americans for Constitutional Action scale.

PLANTS: the number of nuclear plants in the legislator's state, divided by the number of districts in the state.

COMMITTEE: 1 if the legislator is a member of the subcommittee (Public Works) of the Appropriations Committee or the authorization committee responsible for Clinch River (1975 and 1976: the Joint Committee on Atomic Energy; 1977 and 1978: the Subcommittee of Fossil and Nuclear Energy Research; after 1979: the Subcommittee on Energy Research and Production).

HIGH PORK: 1 if total contracts for Clinch River–related work to the legislator's district exceeded $10 million, 0 otherwise. Most of the contracts were signed in 1976 and 1977, and paid out up to 1983. Expenditures on the project rose dramatically in fiscal 1977, and were relatively stable through 1982 (see table 9-5).

MED PORK: 1 if contract expenditures for Clinch River–related work to the legislator's district were between $1 million and $10 million; 0 otherwise.

LOW PORK: 1 if contracts for Clinch River–related work to the legislator's district were between $100,000 and $1,000,000; 0 otherwise.

RETURNS: $\exp[-r(T - t)]/E(C)$, where t is the current year, T is the estimated date of breeder commercial introduction in year t, $r = 0.1$, and $E(C)$ are estimated real costs of the Clinch River project in year t (see table 9-4).

PRES: the ratio of the president's request for Clinch River to the House bill (see table 9-6) for Republicans in 1975–76 and 1981–82, -1 plus the same ratio for Democrats in 1977–79, and 0 for legislators not of the president's party.

the right sign only in the high group (table 9-9). An alternative specification, with a single continuous variable for district expenditures, yields a weak positive relationship. The coefficient for pork (defined as the logarithm of the value of contracts) is positive in all years except 1981, but is significant only in 1978. The dummy variable approach reported here yields significantly better fits than the continuous approach, with likelihood statistics from 0.2 to 3.7 percent lower.

The coefficient for *HIGH PORK* is positive in all years, but not significant until 1977, which is the first vote after the contracts were signed and the first year following big increases in project expenditures. In the pooled time-series regression, replacing the three *HIGH PORK* variables, one for each time period, with a single variable for the entire history of the project results in a significant loss of explanatory power (table 9-11). This strongly

Table 9-8. *Summary Statistics for Variables Used in House Regressions, 1975–82*[a]

Variable	*1975*	*1976*	*1977a*	*1977b*	*1978*	*1979*	*1981*	*1982*
y	0.619	0.545	0.512	0.601	0.563	0.564	0.525	0.475
	(383)	(400)	(420)	(414)	(355)	(427)	(402)	(413)
y Rep	0.780	0.720	0.709	0.826	0.816	0.821	0.605	0.553
	(118)	(132)	(141)	(138)	(114)	(156)	(177)	(179)
y Dem	0.547	0.549	0.412	0.489	0.444	0.417	0.462	0.415
	(265)	(268)	(279)	(276)	(241)	(271)	(225)	(234)
PARTY	0.308	0.330	0.336	0.333	0.321	0.365	0.440	0.433
	(0.462)	(0.471)	(0.473)	(0.472)	(0.468)	(0.482)	(0.497)	(0.496)
ACA	43.272	40.890	42.998	43.082	47.462	44.689	49.706	48.620
	(31.863)	(33.221)	(31.687)	(31.569)	(31.972)	(35.382)	(30.118)	(31.935)
ACA/R	74.831	69.348	73.865	73.710	74.763	80.686	75.232	75.268
	(20.68)	(24.769)	(20.184)	(20.381)	(23.624)	(18.616)	(15.536)	(19.969)
ACA/D	29.219	26.873	27.398	27.768	34.548	23.967	29.627	28.235
	(25.222)	(27.398)	(24.097)	(24.159)	(26.897)	(24.424)	(22.679)	(23.149)
PLANTS	0.550	0.548	0.533	0.529	0.480	0.441	0.371	0.331
	(0.357)	(0.334)	(0.339)	(0.331)	(0.326)	(0.309)	(0.315)	(0.205)
COMMITTEE	0.042	0.040	0.083	0.085	0.090	0.059	0.052	0.058
	(0.200)	(0.196)	(0.277)	(0.279)	(0.287)	(0.235)	(0.223)	(0.234)
HIGH PORK	0.034	0.025	0.033	0.034	0.037	0.033	0.032	0.031
	(0.181)	(0.156)	(0.180)	(0.181)	(0.188)	(0.178)	(0.177)	(0.175)
MED PORK	0.050	0.048	0.050	0.051	0.051	0.049	0.045	0.051
	(0.217)	(0.213)	(0.218)	(0.220)	(0.220)	(0.216)	(0.207)	(0.220)
LOW PORK	0.052	0.053	0.048	0.048	0.056	0.054	0.052	0.051
	(0.223)	(0.223)	(0.213)	(0.215)	(0.231)	(0.226)	(0.223)	(0.220)

a. Numbers not in parentheses are means. Numbers in parentheses for *y*, *y Rep*, and *y Dem* are total numbers voting, and for all other variables are sample standard deviations. See figure 9-1 for an explanation of the variables.

supports the retrospective voting hypothesis for distributive benefits: pork in hand counts, but prospective pork is heavily discounted.

The decline in both the magnitude and significance of the *HIGH PORK* coefficient in the third period is due entirely to a decline in pork barrel support among Republicans (columns 2 and 3 in table 9-11). Hence, though pork barrel politics may be important, it is not a sufficient condition for support. For at least some Republican members of the House, the deterioration of program efficiency and the emphasis of the Reagan administration on cutting government expenditures offset district pork barrel considerations.

The importance of the pork barrel effect can be estimated by using the

Table 9-9. *Annual Regressions for House Votes, 1975–82*[a]

Variable	*1975*	*1976*	*1977a*	*1977b*	*1978*	*1979*	*1981*	*1982*
Constant	−1.945	−1.761	−2.900	−2.119	−2.401	−1.973	−2.710	−2.222
	(0.339)	(0.294)	(0.347)	(0.332)	(0.345)	(0.285)	(0.332)	(0.284)
PARTY	−2.791	−1.146	−1.310	−1.011	0.134	−0.534	−2.707	−1.906
	(0.561)	(0.382)	(0.389)	(0.417)	(0.368)	(0.405)	(0.472)	(0.394)
ACA	0.086	0.055	0.061	0.064	0.049	0.047	0.071	0.052
	(0.010)	(0.006)	(0.007)	(0.007)	(0.006)	(0.006)	(0.007)	(0.006)
PLANTS	−0.022	0.278	1.353	0.590	0.537	0.802	1.316	0.807
	(0.449)	(0.403)	(0.399)	(0.417)	(0.438)	(0.404)	(0.455)	(0.433)
COMMITTEE	2.958	2.193	1.136	0.819	0.643	1.513	1.356	1.427
	(1.105)	(0.833)	(0.475)	(0.500)	(0.500)	(0.586)	(0.575)	(0.534)
HIGH PORK	0.087	0.622	1.403	2.465	2.325	1.629	0.757	1.155
	(0.840)	(0.887)	(0.793)	(1.107)	(1.119)	(0.861)	(0.732)	(0.752)
MED PORK	0.488	0.454	0.218	0.273	0.855	−0.179	−1.832	−0.718
	(0.605)	(0.581)	(0.555)	(0.535)	(0.576)	(0.538)	(0.724)	(0.603)
LOW PORK	0.410	−0.131	0.047	−0.116	0.933	0.760	1.250	0.720
	(0.625)	(0.554)	(0.577)	(0.602)	(0.582)	(0.552)	(0.573)	(0.517)
−2*L*	321.7	391.9	404.6	377.4	331.4	418.0	403.3	452.3
PCP	0.804	0.773	0.769	0.763	0.761	0.768	0.766	0.734
y	0.619	0.545	0.512	0.601	0.563	0.564	0.525	0.475
Degree of freedom	375	392	412	406	347	419	394	405

a. Numbers in parentheses are standard errors. See figure 9-1 for an explanation of the variables.

equations in table 9-9 to predict vote outcomes in the absence of distributive benefits. This procedure reveals that the amendment to limit federal liability for cost overruns passes in 1977, and the amendment to delete funds in 1978 results in a tie. Either amendment would have killed the program. Because the effect of distributive benefits is likely to be understated, as discussed in chapter 5, we conclude that, in the absence of pork barrel benefits, the program would have been canceled roughly when it should have been on economic grounds.

NUCLEAR POWER BENEFITS. Legislators from states with a large per capita nuclear presence were more likely to favor the project. This result has three possible interpretations. First, it may reflect the distribution of the expected commercial benefits, assuming that states with a heavy nuclear presence in the 1970s are likely to benefit disproportionately from a successful breeder industry. Second, it may reflect the extent of public acceptance and support of nuclear power options. Third, nuclear utilities

Table 9-10. *Pooled Regressions for House Votes, 1975–82*[a]

Variable	*All years*	*1975–76*	*1977a, 1977b, 1978, 1979*	*1981–82*
Constant	−2.545	−1.771	−2.262	−2.424
	(0.131)	(0.218)	(0.159)	(0.216)
PARTY	−1.426	−1.789	−0.656	−2.228
	(0.152)	(0.316)	(0.192)	(0.300)
ACA	0.056	0.067	0.053	0.060
	(0.002)	(0.005)	(0.003)	(0.005)
PLANTS	0.781	0.123	0.819	1.083
	(0.146)	(0.298)	(0.202)	(0.311)
COMMITTEE	1.257	2.537	0.981	1.381
	(0.198)	(0.669)	(0.252)	(0.391)
HIGH PORK	1.189	0.367	1.845	0.970
	(0.292)	(0.599)	(0.450)	(0.528)
MED PORK	−0.026	0.474	0.228	−1.217
	(0.198)	(0.417)	(0.271)	(0.464)
LOW PORK	0.449	0.078	0.402	0.960
	(0.194)	(0.412)	(0.284)	(0.378)
RETURNS	1.715	. . .	. . .	. . .
	(0.378)			
PRES	0.093	. . .	. . .	. . .
	(0.062)			
$-2L$	3,225.1	724.2	1,564.8	862.6
R^2	0.33	0.39	0.35	0.30
PCP	0.762	0.785	0.761	0.751
y	0.550	0.581	0.560	0.499
Degree of freedom	3,204	775	1,608	807

a. Numbers in parentheses are standard errors. See figure 9-1 for an explanation of the variables.

may have been more active in lobbying members of Congress from their states. Utilities stood to benefit from the breeder program not only because it affected fuel costs but also because they participated in financing the Clinch River program and wanted some return to their investments. Utilities with nuclear investments also saw the program in the same light as Secretary of Energy Edwards: a sign of commitment to the nuclear industry by the U.S. government.

The data support the third interpretation. With the exception of the first 1977 vote, the magnitude and significance of the *PLANTS* coefficient increases over time, while the program's economic benefits were disappearing into the future. Had economic benefits to consumers or general

Table 9-11. *House Regressions by Party, with Data Pooled for All Years, 1975–82*[a]

Variable	*Both parties*	*Republicans*	*Democrats*
Constant	−1.811	−1.585	−2.137
	(0.142)	(3.742)	(0.186)
1 if Republican	−1.459	...	...
	(0.151)		
ACA			
1975–76	0.062	0.051	0.078
	(0.004)	(0.006)	(0.004)
1977–79	0.059	0.057	0.050
	(0.003)	(0.004)	(0.004)
1981–82	0.046	0.042	0.071
	(0.003)	(0.005)	(0.007)
COMMITTEE			
1975–76	2.452	1.438	3.332
	(0.655)	(0.899)	(1.074)
1977–79	1.068	0.710	1.079
	(0.254)	(0.544)	(0.276)
1981–82	1.119	2.394	0.613
	(0.370)	(0.829)	(0.478)
HIGH PORK			
1975–76	0.326	−0.110	0.398
	(0.580)	(1.096)	(0.708)
1977–79	1.911	1.362	1.969
	(0.452)	(0.920)	(0.51)
1981–82	0.870	0.550	1.912
	(0.502)	(0.575)	(1.151)
RETURNS	0.250	1.222	1.188
	(0.579)	(1.679)	(0.787)
PRES	0.209	−1.388	−0.224
	(0.068)	(3.709)	(0.111)
$-2L$	3,218.810	1,101.819	2,076.985
R^2	0.33	0.23	0.33
PCP	0.760	0.776	0.758
Degree of freedom	3,201	1,143	2,047
y	0.550	0.717	0.456

a. Numbers in parentheses are standard errors. See figure 9-1 for an explanation of the variables.

support for nuclear power been the motive for the positive relationship, an opposite pattern would be expected. If the relationship was due to lobbying by utilities, it would be more pronounced as the political prospects for the program became questionable. Thus the 1977 vote on cost overruns (an issue especially important to the utilities, which were expected to pick up the cost) would be expected to generate intense lobbying, especially after the extremely close vote on the same proposal in 1976. Similarly, in the later years of the program, general support had declined in Congress. Because the interest of the utilities in the Clinch River project was not based solely on economic efficiency, their activities could be expected to pick up in the later years as they became more isolated as supporters of the program.

Committee Support

During the life of the program, the strength of subcommittee support varied considerably. One explanation of this result is that responsibility for oversight of the program changed during the program's life. The strongest support is found in 1975 and 1976, when the Joint Committee on Atomic Energy had authorizing responsibility. With the breakup of the Joint Committee, project responsibility was transferred to the less pronuclear Fossil and Nuclear Research Subcommittee, and support among subcommittee members became insignificant. In 1979 committee responsibility shifted once more, to the Energy Research and Development Subcommittee. Coefficients once again become significant, although support cannot be predicted as consistently as for legislators sitting on the Joint Committee.

Alternatively, the *COMMITTEE* variable may reflect presidential influence through the institutional power structure in Congress. Indeed, pooling the committee coefficients across presidencies is rejected (table 9-11, col. 2). However, the evidence contradicts a hypothesis of consistent presidential influence. First, the increase in the *COMMITTEE* coefficient occurred in 1979 (see table 9-9) with the second change in committee structure but while Carter was still president. (Of course, the change in committee structure in 1977 may have been due to administration influences or to the same political pressures that brought Carter to power in 1977.) Furthermore, the reduction in committee support in the Carter period is disproportionately due to Republican committee members (table 9-11, cols. 2 and 3).

In the last period Republican committee members were likely to support the program while Democratic members were not, which suggests a presidential effect. The timing, however, coincides with the election of

Senator Howard H. Baker, Jr., of Tennessee, a strong supporter of the project, to Senate majority leader. Presidential influence consequently is confounded by a potential Baker effect.

Party and Ideology

The variables measuring party affiliation and *ACA* (Americans for Constitutional Action) score cover overlapping ground in identifying coalitions that support or oppose the program. The separate inferences we draw for the two variables are consistent with the other energy programs and the discussion in chapters 4 and 5. Conservatives are more likely to favor hard technology energy options like the breeder reactor. But the conservative wing of the Republican party usually opposes (to a relatively small degree in this case) large-scale government commercialization programs.

Our conclusions are unavoidably confounded because of limitations in the use of ideology rankings (as discussed in chapter 5). The simple model reported here may correct for this problem by assigning a negative value to the Republican coefficient. This is borne out in part by the results reported in table 9-11. The separate equations for the two parties reveal lower *ACA* coefficients for Republicans than for Democrats in the first and last time periods. These equations predict that Republicans with *ACA* scores that would put them among the top third of Democrats in terms of conservatism (that is, with *ACA* scores of about 50) are likely to vote against the Clinch River project, along with liberal Democrats. Democrats with an *ACA* score of 50 are almost sure to vote for the project.[52]

The pooled time-series regressions reported in table 9-11 show a presidential effect. The party differential closes in the second period, corresponding to the Carter administration's opposition. The effect is strongest for conservative Democrats, as reflected in the lower *ACA* coefficient for Democrats in this period. Because the liberal wing of the Democratic party was already opposed to the program, Carter's potential influence was higher among conservatives. Liberals who supported the program did so despite ideological considerations (that is, for distributive reasons), and were thus less likely to be influenced by the president.

A similar effect is not seen in the final period: Reagan, who supported the program, did not close the party differential. Indeed, it widened during

52. A second problem can arise if support for the Clinch River project does not increase monotonically with the ACA scale: for example, if some maximum level of support resides with legislators with ACA statistics less than 100. There is no evidence to support this view: support for Clinch River increases monotonically with the ACA score among each party.

the final period, when the project was supported by both the Republican administration and the Republican Senate majority leader, Howard Baker. This trend is tied to program efficiency.

Longitudinal Variables: Efficiency and Presidential Effects

In the pooled time-series regressions (table 9-10, col. 1) two variables are added. The efficiency variable, *RETURNS*, has a high positive coefficient and is highly significant. In the equations reported in table 9-11, however, the coefficients for *RETURNS* lose their significance. When a single variable is inserted for *ACA*, it regains significance. Thus the decline in the *ACA* coefficient among Republicans captures the decline in support that was ascribed to efficiency in table 9-10. The implication of this is that programmatic efficiency was most relevant to conservative legislators. They were the most likely to favor Clinch River in the first place, and the results show that their early support and later defection were influenced by the changing economic fortunes of breeder technology.

The presidential support variable is positive, as expected, but is only significant at the 90 percent level (using a one-tailed test). Presidential support probably helps the project, but the effect will probably not be consistent. Presidents are likely to have different degrees of influence on Congress and to accord different degrees of importance to specific projects. A Carter effect exists in the decreased support during his administration among conservative Democrats, but no specific Reagan effect could be distinguished in the data. In short, "support" as measured here is too crude to capture the effectiveness with Congress and the priority ascribed to the program by different administrations.

Senate Votes

An analysis of votes in the Senate gives further support to the conclusions of the House study. Votes on Clinch River in the Senate occurred between 1975 and 1983. Because there were no votes in the Senate between 1978 and 1980, the middle period identified in the House is represented by a single vote in 1977. The votes in the Senate in all years except 1976 were amendments to delete funds for the project. Some votes deal with more convoluted issues than others do (for example, motions to table motions to

Figure 9-2. *Senate Votes on Clinch River Project*

1975: S598. Energy Research Authorization. Motion to table amendment to delete $94.1 million in authorizations for procurement of long-lead-time components for Clinch River. Agreed to, 66–30 (R: 30–7; D: 36–23).

1976A: S3105. ERDA authorization. Amendment to require private utilities to pay half of cost overruns above $2 billion. Rejected, 31–50 (R: 6–22; D: 25–28).

1976b: S3105. ERDA authorization. Amendment to require the National Research Council to make safety ruling before granting construction permit for Clinch River. Rejected, 30–53 (R: 7–22; D: 23–31).

1977: S1811. ERDA authorization. Substitute amendment to limit Clinch River spending to $33 million instead of $75 million specified in the amendment. Rejected 38–49 (R: 6–27; D: 32–22).

1981a: HR4144. Energy and Water Development Appropriation. Motion to table the amendment to reduce to $90 million the Clinch River appropriation. Agreed to, 48–46 (R: 36–14; D: 12–32).

1981b: HR4144. Energy and Water Development Appropriation. Motion to table the motion to reconsider the vote by which the amendment to reduce Clinch River appropriations was tabled. Agreed to, 50–45 (R: 38–12; D: 12–33).

1982a: HJ Res 599. Continuing Appropriation. Amendment to bar use of funds for Clinch River. Rejected 48–49 (R: 18–36; D: 30–13).

1982b: HJ Res 631. Continuing Appropriation. Amendment to drop House-approved provision eliminating construction funds for Clinch River. Adopted, 49–48 (R: 38–14; D: 11–34).

1983: HR3959. Supplemental Appropriation. Motion to table the amendment to add $1.5 billion to complete Clinch River. Agreed to, 56–40 (R: 23–30; D: 33–10).

reconsider amendments to delete funds), but all seem to measure support for the project (figure 9-2).

The independent variables (table 9-12) are modified from the House study to take into account the difference sizes of constituencies served by different senators. The pork variables measure per capita contract dollars. *HIGH PORK* takes positive values for the top 10 percent of the states in per capita contracts, *MED PORK* takes positive values for the 80th to 90th percentile, and *LOW PORK* takes positive values for the 65th to 80th percentile.[53] Finally, *COMMITTEE* takes on a value of 1 only for members of the Appropriations Subcommittee. In regressions not reported here,

53. The best fit was obtained by interacting the dummy variables with the log of per capita expenditures: *HIGH PORK* = *D*1 · log [(contracts to state/state population) · 100]; *MED PORK* = *D*2 · log [(contracts to state/state population) · 100]; *LOW PORK* = *D*3 · log [(contracts to state/state population) · 100].

Table 9-12. *Summary Statistics for Senate Data, 1975–83*[a]

Variable	*1975*	*1976a*	*1976b*	*1977*	*1981a*	*1981b*	*1982a*	*1982b*	*1983*
y	0.688	0.617	0.635	0.563	0.511	0.526	0.505	0.505	0.418
	(96)	(81)	(85)	(87)	(94)	(95)	(97)	(99)	(98)
PARTY	0.385	0.346	0.341	0.368	0.543	0.526	0.557	0.545	0.561
	(0.489)	(0.479)	(0.477)	(0.485)	(0.501)	(0.502)	(0.499)	(0.500)	(0.499)
ACA	34.521	35.765	35.365	40.471	52.660	51.737	53.722	53.0	51.76
	(30.799)	(34.516)	(34.966)	(33.108)	(25.743)	(25.823)	(24.996)	(25.22)	(30.3)
PLANTS	0.558	0.561	0.566	0.530	0.492	0.498	0.446	0.437	0.362
	(0.516)	(0.505)	(0.498)	(0.458)	(0.424)	(0.419)	(0.405)	(0.406)	(0.36)
COMMITTEE	0.115	0.123	0.129	0.126	0.117	0.116	0.113	0.111	0.102
	(0.320)	(0.331)	(0.338)	(0.334)	(0.323)	(0.322)	(0.319)	(0.316)	(0.30)
HIGH PORK	0.355	0.328	0.352	0.392	0.363	0.359	0.351	0.344	0.348
	(1.054)	(1.001)	(1.035)	(1.101)	(1.064)	(1.059)	(1.049)	(1.04)	(1.04)
MED PORK	0.259	0.153	0.209	0.259	0.239	0.261	0.256	0.226	0.258
	(0.765)	(0.604)	(0.706)	(0.769)	(0.743)	(0.769)	(0.762)	(0.72)	(0.76)
LOW PORK	0.333	0.299	0.285	0.332	0.325	0.306	0.313	0.323	0.326
	(0.627)	(0.595)	(0.585)	(0.625)	(0.623)	(0.610)	(0.612)	(0.620)	(0.623)

a. Numbers not in parentheses are means. Numbers in parentheses are sample standard errors except for *y*, where they are total votes cast.

membership on Senate authorization subcommittees did not systematically influence support for Clinch River.

The pooled equations (table 9-13) sort into four groups, corresponding precisely to the groups found in the House: 1975 and 1976; 1977; 1981 and 1982; and 1983. The pooled equation across all time periods was also estimated, with selected independent variables that do not pool interacted with dummy variables for each time periods. Problems with the small sample made it impossible to estimate this equation for Republicans, so the equation is reported for the entire group and for Democrats.

Distributive Politics

CONTRACT AWARDS. The coefficients for the expenditure variables support the threshold hypothesis. Coefficients are positive, but in general not significant for the two lower brackets. The pattern for *HIGH PORK* is slightly different from the House results, but supports the theory of retrospective voting. A pork barrel effect is not apparent in 1977, but is in 1981 and 1982. As discussed in chapter 5, because senators serve large and diverse constituencies and serve longer terms, we expect distributive politics to take longer to emerge and to require larger expenditures (even correcting for population) to be detectable in the Senate than in the House. By the same argument, a better measure of pork barrel stakes in the Senate is cumulative outlays rather than contract finalization. This hypothesis is consistent with the insignificant coefficient for *HIGH PORK* in the 1977 equation.

NUCLEAR POWER BENEFITS. The nuclear plant variable is not significant in any year, and indeed is negative in all years but two (1975 and 1983). In the pooled equation, the coefficient for *PLANTS* is negative and significant. The source of this relationship is unclear. The variable is not highly correlated with the other independent variables, nor with state size or geographic region. Because pooling across time periods cannot be rejected for the *PLANT* variables, we cannot profitably speculate about why a large nuclear presence in a state diminishes support for Clinch River by the state's senators.

Committee Support

The *COMMITTEE* variable is positive and significant in all years after 1976. A pattern like that in the House is not observed in the Senate; however, the variable includes only the Appropriations subcommittee and thus would not show variations based on shifting committee responsibilities. Pooling is not rejected for the committee coefficients across time periods. That it does not vary with changes in administrations lends further support to the institutional interpretation of variations in committee support for the House.

Party and Ideology

As in the House, conservatives were more likely to support the program. The party coefficient differs between the Senate and the House. Although President Carter's influence is not detected, the data are not adequate to rule it out. Indeed, the absence of votes during the Carter administration suggests that the Senate leadership may have avoided a vote because it feared either that the administration might be powerful enough to cancel the project, or that a vote would place too many senators in conflict with their president.

A strong partisan influence existed during the Reagan administration. The data suggest evidence of logrolling by Senator Baker, rather than an administration effect. This interpretation is based on two factors. First, the coefficient for *ACA* declines in the later years, showing a loss in support among conservatives, the group expected to respond most strongly to Reagan. Second, Republican support declined in 1983. At that time, the

Table 9-13. *Pooled Regressions for Senate, 1975–83*[a]

Variable	*All variables*	*Test for party*	*Test for ACA*	*Test for committee*	*Test for pork*	*Democrats only*
Constant	−2.828	−2.783	−3.225	−2.691	−2.630	−3.844
	(0.322)	(0.316)	(0.284)	(0.309)	(0.309)	(0.508)
PARTY						
All	. . .	0.846	. . .	. . .	. . .	. . .
		(0.206)				
1975–76	−0.054	. . .	0.092	−0.054	−0.312	. . .
	(0.389)		(0.353)	(0.390)	(0.380)	
1977	−0.013	. . .	0.743	−0.014	−0.142	. . .
	(0.683)		(0.524)	(0.681)	(0.678)	
1981–82	1.420	. . .	1.165	1.372	1.419	. . .
	(0.283)		(0.253)	(0.278)	(0.276)	
1983	1.019	. . .	0.619	0.897	0.989	. . .
	(0.669)		(0.381)	(0.651)	(0.654)	
ACA						
All	. . .	. . .	0.036	. . .	. . .	. . .
			(0.004)			
1975–76	0.046	0.041	. . .	0.046	0.050	0.053
	(0.008)	(0.007)		(0.008)	(0.008)	(0.012)
1977	0.054	0.048	. . .	0.053	0.053	0.053
	(0.010)	(0.009)		(0.010)	(0.015)	(0.012)
1981–82	0.027	0.033	. . .	0.026	0.028	0.050
	(0.005)	(0.005)		(0.005)	(0.005)	(0.009)
1983	0.025	0.026	. . .	0.025	0.024	0.058
	(0.009)	(0.006)		(0.009)	(0.009)	(0.014)

administration still ostensibly supported the project, but Senator Baker had announced his resignation.

The results follow the pattern in the House. The *ACA* coefficients for Democrats are higher than for the entire group except in 1977, the one Carter year. They remain high at the end. Thus the decline at the end in *ACA* coefficients for the entire sample is due entirely to a decline among Republicans.

Program Efficiency

Two measures of program efficiency—*RETURNS* and *SUCCESS*—were included in pooled equations reported in table 9-13. The first is the discounted returns to investment, and the second is the expected year of breeder commercialization. The variables are highly significant in all

Table 9-13 *(continued)*

Variable	*All variables*	*Test for party*	*Test for ACA*	*Test for committee*	*Test for pork*	*Democrats only*
COMMITTEE						
All	. . .	. . .	. . .	1.492	. . .	. . .
				(0.312)		
1975–76	0.371	0.400	0.398	. . .	0.800	0.800
	(0.519)	(0.508)	(0.497)		(0.502)	(0.70)
1977	1.446	1.501	1.706	. . .	1.173	2.554
	(0.934)	(0.933)	(0.879)		(0.865)	(1.24)
1981–82	1.900	1.751	1.848	. . .	1.898	1.722
	(0.460)	(0.447)	(0.461)		(0.452)	(0.52)
1983	2.669	2.617	2.794	. . .	2.605	4.019
	(0.923)	(0.916)	(0.935)		(0.904)	(1.24)
HIGH PORK						
All	. . .	. . .	. . .	. . .	0.193	. . .
					(0.082)	
1975–76	−0.224	−0.313	−0.242	−0.249	. . .	−0.376
	(0.150)	(0.148)	(0.142)	(0.150)		(0.23)
1977	−0.038	−0.108	−0.011	−0.045	. . .	−0.114
	(0.255)	(0.257)	(0.244)	(0.241)		(0.36)
1981–82	0.478	0.480	0.499	0.468	. . .	0.227
	(0.133)	(0.131)	(0.135)	(0.130)		(0.186)
1983	0.386	0.380	0.409	0.391	. . .	0.208
	(0.233)	(0.231)	(0.340)	(0.224)		(0.445)
RETURNS	5.816	5.545	7.514	5.321	4.96	8.409
	(1.076)	(1.062)	(0.846)	(1.041)	(1.56)	. . .
−2*L*	844.085	855.264	853.715	850.779	858.375	436.2
PCP	0.789	0.745	0.746	0.740	0.733	0.744
Degree of freedom	814	817	817	817	817	428

a. Numbers in parentheses are standard errors. See figure 9-1 for an explanation of the variables.

specifications of the pooled regressions: Senate support declines over the period, and all the measures of program efficiency decline as well.

Conclusion

The Clinch River breeder reactor program was the quintessential example of a technological turkey by the time it was mercifully put to rest in 1983. Although many predicted as much from the beginning, from the perspective of a government official Clinch River looked like an important component of a very good program in the early 1970s, especially after the increase in the price of oil and other hydrocarbon fuels made the nuclear option seem especially attractive.

Bad luck on forecasts of economic demand accentuated and exacerbated the bad decisions that arose from the optimistic early evaluations by AEC technologists. The bad news began to appear in 1975: the reluctance of electric utilities to participate further and the drop in orders for new nuclear power plants. In 1976 and 1977 the collapse of the nuclear industry was under way, and by 1977 the benefit-cost study of Clinch River clearly showed that the time was no longer ripe for a commercial demonstration program.

The roll call analysis identifies a host of factors contributing to inappropriate program stability in the late 1970s. Such support is not automatic; for example, political parties, ideological proponents, and committees may all defect from a program (though with lags) once that program loses its economic rationale. For Clinch River the lag was six years.

The key to the political success of the Clinch River project after the mid-1970s appears to be the result of a coincidence: the big rise in annual expenditures took place just before the bottom fell out of the economic justification for the program. Between fiscal 1973 and 1975 appropriations doubled; by fiscal 1977 they had nearly doubled again. The key contracts were signed in 1976.

Unfortunately, the bad economic news had arrived a year or two too late. The distributive aspects of the program were, by then, in full force. As the nuclear power controversy swirled around Congress, and the votes for financing Clinch River came perilously close to terminating it, the distributional effects proved decisive in keeping the program going after it was a clear mistake.

Although Carter could suspend licensing for Clinch River, so that construction could not begin at the plant site, expenditures continued for purchasing plant components. It is interesting to speculate on what might have happened had the disagreement between Congress and the administration not been so profound. Clearly Clinch River was not the breeder of choice in the late 1970s, either in the details of the technology or in its licensing requirements. Had more agreement existed in government, possibly a substitute breeder program—one that might have required component purchases from the same companies—could have been approved in new legislation, which would require approval of both Congress and the administration. Program continuation, in contrast, is possible without such agreement. Thus controversy may well have contributed to rigidity in the program.

The cost of continuing the program was to cut back other breeder research and development projects. These alternatives were now more appropriate, for a longer lead time until commercial readiness had become a less troubling feature of a technology. Yet the long-run attractions of a broad-based research program lost out to the short-run forces of distributive politics. Not until the budgetary crises of the Reagan administration, amidst a massive cost overrun, could the program be terminated.

Chapter 10

Synthetic Fuels from Coal

LINDA R. COHEN & ROGER G. NOLL

THE FEDERAL program to make synthetic fuels from coal promised an economic and political bonanza. The program would develop an enormous domestic energy resource that would satisfy the political goal of energy independence while minimizing air pollution. It would increase demand for the perennially depressed and declining eastern coal industry.

Federal support for research on coal-based synthetic fuels was minimal until the 1970s, when it experienced a substantial upsurge, only to flounder in the early 1980s. The obvious interpretation is to equate the increased interest in synfuels with the energy crises of the 1970s—and, we will argue, the obvious interpretation is correct. However, our examination of the political economy of the synfuels-from-coal program identifies other, less obvious phenomena.

The synfuels research and development (R&D) program was more diverse and complex than most of the other cases considered in this book. The program pursued a "parallel development effort," or mixed strategy. Its scope was broad in that it investigated multiple technical options and carried the investigations through multiple R&D stages. The program included small-scale research and pilot plants, as well as plans for several commercial demonstrations, although only two of these were completed. The short-lived U.S. Synthetic Fuels Corporation (SFC) attempted to take development a step further by subsidizing commercial introduction of the technology. Between 1970 and 1984 cumulative expenditures by the government were about $2 billion.

This chapter examines the entire coal-based synfuels program, which necessarily risks oversimplifying by ignoring the details of individual projects. The compensating advantage is that it permits an analysis of the economics, politics, and management of a broad program. A mixed strategy of research, and especially a program that avoids large projects with concentrated pork barrel benefits, conforms more to recommendations for an ideal approach to government-supported R&D than do single-project programs. However, research without immediate distributive benefits is less likely to have strong political support. This is indeed apparent in the synfuels program. The political viability of the program required the creation and maintenance of a rather complex, diverse, and ultimately unstable coalition of regional and national interest groups.

The regional support for the program was eastern coal. In the 1970s the program enjoyed enhanced support that depended on its addressing near-term national energy issues—namely, the shortages of natural gas and oil. Most of the resources in the synfuels program went to technological options that could use eastern coal. Within this set, the program concentrated on options that had the greatest potential for producing near-term substitutes for natural gas and oil. Unfortunately, these goals were not compatible. Much more is known about making synthetic fuels (especially "high-grade" fuels like natural gas and gasoline) from western coal. Nevertheless, the technologies chosen for development concentrated on products that addressed the energy crises of the day by using eastern coal rather than on technologies and coal supplies that were most likely to have commercial applications. Consequently, the synfuels program was seen by westerners as a distributive cost. The antiwestern, pro-eastern orientation thus created both technical and political problems.

Synthetic Fuels Technology

The cornerstone of the U.S. synthetic fuels program has been the production of gas and liquids from coal.[1] To understand the implications of these goals requires some knowledge of synthetic fuels technology.

Coal Gasification

The most important distinction among coal gas products is heating value.[2] High-Btu gas (also called substitute natural gas, or SNG) is composed

1. The program has also investigated the production of synthetic fuels from shale, peat, garbage, waste products, and agricultural products. In addition, the program supported other coal research issues, including improved combustion and advanced electricity production techniques. We do not treat these parts of the program unless they are important to coal synfuels R&D and commercialization (for example, oil shale development, which became a critical component in the mid to late 1970s).

2. The following discussion of coal gasification is based on Harry Perry, "Clean Fuels from Coal," in Peter Auer, ed., *Advances in Energy Systems and Technology*, vol. 1

almost entirely of methane, and it has a heating value of nearly 1,000 British thermal units (Btu) per standard cubic foot (scf). Because this is equivalent to natural gas, SNG can be mixed with natural gas in pipelines.

Medium-Btu gas, rated from about 250 to 500 Btu/scf, consists of methane, carbon monoxide, hydrogen, and carbon dioxide. While used for many of the same applications as high-Btu gas, it cannot be piped economically more than about a hundred miles from its source. Medium-Btu gas that is mainly carbon monoxide and hydrogen is called synthesis gas. It is the preferred input for indirect liquefaction (as discussed later) and certain chemical feedstock applications.

Low-Btu gas has a heating value of less than 150 Btu/scf, and so it cannot be burned in equipment designed for medium-Btu or natural gas. Low-Btu gas cannot be piped economically.

The type of gas produced in a gasification process depends on the temperature, pressure, and physical process of the gasifier. Modern coal gasification techniques are usually classified by the method in which the coal comes into contact with reaction gases. There are three kinds of reactor "beds": the fixed bed, the fluidized bed, and the entrained bed. Each uses different kinds of coal and produces different outputs.

The *fixed bed reactor*, of which the Lurgi reactor is the best known, is the most widely used commercial gasification process. Developed in prewar Germany, it is based on processes used in Europe and the United States in the nineteenth and early twentieth centuries. Although called a fixed bed, the reactor actually contains moving columns of coal particles. The particles enter the top of the reactor and pass through three zones. In the top two zones, the coal undergoes reactions that turn part of it into gas. A fraction of the coal is not gasified and falls to the bottom (the "combustion zone"), where it is burned to provide heat for the reactions in the top two zones. In the top ("distillation") zone, the coal is heated and volatile materials (mainly methane, oil, and tars) are released. After the coal is distilled, a char, or carbon residue, is left, which falls into the second ("gasification") zone, where part of it reacts with steam and heat to form

(Academic Press, 1978), pp. 243–325, especially pp. 253–72; Christopher F. Blazek, Nathaniel R. Baker, and Raymond R. Tison, *High-Btu Coal Gasification Processes*, Report prepared by the Institute of Gas Technology for Argonne National Laboratory, ANL/CES/TE 79-2 (1979); Nathaniel R. Baker, Christopher F. Blazek, and Raymond R. Tison, *Low- and Medium-Btu Coal Gasification Processes*, Report prepared by the Institute of Gas Technology for Argonne National Laboratory, ANL/CES/TE 79-1 (1979); and J. G. Patel, "Low, Medium BTU Gas from Coal Lead Conversion Routes," *Oil and Gas Journal*, June 29, 1981, pp. 94–113.

carbon monoxide and hydrogen. The remaining char falls to the combustion zone, where it is burned to ash and removed. Fixed bed reactors produce either medium- or low-Btu gas, depending on the process gas used in the combustion zone. If the char is burned with oxygen, medium-Btu gas is produced. If air is used, the product gas is diluted with nitrogen, producing low-Btu gas.

Among fixed bed technologies, Lurgi reactors have excellent thermal efficiency. Because they operate at moderate temperatures, they produce a large amount of methane (table 10-1). The main technical disadvantages of the Lurgi reactor are that the coal particles must be a specific size and must be evenly distributed through the reactor. A typical Lurgi reactor requires coal particles from 0.5 to 1.5 inches in diameter. Most eastern U.S. coals have a property called "caking": they swell and stick together when heated. (Western coal is noncaking.) Consequently, eastern coals cannot be used in standard Lurgi reactors. In addition, eastern coals have a high proportion of "coal fines" (essentially, very small pieces of coal). Coal fines cannot be used in Lurgi reactors unless they are briquetted. Finally, the even distribution requirement and the slow reaction rate (the particles remain in the reactor for one to three hours) limit the capacity of Lurgi reactors. In 1979 the largest available Lurgi reactor had a 12.5-foot diameter, and it could process 108 tons of bituminous coal a day.[3]

Fluidized bed reactors use fine particle coal. Steam and oxygen enter the reactor below the coal. As the reaction gases move upward and expand, the coal particles move about randomly, like a fluid, thoroughly mixing the coal and reaction gases. Reaction time is between 20 and 40 minutes. Unlike fixed bed reactors, fluidized bed reactors do not contain moving parts. The Winkler gasifier is available in commercial sizes up to 18 feet in diameter, and it has a capacity of up to 1,000 tons a day of bituminous coal. The temperature in a Winkler reactor ranges from 1,500 to 1,900 degrees Fahrenheit, and exit gas temperature is between 1,300 and 1,600 degrees Fahrenheit.

Winkler reactors have a number of drawbacks for U.S. applications. Like fixed bed reactors, they cannot use caking coal or coal fines because either destroys the fluid properties; however, this problem is less severe than in Lurgi reactors, so that "weakly caking" coal can be used. Second, the distillation products are consumed in the reaction. The Winkler reactors

3. Baker, Blazek, and Tison, *Low- and Medium-Btu Coal*, p. 11. The size of these reactors keeps growing. The South Africans recently tested a sixteen-foot-diameter reactor at their Sasolburg facility. Perry, "Clean Fuels," p. 264.

Table 10-1. *Product Composition from Commercial Gasifiers*
Percent of dry volume unless otherwise specified

Product	*Fixed bed (Lurgi)*	*Fluid bed (Winkler)*	*Entrained bed (Koppers-Totzek)*	*Substitute natural gas (Lurgi)*
Hydrogen	37.4	36.0	34.0	9.70
Oxygen	0.0	0.0	0.1	0.0
Carbon monoxide	26.0	44.4	51.1	0.05
Carbon dioxide	26.1	15.7	12.6	3.18
Methane	9.1	1.6	0.1	85.99
Other hydrocarbons	0.6	0.0	0.0	0.0
Nitrogen	0.8	0.8	1.9	1.07
Hydrogen sulfide	0.0	1.5	0.2	0.0
Heating value[a]	310	270	275	902.6
Thermal efficiency (percent)	70–80	70	70	71.7

Sources: Harry Perry, "Clean Fuels from Coal," in Peter Auer, ed., *Advances in Energy Systems and Technology* (Academic Press, 1978), pp. 260–62; and two 1979 reports prepared by the Institute for Gas Technology for Argonne National Laboratory: Nathaniel R. Baker, Christopher F. Blazek, and Raymond R. Tison, *Low- and Medium-Btu Coal Gasification Processes*, ANL/CES/TE 79-1; and Blazek, Baker, and Tison, *High-Btu Coal Gasification Processes*, ANL/CES/TE 79-2.

a. Btus per standard foot.

produce little methane, although they also produce little environmentally undesirable tar and soot. Third, about 70 percent of the ash produced in a Winkler reactor leaves with the product gases and must be cleaned. Fourth, carbon usage is not as efficient as in fixed bed reactors: up to 30 percent of the carbon is unreacted in its first pass through the reactor, and it exits with the ash. Fifth, the high temperature in the reactor requires large quantities of oxygen.

Most of the research gasifiers have been fluidized bed reactors, considered to have the most potential for economic production of SNG from eastern coals. If pretreated, which involves extra complications and expense, caking coals will not cake. Coal pretreatment has been a major focus of research. Another line of research has been hydrogasification: use of hydrogen instead of steam in the reactor. Under appropriate conditions, hydrogen and carbon react to form methane. Hydrogen is expensive, so several processes for efficient hydrogen production have been investigated. Another focus of research has been the use of a series of fluidized beds operating at different temperatures to reduce carbon consumption and to produce distillation gases more efficiently.

Methods to conserve or eliminate oxygen use by using catalysts also have been researched. Catalytic reactions vary by type of coal. Western lignite is generally more reactive than eastern coals. Another area of research has been to agglomerate ash so that it can be removed from the bottom of the reactor, instead of exiting with the product gases. Because the reactor bed must remain fluid, agglomeration is especially difficult.

In the *entrained bed reactor*, pulverized coal is rapidly fed into the reactor with steam and oxygen or air. The coal is suspended (entrained) in the gas, heated to more than 3,000 degrees, and converted into product gases, which exit at 2,000 to 2,700 degrees. Coal residence time is less than one second, and all distillation products are destroyed. Some ash leaves with product gases, and some of it agglomerates and is removed as slag. Entrained bed reactors can use any kind of coal and coal fines. Commercially available entrained bed reactors have capacities of 660 to 850 tons per day.[4] In theory, entrained bed reactors have the highest capacity of all coal gasifiers, but in practice, scaling up this kind of facility has been difficult. The major technical problem of entrained bed reactors is that their very high operating temperature requires special materials and techniques to ensure safe, reliable operation.

Hot product gases are a disadvantage if the end product is SNG for pipeline distribution, but high temperatures are attractive for electricity generation. Unlike the corrosive hot gas from coal combustion, scrubbed synthesis gas can be run through a gas turbine to produce electricity. The heat from the reactor and the hot gas exiting the turbine can also be used to create high pressure steam for a steam generator. The "integrated gas combined cycle" plants, now near commercialization, have much higher thermal efficiencies than do direct coal combustion plants, and they produce minimal sulfur and nitrogen emissions.

The product gas from all gasifiers contains pollutants. The most important, hydrogen sulfide, can be removed from the product gas stream. However, commercially available processes to remove hydrogen sulfide operate at low temperatures, so product gases have to be cooled for cleanup (especially for entrained reactors). Hot gas cleanup has been one component of the synfuels R&D program.

The final step in an SNG facility is methanation. In the methanation unit, the composition of synthesis gas is "shifted" to contain the correct proportions of carbon monoxide and hydrogen by either adding hydrogen or removing carbon monoxide. The gas is then reacted with catalysts to

4. A 1,000-ton-per-day entrained bed gasifier has operated successfully for several years at the Cool Water Integrated Gas Combined Cycle facility, a "demonstration" plant partially supported by the Synthetic Fuels Corporation. This project is discussed later.

produce methane and water. This reaction takes place at low temperatures and produces some heat, but not enough to be commercially interesting.

Liquefaction

Coal liquefaction involves transforming coal into clean liquids: gasoline, methanol, synthetic crude oil, and fuel oil.[5] Some products are actually solid at room temperature, but are relatively clean fuels suitable for chemical feedstocks or boilers. Coal liquids are substitutes for oil for transportation, industrial processes, chemical feedstocks, or electricity production. An advantage of coal liquids over coal gas for electricity generation is that they can substitute directly for boiler fuel (no. 6 oil) in existing oil-fired generators.

All coal liquefaction requires breaking down the coal, increasing its hydrogen-to-carbon ratio, and catalytically recombining it to form oil and distillate products. The basic liquefaction methods are direct liquefaction, which transforms coal to liquids, and indirect liquefaction, which gasifies coal and then converts the gas to liquids. In both cases the product mix depends on the type of coal, the temperature and pressure in the reactor, the quantities of air, oxygen, hydrogen, and steam added to the reactor, and the quantities and types of catalysts and other additives.

Both approaches date to prewar Germany. Friedrich Berguis invented a direct coal liquefaction process in 1911, and Fischer and Tropsch developed indirect liquefaction in the 1920s. During World War II coal liquids provided as much as 90 percent of Germany's aviation and motor fuel. Nearly 90 percent of the synthetic oil manufactured in Germany during the war came from the Berguis process and the remainder from Fischer-Tropsch plants. After the war the Berguis process was abandoned, and, in a move that inspired thriller novels, the formula was lost.[6]

German wartime production was not at commercial scale by modern standards. It peaked at 18 plants producing 70,000 barrels of liquids per day,[7] including 12,000 barrels of gasoline. The largest of the German plants

5. The following discussion of liquefaction is drawn from Perry, "Clean Fuels"; Paul F. Rothberg and others, *Synfuels from Coal and the National Synfuels Production Program: Technical, Environmental, and Economic Aspects*, printed at the request of the Senate Committee on Energy and Natural Resources, 97 Cong. 1 sess. (Congressional Research Service, January 1981), pt. 2; and General Accounting Office (GAO), *Liquefying Coal for Future Energy Needs*, Report to the Congress, EMD-80-84 (August 12, 1980).

6. Rothberg and others, *Synfuels from Coal*, p. 80.

7. J. B. O'Hara, "State-of-the-Art Coal Liquefaction," *Hydrocarbon Processing*, vol. 55 (November 1976), p. 222.

produced 4,000 barrels of gasoline a day. A commercial-scale plant in the United States is expected to produce 50,000 barrels a day.

Coal liquefaction was pursued briefly after the war. The Bureau of Mines of the U.S. Department of the Interior financed two small plants to test German processes; however, the development of far cheaper oil resources in the Middle East caused interest in coal liquefaction to wane in both the United States and Europe. The Bureau of Mines program ended in 1954.[8]

Development of coal liquefaction continued in South Africa, which began operating SASOL-I, a 10,000 barrels-a-day Fischer-Tropsch plant, in 1955. This plant has run continuously since then, and it has been joined by SASOL-II and SASOL-III, each rated at about 60,000 barrels a day. South Africa has the only commercial liquefaction industry in the world.

For the U.S. market, SASOL has two main technical drawbacks. First, it is not suited to eastern U.S. coals because of caking. Second, SASOL synthetic gasoline has a lower octane rating than does U.S. gasoline. In addition, until the late 1970s, indirect liquefaction was thought to be more expensive and considerably more polluting than direct liquefaction (tables 10-2 and 10-3).

U.S. research on indirect liquefaction has been aimed at improving the product mix by altering reactor conditions and adding catalysts. Indirect liquefaction has also been investigated for the production of methanol, which is valuable for chemical processes and as a gasoline additive. Methanol cannot be used in standard automobiles in concentrations greater than 10 percent, but specially designed engines can use pure methanol. Another line of research has been the conversion of methanol to high-octane gasoline.

Eastern coals are suitable for all direct liquefaction processes, although the results vary with the specific kind of coal used. Three chemical processes are used for direct liquefaction: pyrolysis, solvent extractions, and direct hydrogenation.

Pyrolysis, the heating of coal to produce gaseous and liquid fuels, is the first stage of many gasification processes (the distillation stage). Liquefaction requires maximizing the production of light distillates. The main product from pyrolysis plants is char. Research on pyrolysis has been directed at increasing oil products and producing an attractive char product. In *solvent extraction,* coal is dissolved in a solvent, then the components react to form oil products and a hydrogen-depleted residue. The solvent can also contain

8. GAO, *Liquefying Coal,* p. 2.

Table 10-2. *Product Composition from Liquefaction Plants*
Barrels per day unless otherwise specified

		Direct liquefaction		
	Indirect liquefaction	*Solvent extraction*		*Hydrogenation*
Product	*(Fischer-Tropsch)*[a]	*SRC-II*[b]	*EDS*[c]	*H-coal*
Gasoline	18,200	*	*	*
Liquid propane gas	18,800	5,500	*	*
Middle distillate	1,200	*	*	*
Propane	*	*	3,270	*
Butane	*	*	3,500	*
Naphtha	*	10,700	19,900	18,200
Fuel oil	2,000	45,300	28,700	42,200
Gas[d]	127.9	23.1	41.9	19.7
Size (tons per day)	25,000	25,000	25,000	25,000
Thermal efficiency (percent)	45–60	65–70	65–70	65–70

Source: General Accounting Office, *Liquefying Coal for Future Energy Needs*, Report to the Congress, EMD 80-84 (August 1980).
* Very small or none.
a. Of the processes in this table, only Fischer-Tropsch has been run at a commercial scale. The other data should be regarded with skepticism.
b. A solvent extraction process.
c. Exxon Donor Solvent, a hydrogen donor solvent process.
d. Millions of standard cubic feet.

hydrogen, which is transferred to the dissolved coal to increase production of light distillates. (This is called a hydrogen donor solvent process.) *Direct hydrogenation* adds hydrogen to coal in the presence of catalysts, creating liquid products. Liquefaction plants frequently combine features of all three processes. All processes require high temperatures and pressures.

The development of direct liquefaction technology focuses on very basic issues because it is at a much more primitive stage than coal gasification. The main research issues are common to all three processes.

A primary research challenge is scaling up from laboratory or pilot plant facilities to commercially attractive sizes. Proposed commercial plants will handle 30,000 tons of coal a day. By the end of the synfuels program in 1981, the largest plants handled several hundred tons a day.[9] In 1970 only units processing less than a dozen tons per day had been successfully operated. Examples of major scale-up problems are moving large quantities of coal through a pressurized reactor, building large reactors that operate at high temperatures and pressures, and handling large quantities of gritty material.

9. Rothberg and others, *Synfuels from Coal*, p. 120.

Table 10-3. *Major Environmental Impacts of Synfuel Processes*[a]
Tons per year unless otherwise specified

	Gasification		*Liquefaction*	
Item	*High Btu (Lurgi)*	*Low Btu*	*Indirect (Fischer-Tropsch)*	*Direct (EDS)*
Air emissions				
Particulates	460	360	1,270	740
Sulfur oxides	8,530	6,630	20,860	4,770
Nitrogen oxides	5,730	4,450	15,720	4.030
Carbon monoxide	425	330	1,160	350
Water requirements[b]	9,000	5,000[c]	11,500	8,000
Solid wastes generated	438,000	352,000	755,000	412,000

Source: Congressional Research Service, *Synfuels from Coal and the National Synfuels Production Program: Technical, Environmental, and Economic Aspects* (January 1981), pp. 155, 160, 175, 227.

a. Data refer to facilities of approximately the same net energy production. Emissions from use of the final product are excluded. All figures are approximate. Not only do they differ from other references, but the source from which this table is derived contains internal inconsistencies. Thus the table should be used to compare technologies rather than to estimate the environmental impact of any given technology.

b. Acre-feet a year.

c. Calculated from proportions reported on p. 227 of the source.

A second problem is to separate the liquids from unreacted solids. Although gas and solids separate somewhat naturally, separation has been a major stumbling block for direct liquefaction.

A third research area has been upgrading products both commercially and environmentally. Direct liquefaction processes present a range of environmental problems. Pollutants in coal cannot be removed from direct liquefaction products as readily as from coal gas, and product liquids retain much of the sulfur, nitrogen, and toxic chemicals in coal. Among the direct liquefaction products are "aromatics": hydrocarbons that are carcinogenic. Moreover, waste products from coal liquefaction plants (water, slag, ash, used solvents, and so on) contain toxic and carcinogenic material. Finally, toxic and carcinogenic materials raise issues of worker safety.

Cost Estimates of Synfuels Processes

Actual and estimated costs for oil, natural gas, and synthetic fuels are reported in appendix tables 10A-1 to 10A-5. The synfuels estimates are from a variety of sources, including industry, government, and academia. Virtually all estimates are accompanied by a disclaimer as to accuracy and so give only a crude picture of relative costs. With this proviso, several issues are apparent.

First, the established European processes (indirect liquefaction and

Lurgi gasification) are more expensive than the estimated costs of "second-generation" U.S. plants. Of course, the latter cost estimates, especially in the early years, were not based on operating data and proved optimistic. Consequently, the cost differentials narrowed or even switched by the end of the program.

Second, among gasification processes, the CO_2 Acceptor using western coal has consistently lower estimates of fuel and capital costs than does the SNG process using eastern coal. Otherwise, no case can be made on the basis of costs for choosing among the second-generation coal gasification processes. The cost estimates for liquefaction processes are even less conclusive, for they are considered to be more speculative and pertain to processes producing different products.

Third, commercial competitiveness depends on the prices and availability of oil and gas. A 1979 study by Saman Majd estimates the net present value of four direct liquefaction processes and a shale oil process, given various oil price scenarios.[10] The oil price scenario that reflects the mid-1970s expectations assumes an annual 2 percent rise from 1977 to 1991, and 3 percent thereafter. For this case, assuming 8 percent discounting, the net present values are positive for SRC-II, H-Coal, and Shale ($32 million to $86 million), and negative for Synthoil, a low-sulfur boiler fuel process previously investigated by the Bureau of Mines, and Exxon Donor Solvent (EDS), a boiler fuel process (−$123 million to −$1,179 million). Majd states that the cost estimates for EDS are the most reliable, so that net present values are likely to be closer to the EDS estimate than the others. For higher discount rates, none of the synthetic fuel processes has a positive net present value.

Majd's oil price scenario 1 assumes a 50 percent price jump in 1986 and a 2 percent annual rise in other years. This scenario fits expectations following the 1979 oil price increase. At these prices, net present values become positive for all processes ($382 million to $591 million) except EDS (−$573 million). However, assuming 40 percent cost overruns, which approximate the increases in synfuels cost estimates following the 1979 oil crisis, the net present value for all processes is negative: −$61 million (oil shale), −$99 million (H-Coal), and −$1,909 million (EDS). With the higher cost estimates and a more optimistic oil price scenario, which fits expectations after 1982, the net present values plunge further.

10. Saman Majd, "A Financial Analysis of Selected Synthetic Fuel Technologies," MIT Energy Laboratory Working Paper, MIT-EL 79-004WP (MIT, Center for Policy Research, January 1979).

Perhaps the most striking feature of the cost estimates is that they rose more rapidly than did the increases in oil and natural gas prices. Indeed, it has been said that estimating the cost of synthetic fuels is easy: add five dollars to the current price of a barrel of imported oil. This method worked well until 1982, when oil prices started falling, but synfuels cost estimates continued to rise.

The Federal Synfuels R&D Program

With the formation of the Office of Coal Research (OCR) in 1960, the federal government resumed research on coal liquefaction and gasification. The office's purpose was to find new markets for the depressed coal industry by undertaking research on synthetic fuels from coal. The goals were grandiose, as this retrospective look suggests:

> By the mid-60s evidence of an impending shortage of petroleum and natural gas began to appear. The prevention of pollution and protection of the environment also became matters of national concern. As a matter of evolution and as documented in OCR's budget justification, the role of the Office of Coal Research was broadened from the initial concept to meet the wider and more imperative national objective of insuring that coal, as our greatest fossil fuel resource, would be utilized in providing clean and convenient supplements for dwindling supplies of indigenous natural gas and petroleum and reduce the dangers to our economy and national security resulting from becoming overly dependent on foreign sources for these vital energy forms.[11]

This ambitious task was to be accomplished on a minuscule budget that averaged $6 million dollars annually. The OCR's appropriations remained small throughout the 1960s (table 10-4). By 1970 the OCR still had fewer than two dozen employees.[12] The budgetary history suggests that the OCR's goals, however laudable, were not particularly compelling as budget priorities. Dwindling supplies notwithstanding, energy prices dropped

11. *Department of the Interior and Related Agencies Appropriations for 1973*, Hearings before a Subcommittee of the House Committee on Appropriations, 92 Cong. 2 sess. (GPO, 1972), pt. 1, p. 717. (Hereafter Fiscal 1973 House Appropriations Hearings.)

12. Fiscal 1973 House Appropriations Hearings, p. 883. The OCR had twenty-one employees in fiscal 1971.

Table 10-4. *Office of Coal Research, Appropriations and Expenditures, Fiscal Years 1961–69*
Current dollars

Fiscal year	*Appropriation*	*Expenditure*
1961	1,000,000	46,677
1962	1,000,000	372,787
1963	3,450,000	1,470,232
1964	5,075,000	2,626,814
1965	6,836,000	3,822,000
1966	7,220,000	7,124,000
1967	8,220,000	9,989,000
1968	10,980,000	11,856,000
1969	13,700,000	8,427,000

Source: *Department of the Interior and Related Agencies Appropriations for 1973*, Hearings before a Subcommittee of the House Committee on Appropriations, 92 Cong. 2 sess. (GPO, 1972), pt. 1, p. 797.

during the 1960s. Therefore, security arguments lacked the force they carried a decade later, while synfuels technology as savior of the depressed coal industry was clearly a longshot.

Between 1963 and 1967 the OCR supported research on five liquefaction processes (table 10-5). Three involved pilot plants, one was limited to bench-scale R&D, and one was expected to be a pilot plant but was canceled after some bench-scale R&D.

The OCR's priorities and political support can be deduced from its choices of projects. All five OCR liquefaction projects were intended to produce gasoline. The Project Gasoline pilot in Cresap, West Virginia, used direct liquefaction and hydrogenation. The Solvent Refined Coal (SRC) Project was a pilot direct liquefaction process, which yielded clean, sulfur-free boiler fuel for power plants that is solid at room temperature. The SRC process can use any coal; however, the facility was to be built in Tacoma, Washington, a long way from any significant coal reserves, but in the home state of Julia Hansen, chair of the House Appropriations subcommittee overseeing the OCR. Although deferred due to budget constraints, the SRC project finally had funds appropriated in 1967.

The third project was Project COED (Char-Oil-Energy-Development) in Princeton, New Jersey, a pyrolysis pilot that yielded gas, char, and liquids. The initial hope was that the char would be valuable for direct combustion. The fourth project was H-Coal. Originally a direct hydrogenation process that was designed to produce gasoline, it was altered to yield mainly fuel oil. Eventually the project was canceled, only to reemerge in

Table 10-5. *Major Industrial Contracts for Coal Liquefaction and Gasification, Office of Coal Research, Fiscal Years 1961–70*
Value in thousands of current dollars

Year	*Value*	*Company*[a]	*Process*	*Product*
1961	...	...	...	...
1962	1,599	FMC	COED[b]	Boiler fuel
	1,240	Spenser	SRC[c]	Boiler fuel
1963	20,377	Con.Coal	Project Gasoline	Gasoline
	4,000	BCR	(BIGAS)	Substitute natural gas
1964	16,506	IGT	HYGAS	Substitute natural gas
	19,336	Con.Coal	CO_2 Acceptor	Substitute natural gas
	1,710	Kellogg	Molten salt	Substitute natural gas
	968	ARCO	Liquefaction	Gasoline, liquids
1965	2,064[d]	HRI	H-Coal	Gasoline, naphtha
1966	11,200	FMC	COED[b]	Boiler fuel
	28,416[d]	P&M	SRC[c]	Boiler fuel
1967	...	...	...	...
1968	...	...	...	...
1969	...	...	...	...
1970	...	...	...	...

Source: *Special Energy Research and Development House Appropriation Bill for 1975*, Hearings before a Subcommittee of the House Committee on Appropriations, 93 Cong. 2 sess. (GPO, 1974), pt. 1, pp. 616–30. (Hereafter Fiscal 1975 House Appropriations Hearings.)

a. Company abbreviations: Con.Coal: Consolidated Coal Co.; BCR: Bituminous Coal Research (research division of the National Coal Association); P&M: Pittsburg and Midway Mining Co. (a subsidiary of Gulf Oil Co., purchaser of Spenser Chemical Co.); HRI: Hydrocardon Research Inc.; ARCO: Atlantic Richfield Co.; IGT: Institute of Gas Technology.

b. Char-Oil-Energy Development.

c. Solvent refined coal.

d. The H-Coal and Kellogg contracts were terminated and the SRC contract deferred in 1967 due to budget constraints. By 1970 less than $1 million had been expended on the P&M contract for the SRC. GAO, *Liquefying Coal*, pp. 10, 12; and Fiscal 1975 House Appropriations Hearings, p. 795 (expenditures). In each case the stated reason for cancellation was lack of funds rather than disappointing results.

1975 when a pilot plant was built. Finally, OCR provided partial support to investigate the use of coal along with crude oil residuum to produce gasoline.

The OCR's gasification program was commensurate with its liquefaction program. All projects sought to produce high-Btu substitute natural gas. Most were cofinanced by the American Gas Association (AGA), a nonprofit organization of gas utilities, wholesalers, and pipelines.[13] OCR also collaborated with the Institute of Gas Technology for a fluidized bed reactor in Chicago (HYGAS) that would use eastern coals and with Consolidated Coal Company for a pilot fluidized bed plant in South Dakota (CO_2 Acceptor) that would use lignite. Two bench-scale projects also were supported: the

13. A list of 1971 members appears in Fiscal 1973 House Appropriations Hearings, pp. 62–64.

Kellogg process, using western coal to produce SNG, and BIGAS, an SNG gasification process in Pennsylvania for eastern coal.

The OCR's selection of projects—concentrating on pipeline quality gas, gasoline, and the use of eastern coals—shows that the program served purposes beyond simply expanding the use of eastern coal. Alternatively, had the sole objective been the production of SNG and gasoline at lowest cost, most funds would have been allocated to fixed bed and indirect liquefaction processes using western coal. Neither type of project received any financial support during the 1960s. By focusing on top-line products, the OCR ignored potentially easier industrial applications but paid heed to potentially undermining the markets of its collaborators. For example, an attractive commercial process for medium-Btu gas would have competed for AGA's industrial gas customers.

Between 1970 and 1974 the OCR's program underwent two transformations. From 1970 until the Arab oil embargo in 1973, high-Btu gasification dominated the program, liquefaction languished, and low-Btu research was nonexistent. The final budget for fiscal 1974 quadrupled the appropriation for coal liquefaction, while research on high-Btu gasification remained constant. Thus by 1974 SNG had been given a substantially lower priority than liquefaction, while low-Btu gas rose to prominence.

In 1970 the Office of Management and Budget (OMB) directed the OCR to require that one-third of the cost of all new or modified pilot plants come from private sponsors. The following year the OCR and the American Gas Association signed an agreement to cosponsor high-Btu gas research for four years. The OCR agreed to provide $20 million a year; AGA, $10 million. Projects would be chosen by a joint board. In the fourth year a demonstration plant would be designed based on the results from three pilot plants (HYGAS, CO_2 Acceptor, and BIGAS). By 1971 gas pipelines were curtailing deliveries, and the outlook for natural gas reserves was pessimistic. Substitute natural gas was projected to cost between $0.85 and $1.05 per thousand cubic feet. While not competitive with regulated natural gas (typically about $0.50 delivered) or gas imported from Canada, SNG cost less than liquid natural gas or gas manufactured from naphtha or oil (between $1.00 and $1.80).[14]

14. *Department of the Interior and Related Agencies Appropriation Bill, 1972*, H. Rept. 92-308 to accompany H. Rept. 9417, 92 Cong. 1 sess. (GPO, 1971), p. 427 (hereafter Fiscal 1972 House Appropriations Report); and Fiscal 1973 House Appropriations Hearings, pp. 703, 740–67, 847–48.

The final factor leading to the AGA agreement was the Nixon administration's attention to energy issues. On June 4, 1971, the president gave his "Clean Energy Message," which emphasized SNG as a clean way to use coal and to expand gas supplies.[15] In 1971 House appropriations hearings thus described the energy picture:

> The forecast shortage of natural gas is of real concern, and the supply soon may need to be greatly supplemented. . . . The OCR budget request for FY72 contains major items for coal gasification. The request for coal liquefaction research work, however, is for a reduced amount. This reflects the shift in priority based upon recognition of problems of supply for a high-quality synthetic pipeline gas. This is due partly to the increasing impact of pollution control levels and increased demand for pipeline gas as a clean high-Btu fuel.[16]

The OCR requested a supplemental fiscal 1972 appropriation of $10,280,000 for the SNG program. The AGA agreement, however, was evidently the last straw in an argument between the House Appropriations subcommittee and the administration over cost-sharing. The regular fiscal 1972 appropriation hearings had concentrated on cost-sharing and the committee's frustration over the administration's freezing of funds for coal projects. The administration froze $1 million that was appropriated in 1970 for the CO_2 Acceptor.[17] The House subcommittee expressed two concerns about the cost-sharing requirement. First, project selection was biased (that is, the choice of pilot plant projects was governed by industry's preferences).[18] Second, the subcommittee was concerned about giving companies patent rights in new technologies. In the fiscal 1972 budget the House recommended appropriating the additional funds for SRC and BIGAS to replace private financing, stating: "On the basis of information provided, the committee has serious reservations as to the wisdom of venturing on a combined funding basis at this stage of the research projects. Many ramifications are involved, and it is not improbable that such action could be detrimental to the best interests of the Federal government."[19]

15. *Department of the Interior and Related Agencies Appropriations for 1977*, Hearings before a Subcommittee of the House Committee on Appropriations, 94 Cong. 2 sess. (GPO, 1976), p. 841. (Hereafter Fiscal 1977 House Appropriations Hearings.)
16. Fiscal 1972 House Appropriations Report, p. 394.
17. Fiscal 1972 House Appropriations Report, p. 424.
18. Fiscal 1972 House Appropriations Report, pp. 441–42.
19. Fiscal 1972 House Appropriations Report, pp. 16–17.

When faced with the supplemental request, the House initially rejected the entire amount, following the subcommittee's warning that "previous research funding of almost $90-million through fiscal 1971 had not established the feasibility through fiscal 1971 of the techniques."[20] Nevertheless, the vote reflected discontent with the agreement rather than with the OCR's gasification program. Later the Senate passed the entire supplemental, and $5,120,000 was agreed in conference. Although the House continued to reject cost-sharing, Congress continued to appropriate funds for gasification projects.

High-Btu gas research doubled in fiscal 1972 and again in fiscal 1973 as the AGA program got under way (table 10-6). In addition to the BIGAS pilot, four other high-Btu projects received substantial support, including, for a single year, the Lurgi fixed-bed process. The HYGAS and CO_2 Acceptor pilots were completed in 1972, and successful tests of both were conducted during the early 1970s.[21]

Industrial participation was not forthcoming in the liquefaction program, which for several years consisted of operating support for two pilots from the 1960s: Project Gasoline and the COED Project. Performance information from these projects became available in the early 1970s, and both were disappointing. Project Gasoline, in Cresap, West Virginia, operated after a fashion from 1967 to 1970. The plant had trouble with virtually all components and subprocesses. Management and union problems compounded technological difficulties, and the plant ultimately achieved continual operation for as long as a week just three times. The final stage, intended to produce gasoline, never operated.[22] About $18 million was spent on the project, including $6 million for construction and $2.5 million to $3 million per year for three years of operation. In 1970 the OCR contracted with the National Academy of Engineers and Foster Wheeler Corporation for project evaluations. It hoped they would recommend conditions for the project to go forward, but both groups recommended modification of the project for different use. The National Academy of Engineers estimated that to test the original process would add $3 million

20. *Congressional Quarterly Almanac, 1971* (Washington, 1971), p. 818.

21. Fiscal 1973 House Appropriations Hearings, p. 726.

22. *Department of the Interior and Related Agencies Appropriations for 1971*, Hearings before a Subcommittee of the House Committee on Appropriations, 91 Cong. 2 sess. (GPO, 1970), pt. 2, p. 835 (union difficulties), p. 833 (operating history) (hereafter Fiscal 1971 House Appropriations Hearings); Fiscal 1972 House Appropriations Report, p. 432 (technical difficulties), p. 435 (NAE conclusions); Fiscal 1973 House Appropriations Hearings, p. 794 (budget history).

Table 10-6. *Office of Coal Research Appropriations, Fiscal Years 1970–74*
Thousands of current dollars

Budget category	*1970*	*1971*	*1972*	*1973*	*1974*
Liquefaction					
Direct hydrogenation					8,000
Con. Coal, Cresap facility	2,660	3,000[a]	3,000[a]	. . .	. . .
HRI: H-Coal	. . .	100	. . .	. . .	. . .
Four nonpilot-plant processes[b]	. . .	. . .	. . .	. . .	6,450
Solvent extraction					
P&M: SRC	. . .	590	3,150	6,107	. . .
Delayed coker, chemicals from coal	. . .	. . .	. . .	. . .	. . .
Pyrolysis					
FMC: COED	3,300	3,000	3,035	3,250	3,000
Clean coke and fuels	. . .	. . .	. . .	. . .	1,500
Indirect hydrogenation	. . .	. . .	. . .	. . .	2,000
Other projects and suppport	. . .	. . .	. . .	650	3,000
Prototype design	. . .	. . .	. . .	. . .	7,000
Total, liquefaction	**5,960**	**6,690**	**9,185**	**10,007**	**43,500**
Gasification					
High Btu					
Fixed bed: Lurgi	. . .	. . .	. . .	1,600	. . .
Fluidized bed					
IGT: HYGAS	3,692	3,205	4,670	4,000	4,670
IGT: Steam Iron	. . .	. . .	. . .	5,180	4,670
CO_2 Acceptor	1,000	3,000	3,870	2,730	6,990
Battelle self-agglomerating fluid bed	. . .	. . .	. . .	1,670	670
Entrained bed					
BCR: BIGAS	900	900	7,377	6,400	1,330
General					
Chemical systems methanation	. . .	. . .	. . .	1,010	340
Applied technical molten iron	. . .	. . .	. . .	1,670	. . .
Process selection, support	. . .	. . .	. . .	900	6,730
Subtotal	5,592	7,105	15,917	25,160	25,400
Low, medium Btu					
Pressurized	. . .	. . .	. . .	. . .	16,200
Atmospheric	. . .	. . .	. . .	. . .	3,000
General, support	. . .	. . .	. . .	. . .	500
Subtotal	. . .	. . .	. . .	3,000	19,700
Total, gasification	**5,592**	**7,105**	**15,917**	**38,167**	**45,100**
Other projects and support	1,100	2,850	2,950	2,433	8,600
Total, synthetic fuel contracts	**12,652**	**16,645**	**28,052**	**40,600**	**97,200**
Administration and supervision	448	495	598	885	2,400
Total, OCR	**13,300**	**17,160**	**30,650**	**43,490**	**123,400**

Sources: For 1970: *Department of the Interior and Related Agencies Appropriations for 1971*, Hearings before a Subcommittee of the House Committee on Appropriations, 91 Cong. 2 sess. (GPO, 1970), p. 793 (Fiscal 1971 House Appropriations Hearings). For 1971: Fiscal 1972 House Appropriations Hearings, p. 413. For 1972: Fiscal 1973 House Appropriations Hearings, pp. 861–63. For 1973: Fiscal 1974 House Appropriations Hearings, pp. 557–59, 584. For 1974: Fiscal 1975 House Appropriations Hearings, pt. 1, pp. 638–42; and *Supplemental Appropriation Bill, 1974*, House Committee on Appropriations, 93 Cong. 1 sess. (GPO, 1973), pp. 384–94. The fiscal 1974 liquefaction subtotals are estimates.

a. Funds not expended.

b. Direct (Synthoil), fixed bed (H-Coal), centrifugal, and zinc chloride gasoline conversion. These projects were funded, but subtotals are not available.

c. Includes OCRC programs other than gasification and liquefaction.

in construction costs and $6 million in operating costs. In 1971 the OCR sought appropriations to convert the plant to test the H-Coal process of Hydrocarbon Research Inc. (HRI), which the OCR had supported at bench scale from 1965 to 1967 and which Atlantic Richfield and other oil companies had supported subsequently.[23] Three million dollars was appropriated for this purpose, but the OCR, HRI, and the private cosponsors could not work out a cost-sharing agreement. The conversion never took place, and HRI development remained in private hands until 1975. The Cresap test facility was reopened in 1974 to produce the low-grade liquid "it is capable of producing" and to test the unreliable components that had plagued both gas and liquids projects.[24]

The COED pilot plant ran tests on different types of coal from 1970 to 1974. It processed 25 tons of coal a day and produced sulfur-free and nitrogen-free crude oil, clean gas, and char.[25] Unlike Project Gasoline, the plant worked, and for several years the Office of Coal Research claimed that the project was a success. According to the acting director of the OCR: "Our pilot plant . . . has been a complete technical success in terms of its original objectives. We have met all our scientific, technical requirements. The plant did everything it was supposed to. . . . The engineering design of a demonstration scale based on that process is in the works."[26]

This rosy assessment prompted discussion of transferring development and demonstration of the technology to private industry.[27] By 1976, however, the Energy Research and Development Administration (ERDA) conceded that the plant was a failure. In brief, the plant worked, but the product was lousy. Little oil and gas was produced, and the char from processing high-sulfur coal was so sulfurous that it had no environmental advantages over direct combustion of coal.[28] The pilot plant was dismantled, but the technology eventually became part of the SRC demonstration and the first step in an oil and gas facility called COGAS.[29]

Following President Nixon's "Clean Energy Message," the OCR changed its emphasis in liquefaction from gasoline to fuel oil, a move that,

23. Fiscal 1972 House Appropriations Report, pp. 429–30.

24. *Department of the Interior and Related Agencies Appropriations for Fiscal Year 1975*, Hearings before a Subcommittee of the House Committee on Appropriations, 93 Cong. 2 sess. (GPO, 1974), pp. 688–99. (Hereafter Fiscal 1975 House Appropriations Hearings.)

25. Fiscal 1975 House Appropriations Hearings, p. 650.

26. Fiscal 1975 House Appropriations Hearings, p. 686.

27. Fiscal 1973 House Appropriations Hearings, pp. 924–26.

28. Perry, "Clean Fuels from Coal," pp. 278–79.

29. Fiscal 1977 House Appropriations Hearings, pt. 5, p. 527.

while less glamorous, was justified on the grounds of environmental effects, technical feasibility, and direct substitution for oil.

Synthetic fuel oil substitutes for residual fuel oil in electrical generating plants. This use promised only short-term benefits because oil-fired generation was being replaced by other electrical generation methods. The longer-term justification was environmental: synthetic fuel oil was a promising substitute for coal. George Fumich, director of the OCR, laid out the argument:

> Of the 308 million tons of coal burned in the electric utility market last year, 93 percent was mined east of the Mississippi River and . . . 70 percent of the coal east of the Mississippi River is high sulfur and can't be utilized today in the electric utility market. In other words, we have a lost resource. . . . [I]t was decided that we would have more reason to go to a higher priority target area, the low sulfur fuel, because of the air pollution regulations. This was a more near run priority; and we felt that we should change our target and develop a low sulfur liquid fuel rather than go down the longer road to develop light fuels that could be refined to gasoline.[30]

The policy was implemented in the gasification program by investigating entrained and fluid bed technology for producing medium- and low-Btu gas. The liquefaction program's response was to proceed with the Solvent Refined Coal Project in Tacoma, Washington. After completion of preliminary work in 1969, the SRC project failed to be funded for two years.[31]

In 1971 the OMB authorized the OCR to request $2.1 million for design of the SRC pilot plant, with the requirement that the OCR solicit private cofinancing.[32] Congress appropriated the entire cost of $3.15 million, and the project proceeded without private support. The SRC plant was completed in 1973.[33] Operations went smoothly, and in 1974 the process was considered the best commercialization prospect among liquefaction processes.[34]

30. Testimony of George Fumich, *Department of the Interior and Related Agencies Appropriations for 1972*, Hearings before a Subcommittee of the House Committee on Appropriations, 92 Cong. 1 sess. (GPO, 1971), pp. 430–432.

31. Fiscal 1971 House Appropriations Hearings, pp. 850–51.

32. Fiscal 1972 House Appropriations Report, pp. 439–41.

33. Fiscal 1975 House Appropriations Hearings, p. 649; and Fiscal 1973 House Appropriations Hearings, p. 923.

34. Fiscal 1975 House Appropriations Hearings, p. 644.

The big budgetary change in scale and relative priorities came in fiscal 1974. The administration's initial liquefaction budget request of $9 million, a decrease from the previous budget of $9.4 million, presaged no significant change. The proposal included $8 million for the SRC project and a decrease to $1 million for the COED project. The administration's overall request for the Office of Coal Research was $52.5 million, a modest increase from the previous budget of $43.5 million. The entire increase in the synfuels program was for low-Btu gasification.

In June 1973 the perceived importance of the program began to increase dramatically. The House Appropriations Committee restored the COED funds and recommended $7 million for designing a demonstration plant using the COED and SRC technologies. By including these projects, the committee recommended nearly doubling liquefaction funds. Days later President Nixon proposed a $100 million increase in energy research, half of which was for coal. The Senate Appropriations Committee issued its report in July, stating:

> The Committee is cognizant of the President's June 29, 1973 message. . . . In the absence of a budget estimate, the Committee has elected to provide a substantial increase on the basis of available research priorities established by the Department. The Committee expects an early supplemental budget estimate detailing further needs in coal research.[35]

The Senate committee recommended an overall budget of $95 million, including increased funds in the liquefaction program for virtually every program the OCR had ever investigated: SRC pilot ($11.5 million), SRC demo ($9 million), the Cresap test facility ($1 million), indirect liquefaction ($2 million), direct hydrogenation ($5.75 million, split among H-Coal, Synthoil, and two other gasoline-oriented processes), and three smaller projects ($1.5 million). In gasification, increases were approved for CO_2 Acceptor and high-Btu projects ($4.4 million) and for low-Btu gasification ($7.6 million). The Senate's budget was approved, as was a supplemental request in October for $28 million more, including increases of $1.6 million for high-Btu gasification (support work) and $5.35 million for liquefaction. The liquids supplemental increased funds for the SRC-COED demonstra-

35. *Department of the Interior and Related Agencies Appropriation Bill, 1974*, S. Rept. 93-362 (GPO, 1973), p. 13.

tion, the Cresap test facility, indirect liquefaction, and smaller research projects.[36] This final fiscal 1974 budget is shown in table 10-6.

By 1974 much of the interest in synfuels had shifted to liquids, although to technologies that had a greater near-term chance of success than had technologies for producing gasoline. The following two excerpts express the program's goals:

> The OCR program for coal liquefaction represents a substantial acceleration of effort over previous years, in response to changes in the Nation's priorities. Coal liquefaction will help meet the increasing demand for petroleum and liquid fuels with reduced reliance on imported oil. . . . A vigorous program adequately funded could lead to commercialization in the late 1970s. . . . The OCR program is an across-the-board approach designed to explore and develop all reasonably possible processes for producing clean liquid fuels from coal. Success in this area is one of the few alternatives to a continuing intolerable dependence on imports for at least our vehicular fuels. . . .
>
> There is an underlying strategy for this distribution of funds . . . "Where can we make the biggest impact in the shortest period?" . . . [B]y concentrating on boiler fuels. . . . When we do this we don't downgrade the high-Btu program. We are keeping that going at as rapid a rate as we think that technology is moving, but . . . most of our growth over the last year is not in that program. It is in liquids, it is in direct combustion, it is in low-Btu.[37]

Rapid scale-up of the synfuels program continued in fiscal 1975 (table 10-7). Design for the H-Coal pilot, construction on the Synthoil pilot unit, and increases in other categories doubled the liquefaction budget. In high-Btu gasification, work began on three new fluidized bed processes, as did serious spending for the BIGAS pilot, more than doubling the budget in this category. Low-Btu gasification increased from about $20 million to $50 million for processes identified in the previous year.

Designs for liquefaction, high-Btu gasification, and low-Btu gasification

36. Fiscal 1972 House Appropriations Report, pp. 16–17; *Department of the Interior and Related Agencies Appropriation Bill, 1974*, H. Rept. 93-322 (GPO, 1973), p. 22; *Department of the Interior and Related Agencies Appropriation Bill, 1974*, S. Rept. 93-362 (GPO, 1973); and *Supplemental Appropriation Bill, 1974*, House Hearing, 93 Cong. 1 sess. (GPO, 1973), pt. 2, pp. 387–89.

37. Fiscal 1975 House Appropriations Hearings, pp. 648, 680.

demonstrations began in fiscal 1976. Despite substantial increases in appropriations for demonstrations, spending actually declined slightly. Congress appropriated $35 million less for coal research in fiscal 1976 than the administration requested, citing the "excessive" $100 million balance carried forward from fiscal 1975, which "casts doubt on the wisdom of providing still another major funding increase."[38]

In 1975 the Ford administration made energy policy a high priority, which it was to remain for the next eight years. In his State of the Union address, President Ford called for accelerating the development of synfuels technology in order to accommodate a rapid future expansion of the industry. The rationale was that synfuels would reduce reliance on imports, thereby safeguarding against an oil embargo and providing less expensive supplies should import prices rise. In addition, the program was said to improve the international position of the United States by establishing leadership in the field and showing resolve to oil-exporting countries. Even though no synfuels project had yet shown commercial promise, President Gerald Ford set a goal to produce the equivalent of 1 million barrels a day from synthetic fuels from coal and oil shale by 1985. Established technologies would be used, including Lurgi gasification and indirect liquefaction, rather than those being developed by the OCR and ERDA. This oriented the program toward western resources such as oil shale, lignite, and sub-bituminous coal.

The Synfuels Interagency Task Force was established to recommend a commercialization program that would achieve the president's goal. Its report, issued in November 1975, concluded that the eventual need for synthetic fuels was clear, but that the expected cost and development risks of synfuels were so high that no projects were likely to be built soon without government aid. The report recommended using incentives to induce synfuels investments. Loan guarantees were proposed for projects built by regulated utilities (gasification plants) to overcome Federal Power Commission regulations regarding their access to capital markets. Price guarantees were recommended for projects (such as shale oil) that might be jeopardized by declines in the future price of oil imports.[39]

38. *Department of the Interior and Related Agencies Appropriation Bill, 1975*, S. Rept. 94-462, 93 Cong. 2 sess. (GPO, 1974), p. 35.

39. *Recommendations for a Synthetic Fuels Commercialization Program*, vol. 1: *Overview Report;* vol. 2: *Cost/Benefit Analysis of Alternate Production Levels*, Report submitted by Synfuels Interagency Task Force to the President's Energy Resources Council (GPO, November 1975).

Table 10-7. *Budgets of the Energy Research and Development Administration, Fiscal Years 1975–78*
Thousands of current dollars

Budget category	*1975*	*1976*	*Transition quarter*	*1977*	*1978*
Liquefaction					
Direct hydrogenation					
H-Coal	12,400	27,096	11,530	20,800	26,000
Synthoil	10,600	7,573	3,250	7,900	11,000
Zinc chloride catalyst, pilot	2,250	2,000	860	2,400	2,000
Disposable catalyst	2,800	516	150	601	500
Multistage liquefaction	0	0	0	0	1,000
Subtotal	28,050	37,185	15,790	31,701	40,500
Solvent extraction					
SRC pilot plant	12,700	12,570	3,145	19,635	16,000
SRC utilization	0	2,020	517	1,120	...
Costeam	787	1,000	250	500	500
Chemicals from coal	1,500	300	125	450	400
Solvent extraction lignite	1,500	500	0	1,000	0
Donor solvent	0	2,000	500	2,000	30,300
Subtotal	16,487	18,390	4,537	24,705	47,200
Pyrolysis					
Clean coke pilot	4,000	3,400	200	2,400	0
COED	3,000	...	...	...	...
Hydrocarbonization	3,600	800	100	900	0
Entrained pyrolysis	0	200	50	600	500
Flash liquefaction	0	500	50	900	500
Fluid coke	0	600	100	600	1,000
Subtotal	10,600	5,500	500	5,400	2,000
Indirect liquefaction	4,800	1,700	750	1,100	0
Support studies	30,470	27,137	4,866	10,051	16,500
Demonstrations					
Design prototype	4,000	...	...	...	...
Clean boiler fuel, operating costs	0	10,000	2,000	...	...
Design and tecnical support[a]	0	5,900	1,750	...	...
Design and technical, SRC, operating costs	0	0	0	8,400	4,000
CBFD, capital	13,000[b]	20,000	8,000	30,000	0
Total, liquefaction	**107,407[b]**	**125,812**	**38,193**	**111,357**	**110,200**

(continued)

Table 10-7 *(continued)*

Budget category	1975	1976	Transition quarter	1977	1978
Gasification					
High Btu					
Fluidized bed					
HYGAS	3,150	6,370	670	2,800	0
Steam iron process	6,400	7,190	2,047	4,600	0
Synthane	8,600	9,000	1,400	10,500	14,500
CO_2 Acceptor	2,000	5,640	710	2,133	0
Self-agglomerating	3,700	3,340	807	2,227	1,500
Hydrane	1,039	2,981	705	1,800	3,500
Catalytic gasification	0	0	0	0	3,500
Entrained bed					
BIGAS	13,200	12,000	1,200	9,800	3,000
Support	21,716	6,843	1,711	10,194	. . .
Demo, operating	0	8,000	2,000	10,000	5,000
Demo, construction	0	0	0	10,000	29,000
Subtotal	59,805	61,364	11,250	64,054	60,000
Low, medium Btu					
Fixed bed, atmospheric	5,200	2,350	588	2,950	4,550
Fluid bed, pressurized	8,000	8,300	750	1,700	5,000
Entrained bed, atmospheric	3,000	0	0	500	1,490
Combined cycle pilot	11,000	6,530	2,000	16,500	20,000
Molten salt pilot, pressurized	4,000	500	0	1,000	0
Hydrogen from coal	0	0	0	5,000	15,000
Gasifiers in industry	. . .	. . .	. . .	1,000	10,000
Support, miscellaneous	18,793	6,852	3,382	4,402	1,300
Demo, operating	0	8,000	2,000	12,000	11,500
Demo, construction	0	0	0	7,300	39,000
Subtotal	49,993	32,532	8,720	52,352	107,840
Total, gasification	**109,798**	**93,896**	**19,970**	**116,406**	**167,840**
Advanced research and support	23,325	35,393	8,850	37,070	40,000
Total	**240,530**	**225,101**	**67,013**	**264,833**	**318,040**

Sources: For 1975, 1976, and transition quarter: *Department of the Interior and Related Agencies Appropriations for 1977*, Hearings before a Subcommittee of the House Committee on Appropriations, 94 Cong. 2 sess. (GPO, 1976), pt. 5, pp. 381–83, 424–27 (Fiscal 1977 House Appropriations Hearings); and Fiscal 1976 House Appropriations Hearings, pt. 5, p. 2708. For 1977: Fiscal 1978 House Appropriations Hearings, pt. 2, pp. 565–68, 637. For 1978: *Fossil Energy Research and Development Program of the U.S. Department of Energy, FY1979*, DOE/ET-0013(78) (March 1978), pp. 15, 61, 62, 75, 85, 105, 107, 123, 161, 175, 182.

a. Includes gasification demo design and technical support.

b. Includes all appropriations through fiscal 1975. Total excluded from total liquefaction figure for fiscal 1975. As of the fiscal 1978 budget request, only about $20 million total of this had been obligated.

The task force conducted an elaborate cost-benefit study of three scenarios: the president's synfuels goal (1 million barrels a day by 1985), a less intense "information" program (335,000 barrels a day by 1985 from commercial plants in each major technology area), and a "maximum" program (1.7 million barrels a day by 1985). The discounted expected net benefits (including estimated values for embargo protection and environmental and socioeconomic costs but excluding other noneconomic goals) were negative for all three options: −$1.63 billion for the information program (1975 dollars), −$5.41 billion for Ford's proposal, and −$10.98 billion for the maximum program. With more favorable but unrealistic assumptions about future synfuels prices (low) and oil import prices (high), the net benefits of Ford's proposal as well as the information program became positive. Based on these results, the task force gamely recommended, and the administration adopted, the information program.

Despite support for synfuels R&D, the commercialization program ran into opposition in Congress and was not authorized. In the appropriations for ERDA, the Senate included $6 billion for synfuels loan guarantees. The provision was included in the conference bill, but the House deleted the provision.[40] In early 1976 the measure was again scheduled for consideration in the House, but was in the interim hit by a barrage of criticism. In January the Congressional Budget Office issued a report projecting that synfuels would be more expensive than imported fuels for the next ten years. It also examined problems with loan guarantees and other incentives, including perverse effects on the efficiency of synfuels producers and difficulties in maintaining budget control (since actual expenditures would depend on uncertain future prices and, in the case of loan guarantees, would be "off budget").[41]

In March 1976 the General Accounting Office also criticized the program. It restated the conclusions of the Congressional Budget Office and in addition expressed concern over the evident hastiness of the legislation, the weak economic justification provided by ERDA, and the imbalance between energy augmentation and energy conservation.[42] GAO then issued a longer report detailing the advantages of other strategies (such as the

40. *Congressional Quarterly Almanac, 1975*, p. 289.

41. *Commercialization of Synthetic Fuels: Alternative Loan Guarantee and Price Support Programs*, Background Paper 3, January 16, 1976, prepared by W. David Montgomery, Congressional Budget Office, Washington, D.C., 1976.

42. *Comments on the Administration's Proposed Synthetic Fuels Commercialization Program*, Report of the Comptroller General of the United States, RED-76-82, March 19, 1976.

Strategic Petroleum Reserve, conservation, and solar power) and further criticizing loan guarantees for subsidizing synfuels commercialization.[43] In September the House voted 192-193 to reject a rule to allow debate on the loan guarantees bill. This effectively killed the proposal once again.[44]

Although Congress narrowly rejected early commercialization, appropriations for ERDA's coal R&D program continued to increase. The largest increase took place in low-Btu gasification, where ERDA supported demonstration designs and industrial gasifiers. Industry cost-sharing on these projects was about 50 percent, and by 1978 they were thought to involve little technical risk:

> The nature of this demonstration is not so much to prove a new research concept as to prove to a businessman, who is investing the stockholders' money, that he can operate one of these machines and make money at it; that it is reliable; and that it meets the environmental standards. Those are the hurdles that have to be crossed with this kind of demonstration to convince businessmen that it is a legitimate investment.[45]

The liquefaction program supported a broad spectrum of technologies, including three major pilot plants, a demonstration, and several smaller projects. Unfortunately, despite the diversity of the program, its results continued to demonstrate that synfuels would not be a quick fix for petroleum dependence or coal use. Enthusiasm for SRC diminished with operating experience. In 1976 Dr. White, ERDA's assistant administrator for fossil fuel, acknowledged that the SRC process was "an expensive proposition. It gives you a fairly low yield of liquid, and to say it is liquid is stretching a point because it is solid as a rock at room temperature. We don't feel it is a very good choice for our first advanced coal liquefaction demonstration plant."[46]

ERDA undertook two large pilot plant projects: H-Coal and Exxon

43. GAO, *An Evaluation of Proposed Federal Assistance for Financing Commercialization of Emerging Energy Technologies,* Report to the Congress by the Comptroller General of the United States, EMD-76-10 (August 1976).

44. *Congressional Quarterly Almanac, 1976* (Washington, 1976), p. 177.

45. Testimony of Robert Fri, acting administrator, ERDA, *Department of the Interior and Related Agencies Appropriations for 1978,* Hearings before a Subcommittee of the House Committee on Appropriations, 95 Cong. 1 sess. (GPO, 1977), pt. 8, p. 529. (Hereafter Fiscal 1978 House Appropriations Hearings.)

46. Fiscal 1977 House Appropriations Hearings, p. 526.

Donor Solvent. H-Coal took twice as long to design (two years rather than one) and nearly twice as long to build (36 months as opposed 20) than planned. It cost 66 percent more than projected ($296.1 million rather than $178.8). Although the H-Coal process had been studied for a decade before the pilot was built, a 1981 GAO study concluded that the main source of problems was the hastiness with which it was undertaken:

> DOE started the H-Coal project prematurely before sufficiently detailed designs were available. This action was taken as a reaction to the "energy crisis" and despite warnings of the plant designer. The project was not adequately planned and key management functions such as a construction schedule, materials handling, inventory systems, and a quality control program were nonexistent for most of the construction period.[47]

The EDS plant, in which the Department of Energy had a 50 percent interest, had a similar record. Detailed design work was completed five months late and construction was completed another four months behind schedule. Plant costs were 24 percent over the baseline estimate of $225 million, but then an attempt was made to increase yields from unused liquids and unreacted coals through a nonliquefaction process called "bottoms development." This added $65 million, for a total cost overrun of 50 percent.[48]

ERDA started its liquefaction demonstration with the Clean Boiler Fuel Demonstration (CBFD), a $227 million contract. Phase I (conceptual design) was never completed, and after design expenditures of $14 million ($10 million over budget), the project was terminated. The purpose was to adapt Union Carbide's western coal process to use eastern coals. The problems in using caking coals were not solved. The General Accounting Office attributed the failure to insufficient research and development, inadequate control by ERDA, and inflexibility when it became evident that the original concept was not going to work. George Fumich, then ERDA's deputy assistant administrator for fossil energy, claimed that the Interior

47. GAO, *Controlling Federal Costs for Coal Liquefaction Program Hinges on Management and Contracting Improvements*, Report to the Congress by the Comptroller General of the United States, PSAD-81-19 (February 1981), p. 46.

48. GAO, *Controlling Federal Costs*, p. 11. Bottoms development via the redesigned equipment was estimated to cost $79.8 million rather than $15 million. This plus other changes led to a total facility cost of $359.3 million instead of $240 million.

Department, "with the support of incoming ERDA officials, decided to award the contract despite its high risk because of the 1973 oil embargo and congressional concern that alternatives to foreign oil be expedited."[49]

Under ERDA, the high-Btu research program wound down without the dramatic cost overruns of the liquefaction programs, but also without technological breakthroughs. The HYGAS plant, using the steam iron process to produce hydrogen, demonstrated technical feasibility on eastern coals. The CO_2 Acceptor and Synthane programs also demonstrated technical feasibility. BIGAS ran into initial difficulties because of design complexities. It operated for 39 hours in 1977 and ran briefly in February 1978 and again in February 1979. The Department of Energy terminated operations in 1979.[50]

Starting in fiscal 1976, ERDA supported design studies for high-Btu gas demonstration plants. Two processes were evaluated: COGAS, based on the COED process with fluidized bed gasification of the char residue, and a Lurgi process, based on a Scottish pilot plant that had successfully used eastern U.S. coal. In fiscal 1979 funds were appropriated for a design based on HYGAS. However, cost estimates for SNG had risen steadily, while supply projections for natural gas had shifted from pessimism to optimism following partial deregulation. By 1977 ERDA conceded that the second-generation pilot plants were too expensive for near-term development. Dr. White of ERDA explained that cost projections for HYGAS substitute natural gas were in the $3 to $3.50 range, while the deregulated price of natural gas was expected to be about $1.75. The following exchange then took place:

> Congressman Yates: "The impression I have is it may have been the Thomas Alva Edison reply. When he was asked about the 300th test that he had undertaken which had proven unsuccessful. He said, 'At least I know now what won't work.' "
>
> Dr. White: "We will have ample accumulation of that sort of knowledge."[51]

49. The CBFD contract was signed by the OCR two days before the transfer of the program to ERDA, "to avoid additional administrative burdens imposed by the Federal Nonnuclear R&D Act of 1974." GAO, *First Federal Attempt to Demonstrate A Synthetic Fossil Energy Technology—A Failure*, Report to the Comptroller General of the United States, EMD-77-59 (August 1977), p. 9. The quotation by George Fumich is on pp. 10–11.

50. Rothberg and others, *Synfuels from Coal*, pp. 42–44.

51. Fiscal 1978 House Appropriations Hearings, pt. 8, pp. 521–23.

In one of his first acts as president, Jimmy Carter issued the National Energy Plan, which attempted a comprehensive attack on the energy problem, and included more than a hundred specific proposals. The plan emphasized reducing energy demand through price decontrol and other regulatory actions, developing known and new conservation technologies, and exploiting solar and renewable resources. Like President Ford before, Carter advocated a plan whereby coal, including synfuels, was to substitute for oil imports and natural gas, even though several more synfuels projects had failed in the intervening years. For administration, Carter proposed establishing the Department of Energy to coordinate energy policy and technology development. The National Energy Act followed, and ERDA was included in the new department.[52] This legislation was immediately accompanied by further budgetary increases for the program (table 10-8).

In early 1979 oil import prices again began to rise sharply, and all the political hysteria of 1973 returned with a vengeance. In 1979, 37 different subcommittees and committees in the Senate and House held hearings on some aspect of energy policy (29 subcommittees of 17 committees and 8 other full committees).[53] A complicated series of bills and amendments on commercialization of synthetic fuels were introduced.[54] The president's proposal called for the establishment of an independent board to invest $88 billion, financed from the windfall profits tax on oil, in a synthetic fuels industry, including coal synfuels, shale oil, biomass, peat, and unconventional natural gas. The proposed Synthetic Fuels Corporation could issue loans, loan guarantees, price guarantees, construction grants, and government purchase agreements. The bill also would establish the Energy Mobilization Board to speed up environmental regulatory processes for the new industry. Finally, the bill established production goals: 2.5 million barrels per day of crude oil equivalent by 1990.

Congress broadened the program by including funds for biomass, wind, hydroelectric power, solar energy, geothermal power, conservation, and the Strategic Petroleum Reserve. It deleted the windfall profits tax and the Energy Mobilization Board. The program had two stages. In its first five years the Synthetic Fuels Corporation would have $20 billion and a

52. Jack M. Holl, *The United States Department of Energy: A History*, DOE/ES-0004 (Department of Energy, November 1982).

53. *Synthetic Fuels*, Report by the Subcommittee on Synthetic Fuels of the Senate Committee on the Budget (GPO, 1979), p. 51.

54. See Eric M. Uslaner, *Shale Barrel Politics: Energy and Legislative Leadership* (Stanford University Press, 1989), especially chapter 4.

production goal of 500,000 barrels of oil equivalent a day (BOED) by 1987. The SFC was to submit a detailed plan of activities, and Congress could then appropriate the remaining $68 billion dollars. Congress lowered the final production goal to 2 million BOED by 1992.[55]

Despite its many shortcomings the synfuels legislation sailed through Congress in early 1980. The bill was not just bad economic and fiscal policy. It was technically unfeasible because of resource constraints ranging from lack of water and coal mines to lack of architect-engineers, skilled construction workers, and building materials.[56]

The production goals of the Synthetic Fuels Corporation required that it support first-generation synfuels plants: SNG using western coal and indirect liquefaction using western coal, shale oil, and low-Btu gas processes. This left synfuels research to the Department of Energy. However, the department redefined its role as supporting a near-term commercial synthetic fuels industry and emphasizing near-term commercialization of second-generation plants, even though not one pilot plant had yet succeeded. The gasification and liquefaction research programs were consolidated and centered on commercial demonstration. The Department of Energy's fossil fuel policy and strategy are expressed in this 1980 document:

> The primary emphasis of the [liquefaction] program will be on furthering the development of improved coal liquefaction technologies that appear to be closest to commercial readiness. The strategy includes emphasis on improved versions of both direct and indirect liquefaction technology. Criteria for selection are process economics, product slate value and potential applications, process efficiency, environmental impacts, and the ability of the process to accept a broad range of U.S. coals. Emphasis is placed on processes capable of producing liquids for applications unsuitable for the direct burning of coal. . . .
>
> The major [gasification] program emphasis is to be concentrated on early demonstration of a limited number of improved gasification

55. Rothberg and others, *Synfuels from Coal,* p. 122.

56. See, for example, Harry Perry and Hans H. Landsberg, "Factors in the Development of a Major US Synthetic Fuels Industry," *Annual Revue of Energy,* vol. 6 (1981), pp. 233–66; Paul L. Joskow and Robert S. Pyndyck, "Synthetic Fuels: Should the Government Subsidize Nonconventional Energy Supplies?" *Regulation,* September–October 1979, pp. 18–24, 43; and Report of the Consulting Firm of Cameron Engineers, "Overview of Synthetic Fuels Potential to 1990," in *Synthetic Fuels,* Senate Report, app. 2.

Table 10-8. *Budgets of the Department of Energy, Fiscal Years 1979–84*
Thousands of current dollars

Budget category	*1979*[a]	*1980*[a]	*1981*	*1982*	*1983*	*1984*
Liquefaction[b]						
Direct hydrogenation						
H-Coal	33,000	64,500	57,000	22,080	0	0
Synthoil	2,389	0	0	0	0	0
Zinc chloride	2,100	2,500	0	0	0	0
Disposable catalyst	1,011	9,500	2,200	4,800	7,200	5,000[c]
Subtotal	38,500	76,500	59,200	26,880	7,200	5,000
Solvent extraction						
SRC pilot plant	13,500	15,000	47,000	0	0	0
Exxon Donor Solvent	34,600	30,000	32,000	28,800	0	0
Costeam	540	0	0	0	0	0
Subtotal	48,640	45,000	79,000	28,800	0	0
Pyrolysis						
Flash liquefaction	2,000	3,000	5,000	5,184	0	0
Subtotal	2,000	3,000	5,000	5,184	0	0
Support studies	22,986	24,506	19,297	25,728	15,800	16,000
Alternate fuels feasibility study	0	0	0	0	2,000	0
Indirect liquefaction	0	0	12,000	11,520	14,700	7,900
Demonstrations						
SRC-I, operating costs	7,000	7,000	5,000	0	0	0
SRC-I, construction	40,000	40,000	157,500	0	0	0
SRC-II, operating costs	7,000	14,000	5,000	0	0	0
SRC-II, construction	50,000	40,000	170,000	0	0	0
Design and technical support	1,600	0	0	0	0	0
Subtotal, demonstrations	105,600	101,000	337,500	0	0	0
Total, liquefaction	**217,726**	**250,006**	**511,997**	**98,112**	**39,700**	**28,900**

(continued)

technologies. The objective is to bring to commercial readiness as soon as possible (no later than the mid-1980s) gasification processes that are significantly superior, economically and environmentally, to presently available technology and that can effectively utilize the major U.S. coal resources. . . . Beyond these near-term demonstration activities, it is assumed that commercialization (including demonstration) of advanced gasification technologies will be achieved by the private sector with minimal or no Federal support through RD&D funds.[57]

57. *Fossil Energy Policy and Strategy, FY 1982–86*, working draft, distributed by the Office of Plans and Technology Assessment, Assistant Secretary for Fossil Energy, Department of Energy, March 1980 (reissued July 30, 1980), pp. 30, 36, 38.

Table 10-8 *(continued)*

Budget category	*1979*[a]	*1980*[a]	*1981*	*1982*	*1983*	*1984*
Gasification						
High Btu						
Synthane	2,620	0	0	0	0	0
BIGAS	7,900	13,500	10,000	d	0	0
Peat (at HYGAS)	4,500	3,000	13,000	d	0	0
Catalytic gasification	7,228	9,000	6,000	d	0	0
Hydrogasification	6,147	7,000	6,500	0	0	0
Support	9,700	4,000	4,500	0	0	0
Demo, operating costs	10,000	24,000	6,000	0	0	0
Demo, construction	42,000	27,000	44,000[d]	0	0	0
Subtotal	90,095	87,500	90,000	0	0	0
Low, medium Btu						
Fixed bed	4,190	6,000	7,000	d	d	0
Fluid bed	4,000	8,000	9,000	d	d	0
Entrained bed	3,790	3,000	3,000	d	d	0
Gasifiers in industry	1,000	500	0	0	0	0
Demo, operating costs	4,300	7,000	0	0	0	0
Demo construction	23,000	−12,000	45,400[e]	0	0	0
Subtotal	40,280	12,500	64,400	0	0	0
Support, miscellaneous	−14,477	15,350	10,000	0	0	0
Advanced process development	0	0	0	5,520	4,250	6,950
System engineering concepts	0	0	0	37,512	27,050	22,400
Environmental and engineering analyses	0	0	0	9,528	7,500	4,400
Capital	0	0	0	528	0	0
Great Plains Coal Gas Project, monitoring	0	0	0	0	0	3,040
Total, gasification	**115,898**	**115,350**	**164,400**[d]	**53,088**	**38,800**	**36,790**
Total	**333,624**	**365,356**	**676,397**	**151,200**	**78,500**	**65,690**

Sources: For 1979, 1980: Department of Energy, *Congressional Budget Request Program Overview for Fiscal 1981* (1980), pp. 25, 33, 34. For 1981: *Congressional Budget Request for Fiscal 1982*, pp. 25, 38, 39. For 1982: *Congressional Budget Request for Fiscal 1983*, pp. 35, 72. For 1983: *Congressional Budget Request for Fiscal 1984*, p. 39. For 1984: *Congressional Budget Request for Fiscal 1985*, pp. 51, 87.

a. Totals exclude other unobligated operating expenses.

b. Liquefaction totals include capital equipment.

c. An estimate. Funds were included in "support studies" category.

d. These projects continued, but they did not appear as line items in the budget.

e. These funds were taken away, so that the ultimate total appropriation for surface gasification in fiscal 1981 was $70,118,000.

The change in emphasis narrowed the department's focus, but required a substantial increase in appropriations. In 1978 DOE signed contracts for two liquefaction demonstrations, both using eastern coal and technology that had not worked in pilot plants. SRC-I was to produce a low-ash, low-sulfur, solid fuel in Newman, Kentucky. SRC-II was to produce fuel oil near Morgantown, West Virginia. The initial cost estimate for each plant was $700 million. However, by the end of 1980, the first year of design work, cost estimates were $1.9 billion for SRC-I and $1.4 billion for SRC-II. Private contractors agreed to put $100 million into each plant, the state of Kentucky agreed to contribute $30 million to SRC-I, and Japan and Germany each agreed to contribute 25 percent of the construction and operating costs (but not design costs) of SRC-II.[58] SRC-II, however, had a short life. In June 1981 the Reagan administration terminated SRC-II, with a rescission of unexpended funds.

In 1982 cost estimates for SRC-I exceeded $2 billion. The Department of Energy then asked the Rand Corporation to review the estimate, based on its extensive study of pioneer plant costs. The Rand analysts concluded that costs would probably be 26 percent to 32 percent over the last estimate and that extensive performance shortfalls could be expected.[59]

Congress approved termination activities at SRC-I in 1982, although "post-baseline activities" costing $28 million continued through 1984.[60] Only a small fraction of the large appropriations for these demonstrations was actually spent. Both were canceled before any construction or major procurement. Shortly after the decision to concentrate on near-term synfuels development came the Reagan policy to support only long-term, high-risk research. The February 1982 budget request characterized the federal government's "four basic energy responsibilities" as follows:

> Protecting against energy supply disruptions, primarily through planning and maintenance of the Strategic Petroleum Reserve.
>
> Providing support to long-term, high-risk, high-payoff research but not research and development for the purpose of accelerating new technologies into the marketplace.
>
> Supporting national defense needs through civilian control of research, development, production and testing of nuclear weapons.

58. GAO, *Controlling Federal Costs*, p. 3; and GAO, *Liquefying Coal*, p. 11.
59. Edward W. Merrow, "Analysis of the SRC-I Baseline," unpublished paper.
60. Department of Energy, "Solvent Refined Coal (SRC-I) Demonstration Plant," Fossil Energy Technology Fact Sheet, August 1983.

Performing tasks such as operation of Federal resource reserves, power marketing, and utility regulation as required by law.[61]

This policy was not applied consistently (see chapter 9 on breeder reactors), but it caused an upheaval in synfuels research. The budget explicitly ended proof-of-concept programs in coal.[62] The gas demonstration and the pilot programs in both gasification and liquefaction ended by 1982. The Ft. Lewis SRC pilot was dismantled, and the small Wilsonville facility was converted to testing indirect liquefaction processes.[63] Efforts were made to sell the large plants (H-Coal and EDS). HYGAS completed its experimental runs and then was used briefly to test peat gasification. BIGAS was finally completed and operated briefly. In 1982 and 1983 the Department of Energy had attempted to dismantle it, but funds continued to be appropriated as Congress considered the possibility of converting it to test medium-Btu projects. Eventually, BIGAS was terminated in June 1984.

All of these projects were considered by the Department of Energy to be appropriate for consideration by the Synthetic Fuels Corporation, and two DOE projects on medium-Btu gasification did receive SFC support.[64] As with the liquefaction demonstrations, most of the funds appropriated for gasification demonstrations were returned to the Treasury.

The Synthetic Fuels Corporation was technically distinct from the rest of the synfuels program. Because the government's research and development program had not yet yielded any liquefaction or SNG process that was remotely commercial, the SFC was expected to rely on western coal conversion and oil shale development. Some overlap was expected in the medium-Btu area. In 1980 Congress appropriated $18 billion for phase I of the synfuels commercialization plan. Because the SFC was not yet organized, $6 billion of this went to the Department of Energy to start private development incentives, and the remainder was placed in reserve. During 1981 DOE signed a $0.4 billion price guarantee with Union Oil for

61. Department of Energy, *Federal Energy Programs, FY 1983*, Budget Highlights, DOE/MA-0062 (February 1982), p. 1.

62. DOE, *Federal Energy Programs*, p. 1.

63. *Department of the Interior and Related Agencies Appropriations for 1986*, Hearings before a Subcommittee of the House Committee on Appropriations, 99 Cong. 1 sess. (GPO, 1985), pt. 7, pp. 870–900, 950–70.

64. *Department of the Interior and Related Agencies Appropriations for Fiscal 1983*, Hearings before a Subcommittee of the House Committee on Appropriations, 97 Cong. 2 sess. (GPO, 1982), pp. 15–27.

the Parachute Creek Shale Oil Project, and loan guarantees of $1.23 billion to the Oil Shale Corporation (TOSCO) project and $1.5 billion for the Great Plains Coal Gasification Project.[65]

President Carter's interim board for the SFC was not confirmed by Congress and resigned in January 1981. A permanent board was not seated until October 1981. In 1982 the SFC was declared operational, and the reserved funds (less $2 billion from a general budget rescission) and the TOSCO and Union Oil contracts were transferred to the Synthetic Fuels Corporation. The Great Plains project remained under DOE supervision.

In late 1983 the SFC lost a quorum of directors so it could no longer approve projects. This impasse lasted until September 1984, when Reagan agreed to appoint three new directors and Congress agreed to rescind an additional $5.38 billion from the SFC's appropriation and to remove the production goals. The reconstituted SFC board decided to support a few technologies at the demonstration level to obtain learning benefits, but not to augment or to compete with the energy supplies market. The plan was to build small plants of commercial scale in components whose scale up was uncertain—essentially a standard demonstration.[66]

The oversight committees in Congress were adamant that the Synthetic Fuels Corporation was not supposed to be in the research business. While the SFC insisted that it was not, the ambivalence of the new directors was fairly clear. For instance, the corporation stated that it would like to support advanced technologies, assuming that it was given a reasonable business proposition. The corporation would not support low-Btu gasification projects, and it canceled the coal-water solicitation because the technology was deemed commercial. The new strategy failed to save the SFC, and Congress rescinded the rest of the SFC's budget in 1985.

By 1985 the SFC had conducted four general solicitations and five competitive solicitations: oil shale, Gulf Coast lignite gasification, eastern coal gasification, coal or lignite gasification, and coal-water fuel. The initial solicitation, including the three proposals supported by the Department of Energy, had 71 respondents: 22 liquefaction, 17 gasification, 15 oil shale, and 17 tar sands, heavy oil, peat, hydrogen, and other. The SFC did not

65. U.S. Synthetic Fuels Corporation, *Comprehensive Strategy Report* (June 1985), pp. 13–18; and U.S. Synthetic Fuels Corporation, *1984 Annual Report, 1985*, p. 20.

66. See *Corcoran, Reichl, and MacAvoy Nominations*, Hearing before the Senate Committee on Energy and Natural Resources, 99 Cong. 1 sess. (GPO, 1985), especially pp. 22, 45–49, 127.

conclude agreements for any of these projects. The second solicitation had 38 responses (9 liquefaction, 7 gas, 7 oil shale, and 15 other); the third solicitation had 49 (11 liquefaction, 9 gasification, 13 oil shale, and 16 other); and the fourth solicitation received 18 (3 liquefaction, 6 gasification, 4 oil shale, and 5 other). Most of the proposals in later solicitations were modifications of earlier proposals, so that about one hundred distinct proposals were received by the SFC.[67]

Financial assistance agreements were reached for two projects. The Cool Water Coal Gasification Plant, which produces gas for a combined cycle power plant, uses an entrained bed medium-Btu gasifier. The SFC awarded Cool Water $115 million in price guarantees. The SFC also concluded a $620 million price guarantee for Dow Syngas, a gasification project producing medium-Btu gas for multiple purposes (power, heat, steam, feedstock).

The Synthetic Fuels Corporation was criticized from all sides during its entire history. It was accused of moving slowly to obstruct its purpose, in keeping with the Reagan administration's opposition. It was accused of mismanagement and excessive salaries, pensions, and benefits to its employees, directors, and consultants. The SFC's defense was that the proposals did not represent a good investment opportunity and that they were not sufficiently well developed to justify support. As the first projects selected by the Department of Energy progressed, SFC could justifiably point to them as examples of bad decisions made in haste. The TOSCO Project was canceled in 1982 and all outstanding loans repaid. A disaster for its private sponsors, Parachute Creek was completed but could not be made to operate. The U.S. government, however, got off cheaply. The price guarantee was not enforceable until commercial production began, and commercial production never occurred. In all, the SFC spent less than $100 million, so little compared with its grandiose conception and stupendous authorization that one member of Congress chided: "Are you saying you have only spent about a hundred million dollars? . . . Actually, you really have not spent any money, have you?"[68]

There are two postscripts to the SFC story. One is the Cool Water

67. U.S. Synthetic Fuels Corporation, "Solicitation History," May 4, 1984; and U.S. Synthetic Fuels Corporation, "Projects Awarded Financial Assistance," April 26, 1984.

68. *Department of the Interior and Related Agencies Appropriations for 1986*, Hearings before a Subcommittee of the House Committee on Appropriations, 99 Cong. 1 sess. (GPO, 1985), pt. 7, p. 488 (quoting Representative Tom Bevill).

"Integrated Gas Combined Cycle" Project, and the second is the Great Plains Coal Gasification Project. Both were completed, but with wildly different results.

The success story of the entire synfuels program is almost certainly Cool Water.[69] Expected to cost about $300 million, the project was completed at a cost of $263 million. It produces power at the designated capacity of 115 to 120 megawatts, and (except when used for experimental purposes) normally operates more than 70 percent of the time. Both capital and operating costs are equal or below the costs of standard coal-fired electric generation facilities that are equipped to satisfy new source emissions standards. The gasification process produces synthesis gas; both the gasification process and the combustion of the synthesis gas are used to produce steam to drive electric generation facilities.

Because the Cool Water process can be connected to standard gas turbine systems that run on fuel oil or natural gas, the technology is flexible and can be added to a utility's mix of generation plants relatively quickly and easily. Because it was intended to be a commercial demonstration, Cool Water uses western coal from Utah. Nevertheless, reflecting the bias in the synfuels program since its inception, experiments have also been run with eastern coal. The good news is that the plant actually can run on eastern coal; however, the bad news is that the eastern coal produces especially corrosive products that are quite damaging to the system. Further research is required before Cool Water technology can be economically operated with eastern coal.

The other legacy of the Synthetic Fuels Corporation is Great Plains, the success of which is perhaps indicated by the SFC's claim that it was really a project of the Energy Department. Blame aside, Great Plains turned out to be a classic technological turkey. In 1985 its private sponsors demanded a subsidy of $5.00 per million standard cubic feet (Mscf) so that the gas could be sold for $5.00/Mscf rather than what the sponsors claimed to be its actual cost of $10.00/Mscf. Although two to three times the price of gas in the region, the Federal Energy Regulatory Commission (FERC) agreed to let pipelines roll in the gas at this price. After elaborate negotiations, the

69. The information in this paragraph is derived from Wayne N. Clark and Vernon R. Shorter, "Cool Water: Mid-Term Performance Assessment," Sixth Annual EPRI Coal Gasification Contractor's Conference, October 1986; Ralph W. Grover and others, "Preliminary Environmental Monitoring Results, Cool Water Coal Gasification Program," National Meeting of the American Institute of Chemical Engineers, April 1986; and *Cool Water Coal Gasification Program: Fourth Annual Progress Report*, AP-4832 (Palo Alto, Calif.: Electric Power Research Institute, 1986).

price support deal fell through, and the Great Plains associates defaulted on their $1.5 billion guaranteed loan. Nevertheless, the plant was completed, and it did produce gas (albeit malodorous gas).

According to a study by the state of North Dakota, plant closure meant a loss of about 4,000 jobs in Mercer County, North Dakota, which has a population of 12,000, and an annual loss to the state of over $200 million.[70] North Dakota considered and rejected a "coal conversion tax," then led a lobbying effort to keep the facility open. The Department of Energy continued to operate the plant, while suing to force the pipeline companies to honor their contracts to buy the gas at the price previously approved by FERC.

The entire synfuels program had a quality of madness to it. Project after project failed. Cost estimates were connected to the price of substitutes rather than to the program itself. Goals were unattainable from the start. Official cost-benefit studies estimated net benefits in the minus billions of dollars. Even apart from the Synthetic Fuels Corporation, the dogged continuation of the research and development program seems incredible. However, despite the rhetoric, it is clear that the wildest scenarios never came close to happening. The nation's entire construction industry was not pressed into service to build plants that would not work in order to impress the Organization of Petroleum Exporting Countries (OPEC) of the United States' commitment to energy independence. Nevertheless, substantial funds were expended, and several patterns emerge.

First, the synfuels program expanded dramatically in response to the energy crises of 1973 and 1979. Project budgets jumped after both events, and project choices were apparently made in haste. The Clean Boiler Fuel demonstration, the H-Coal and BIGAS pilot plants, the SRC demonstration plants, and TOSCO and Great Plains were blamed on hasty decisionmaking and were regretted at leisure.

The crises also led to concentration on short-term payoffs from a technology, direct liquefaction, that was unlikely to work. The decision to concentrate on fuel oil for existing electrical generators was justified as an attempt to minimize the time before some payoff might take place. The choice of fuel oil may have been without alternative: the gasoline processes

70. Bill Rankin, "Energy Agency Plays the Spread on Great Plains Debt," *Energy Daily*, October 7, 1985, p. 4; Rankin, "American Natural Pays DOE for Great Plains Gas," ibid., February 10, 1986, p. 1; "The FERC Makes Abandonment Easier," ibid., February 27, 1986, pp. 1, 4; and Rankin, "Great Plains: Going, Going, Gone . . .," ibid., July 1, 1986, p. 1.

had not worked at all. Indirect processes received very little support, although they were technically more certain. The eastern coal processes that were the most likely to work technically (albeit not commercially) were for fuel oil. In this choice, and in the choices made earlier in the SNG program, a bad bargain is evident: produce a fuel that is a clear response to a crisis, is (maybe) technically feasible, and uses eastern coal.

Voting Analysis of the Coal R&D Program

Between 1975 and 1985 the House of Representatives cast roll call votes on different aspects of the synfuels program: loan guarantees and the Synthetic Fuels Corporation, the big liquefaction demonstrations (SRC-I and SRC-II), and a proposed oil shale demonstration (table 10-9). In addition to providing tests for the timing and nature of distributive and ideological support, the synfuels votes allow an opportunity to test for commonality and divergence of support for different program components and for the different research stages corresponding to research, development, and commercial demonstration. The variables in the regressions are described in figure 10-1, and the regressions are reported in table 10-10.

Expenditure-based Distributive Support

Expenditure-based distributive support apparently was not an important consideration in the politics of the synthetic fuels program. This conclusion may reflect the data limitations, but the result is not surprising given that expenditures on particular projects were never as large as they were for the supersonic transport, the space shuttle, and the Clinch River Breeder Reactor.[71] Beyond this general assessment, several interesting phenomena can be discerned in the estimates.

71. Our contract variables are less exact than is optimal for the purposes of this analysis. Four dummy variables identify districts with current and previous DOE liquefaction pilot plants and gasification pilot plants. Because of the small number of plants, we cannot further distinguish expenditure levels or the gasification districts by high- and low-Btu gasification projects. Both omissions are expected to reduce the significance of the estimates. The high-Btu versus low-Btu projects distinguish between speculative and more easily commercializable technology, so pooling may weaken results in that area as well. In the liquefaction area, small sample size problems do not make it possible to distinguish one SRC technology project from another.

We identify districts with SFC projects or active proposals, those that will have projects or proposals in the future, and those that had proposals previously, but withdrew them before the date of the vote. Because financial arrangements for each project are complicated, it is difficult to estimate either the expected cost of the project or the expected government subsidy. Consequently, we settle for a dummy variable, losing measurement possibilities for threshold effects.

Table 10-9. *House Roll Call Votes on Synfuels Measures, 1975–85*

Year	*Bill number*	*Substance of measure*	*Vote*
1975a*	HR3474	Motion to delete from conference version of energy research authorization bill $6 billion in loan guarantees for commercial development of synfuels.	Passed, 263–140
1975b*	HR3474	Motion to delete from conference version of energy research authorization bill the authorization for ERDA to conduct an oil shale demonstration program with a private company.	Passed, 183–117
1976	HR12112	Adoption of rule to permit House floor consideration of bill authorizing $3.5 billion in loan authority and $500 million in price supports for synfuels development.	Rejected, 192–193
1978	HR12163	Amendment to authorization bill to add $75 million for initial construction costs for SRC-II.	Passed, 165–132
1979	HR3930	Defense Production Act.	Passed, 368–25
1980	S932	Adoption of conference report on Synthetic Fuels/Defense Production Act.	Passed, 317–93
1981	HR4035	Amendment to Interior Department appropriations bill to delete $135 million for SRC-I.	Rejected, 177–236
1984a*	HR5973	Adoption of rule to consider Interior Department appropriations bill that would not have waived points of order against amendments to rescind funds for Synthetic Fuels Corporation.	Rejected, 148–261
1984b*	HR5973	Amendment (rescinding $5 billion from SFC appropriation) to the amendment to rescind $10 billion from SFC appropriation.	Passed, 236–177
1985	HR3011	Amendment to Interior Department appropriations bill to rescind all but $500 billion from previously appropriated amounts to SFC.	Passed, 312–111

Sources: *Congressional Quarterly Almanac*, various issues.
* The letters a and b are used to distinguish between votes in the same year (see table 10-11, columns 1 and 2).

Table 10-10. *Annual Regression Results, Logit Analysis*[a]

	Votes				
Variable	*Loan guarantees, 1975*[a]	*Oil shale demonstration, 1975*[b]	*Loan guarantees, 1976*	*SRC-II, 1978*	*Defense Production Act, 1979*
Percent voting for synthetic fuels	35.4	30.0	49.9	55.9	93.6
$-2L$	455.42	410.74	460.2	349.1	142.8
Degree of freedom	382	398	395	370	282
Constant	−0.92 (0.70)	−1.60 (0.77)	−0.35 (0.67)	0.38 (0.75)	1.04 (1.55)
Party	2.42 (0.84)	0.95 (0.96)	2.04 (0.65)	0.67 (0.78)	19.05 (6.72)
ACA-R	−0.019 (0.010)	0.003 (0.01)	−0.005 (0.008)	−0.018 (0.010)	−0.21 (0.07)
ACA-D	0.034 (0.006)	0.044 (0.006)	0.025 (0.006)	0.021 (0.007)	0.002 (0.012)
Sci & Tech	1.18 (0.71)	0.745 (0.72)	1.37 (0.71)	0.98 (0.59)	[c]
FSF	. . .	. . .	. . .	. . .	. . .
Approp	0.86 (0.82)	2.15 (0.88)	0.75 (0.79)	0.93 (0.85)	[c]
E-coal	0.52 (0.029)	0.0002 (0.03)	−0.012 (0.028)	0.051 (0.045)	−0.031 (0.061)
Low W-coal	1.12 (0.51)	0.32 (0.53)	0.97 (0.64)	0.50 (0.64)	1.036 (0.84)
Med. W-coal	−0.121 (0.76)	0.41 (0.80)	0.065 (0.87)	1.16 (1.18)	[c]
High W-coal	−0.069 (0.95)	2.314 (0.96)	0.53 (0.97)	0.67 (1.37)	−3.39 (1.38)
Oil res.	0.15 (0.11)	0.046 (0.11)	0.046 (0.11)	0.023 (0.135)	−0.41 (0.16)
Oil con.	−0.011 (0.006)	−0.008 (0.007)	−0.009 (0.006)	−0.006 (0.007)	0.024 (0.016)
Liquid PP	0.46 (0.82)	1.47 (0.77)	1.45 (0.92)	1.62 (1.40)	−0.13 (1.26)
Gas PP	0.18 (0.93)	−0.034 (0.98)	1.66 (1.17)	−0.84 (1.34)	[c]
Past Liq PP	. . .	. . .	. . .	. . .	. . .
Past Gas PP	. . .	. . .	. . .	. . .	. . .
Fut SFC Prop	0.24 (0.34)	−0.61 (0.39)	0.63 (0.35)	0.42 (0.45)	2.02 (1.09)
SFC Prop	. . .	. . .	. . .	. . .	. . .
Past SFC Prop	. . .	. . .	. . .	. . .	. . .

(continued)

Table 10-10 *(continued)*

Votes					
Defense Production Act, 1980	*SRC-I, 1981*	*Adoption of rule, 1984*[a]	*Amendment to rescind $5 billion, 1984*[b]	*SFC rescission, 1985*	*Pool (1975*[a]*, 1976, 1979, 1980, 1985)*[b]
76.9	57.0	36.1	57.1	25.5	. . .
302.0	509.9	409.9	451.0	427.0	2,044
401	400	395	399	406	2,010
3.52 (1.04)	0.21 (0.64)	−1.13 (0.72)	0.38 (0.69)	−1.65 (0.69)	0.83 (0.33)
3.96 (1.58)	0.32 (0.52)	1.46 (1.09)	−0.40 (0.81)	−0.21 (0.83)	1.97 (0.38)
−0.096 (0.019)	−0.001 (0.006)	0.001 (0.01)	0.005 (0.009)	−0.010 (0.010)	−0.026 (0.004)
−0.045 (0.011)	0.03 (0.009)	0.028 (0.007)	0.052 (0.01)	−0.008 (0.008)	0.015 (0.003)
1.81 (1.09)	−0.28 (0.48)	−0.002 (0.58)	0.92 (0.57)	−0.164 (0.60)	0.84 (0.29)
. . .	. . .	−2.80 (0.99)	−0.16 (0.62)	−0.44 (0.67)	. . .
−0.71 (0.99)	1.45 (1.10)	3.41 (1.12)	2.30 (1.09)	1.09 (0.75)	0.45 (0.40)
−0.014 (0.038)	0.17 (0.06)	0.078 (0.032)	0.073 (0.038)	0.11 (0.03)	0.042 (0.014)
−0.033 (0.61)	0.62 (0.54)	1.07 (0.56)	1.00 (0.55)	0.43 (0.52)	0.45 (0.26)
0.66 (0.96)	1.71 (1.13)	1.24 (0.76)	1.17 (0.88)	1.14 (0.63)	0.55 (0.39)
−0.69 (1.05)	−0.29 (0.91)	−1.24 (1.36)	−1.03 (1.12)	0.29 (1.00)	−0.23 (0.48)
−0.30 (0.13)	0.024 (0.11)	0.21 (0.14)	0.32 (0.16)	0.27 (0.13)	−0.015 (0.05)
0.004 (0.009)	−0.007 (0.005)	0.002 (0.006)	−0.011 (0.006)	0.005 (0.006)	−0.003 (0.002)
[c]	0.84 (1.43)	. . .	. . .	. . .	0.50 (0.51)
[c]	−0.70 (1.15)	. . .	. . .	. . .	1.35 (0.59)
−0.45 (1.19)	0.13 (0.96)	0.43 (0.74)	−0.022 (0.76)	0.81 (0.66)	. . .
−0.16 (1.52)	−0.43 (1.51)	−0.57 (0.95)	−0.39 (0.87)	−0.054 (0.87)	. . .
0.60 (0.45)	0.35 (0.76)	−0.33 (0.70)	1.18 (0.85)	. . .	0.38 (0.19)
. . .	0.47 (0.37)	1.43 (0.60)	1.14 (0.69)	0.87 (0.45)	1.13 (0.40)
. . .	. . .	−0.69 (0.52)	−1.29 (0.52)	0.16 (0.42)	. . .

a. The numbers in parentheses are standard errors. For a description of variables and votes, see figure 10-1.
b. Dummy 1976: 0.63 (0.15); dummy 1979: 2.96 (0.24); dummy 1980: −1.66 (0.24); dummy 1985: −2.35 (0.17).
c. All members voted for synthetic fuels.

Figure 10-1. *Description of Variables*

Party: 1 if Republican, 0 otherwise.

ACA-R: Americans for Constitutional Action score if Republican, 0 otherwise.

ACA-D: ACA score if Democratic, 0 otherwise.

Sci & Tech, FSF, Approp: House Science and Technology Committee, Fossil Synthetic Fuel Subcommittee of the House Committee on Energy and Commerce, and House Appropriations Committee: 1 if member of relevant committee or subcommittee, 0 otherwise.

E-coal: 1,000 short tons of coal reserves in state divided by number of districts in state with reserves; 0 for districts in state without reserves. Eastern coal (bituminous) only.

W-coal: per district reserves calculated as for *E-coal* for western coal (lignite, subbituminous) states. Low = 1 if $0 < \text{reserves} < 12$; med. = 1 if $12 < \text{reserves} < 40$; high = 1 if $\text{reserves} > 40$.

Oil res.: estimated oil reserves in state divided by number of districts and scaled: 0 if 0-100 (million barrels); 1 if 100–500; 2 if 500–1,000; 3 if 1,000–2,000; 4 if over 2,000.

Oil con.: index of 1972 per capita consumption of petroleum products by state, aggregate use by all sectors (U.S. Btu per capita set at 100).

DOE–SFC: Department of Energy–Synthetic Fuels Corporation project variables: 1 if district contains, will contain, or did contain pilot plant or demonstration, proposal, or project of given type.

Liq. PP: Department of Energy Liquefaction Pilot Plant variables: 1 if district contained a project; 0 otherwise.

Gas PP: Department of Energy Gasification Pilot Plant variables: 1 if district contained a project; 0 otherwise.

SFC Prop: 1 if a proposal from the district was pending at the Synthetic Fuels Corporation; 0 otherwise.

Fut SFC Prop: 1 if at sometime in the future a proposal from the district would be pending at the Synthetic Fuels Corporation; 0 otherwise.

Past SFC Prop: 1 if at sometime in the past a proposal from the district had been pending at the SFC; 0 otherwise.

Sources: For votes, *Congressional Quarterly Almanac,* various years. For ACA scores, *Congressional Quarterly Weekly Report,* selected issues. For committee membership, Michael Barone and others, *Almanac of American Politics* (Dulton, various years). For coal statistics, *Synthetic Fuels*, Report by the Subcommittee on Synthetic Fuels of the Senate Committee on the Budget (GPO, 1979), p. 159. For oil reserves, Department of the Interior, *Geological Estimates of Undiscovered Recoverable Oil and Gas Resources in the United States,* Geological Survey Circular 725 (1975), pp. 74–75. For oil consumption, Irving Hoch, *Energy Use in the United States by State and Region,* Research Paper R-9 (Washington: Resources for the Future, 1978), p. 164. For Department of Energy project variables, see table 10-8. For SFC project variables, U.S. Synthetic Fuels Corporation, "Solicitation History," May 4, 1984; April 26, 1984; July 9, 1984; and U.S. Synthetic Fuels Corporation, *Comprehensive Strategy Report* (June 1985).

The coefficients of the pilot plant and SFC variables suggest weak to nonexistent coalitional support for separate program components based on a common interest or a synfuels log roll. Districts with liquefaction plants were likely to support the SRC demonstrations, unlike gasification districts or districts with either future or actual SFC solicitations. Districts with gasification plants show more enthusiasm for the SFC program than those with liquefaction plants. This is expected because gasification projects were closer to commercialization, and some were expected to receive support from the Synthetic Fuels Corporation. Districts with past pilot plants have even weaker or negative coefficients. Thus, such support as there was depended on the existence of an active project.

The SFC program fits the retrospective model shown in the other cases. Districts with future proposals are somewhat more prone to support the SFC. Districts with active proposals are much more likely to support SFC budgets. Those with past SFC projects are not apparently interested in the program at all. Because retrospective distributive support is very evident in these votes, the lack of active solicitations is significant in the demise of the corporation.

Actual expenditures on synfuels projects were small, although potential expenditures were enormous. According to some, the programs were later scuttled with unseemly haste. Thus we expect to find a pattern of support that differs from the Clinch River situation: while the potential for expenditure-based distributive benefits to enhance political support may have encouraged haste in project selection, the low rate of expenditures before the programmatic bad news should have mitigated the extent to which distributive support could overcome shifts in program fortunes.

Resource-based Distributive Support

The beneficiaries of successful synfuels development fall into three groups: coal and shale interests, the industrial firms that received subsidies to develop technology, and consumers.[72]

72. Eastern coal (*E-coal*) is a continuous variable. The deposits per district (for both eastern and western coal) are approximated by dividing the total reserves in the state by the number of districts containing coal within the state. We use three dummies for western coal states: those districts with moderate, large, and enormous reserves. As is discussed in chapter 5 with relation to funding variables, this specification is superior to a continuous variable because of the large variance but relatively coarse potential response (that is, vote yes or vote no). Oil resources enter the equation as a step function, described in figure 10-1, for the same reason. Oil consumption data are available on a per capita–state basis; this variable is a 1978 index, with the national average set at 100. New England states tend to be at the high end of the index, followed by southern states and northern and midwestern

COAL. The coefficients of the eastern coal variable are positive as predicted, but not so strong as to suggest that the synfuels program was a major priority among these districts. Support on the first, hasty loan guarantee bill in 1975 is positive and significant. No relation exists between eastern coal and the oil shale vote: such support would exist only if there was a synfuels logroll, for there is neither common technology nor a common resource base between coal and shale. For SFC votes in 1976, 1979, and 1980, the coefficients are all negative but not significant. These results are consistent with the view that early commercialization would focus on western coal. The SRC votes both get positive response, and the SRC-I coefficient is significant. Interestingly, the coefficients reverse for the SFC votes in 1984 and 1985: eastern coal districts show significant support for the program in its later years. This change coincides with the demise of the coal research program at the Department of Energy and the reorientation of the SFC program toward more developmental investments, which are more likely to use eastern coal.

The western coal group displays ambivalence toward synfuels. Support cannot be detected prior to the establishment of the SFC. Some retrospective support is shown in the votes in the 1980s; however, the districts with the largest reserves come out (weakly) opposed to the program. Thus the persistent focus of synfuels programs on less attractive eastern coal is broadly consistent with positive political support in the East but ambivalence in the West.

INDUSTRY. Industrial developers of synthetic fuels technology are difficult to identify. We include oil resources as an independent variable in order to detect the influence of one potential beneficiary. Many oil companies were involved in the development of synfuels from either coal or shale, and presumably their influence is greater in areas with oil reserves. In addition, the loan guarantee program was denounced in both the 1975–76 and 1979–80 debates as a boondoggle for oil companies, who were at that time characterized as the bad guys in the energy crisis. Of course, synfuels may be regarded as competitors to oil and so opposed by oil state representatives. The sign and magnitude of the coefficient on oil reserves thus indicate the actual resolution of these conflicting effects.

Oil resources do not correlate with SFC or DOE support until the last votes. These results parallel the support for the Clinch River program by

states with coal reserves. Southern and southwestern states with natural gas reserves are at the bottom of the index. The variable does not correlate highly with oil reserves.

nuclear utilities. Consequently, the pattern can be similarly interpreted: resource competition was not an issue, and lobbying efforts intensified when the program lost general support. In addition, the pattern of coefficients supports the general retrospective voting hypothesis. While the SFC loan guarantee program may have benefited oil companies, they did not constitute a source of support before the program began. Once the oil companies became involved with the government in planning projects, their interest in maintaining the program increased.

CONSUMERS. The economic rationale of synfuels was the displacement of petroleum. Consequently, geographic regions that consume more oil per capita disproportionately benefit from commercial availability of synfuels. For the reasons given in chapter 4, we ordinarily would not expect consumers to be a significant source of program support, especially for the more research-oriented parts of the program. To the extent that the commercialization program was considered a reasonable short-term strategy for energy supply enhancement, consumers might contribute to political support for it. Failing an optimistic assessment of the program's commercial potential, this group still might influence votes if OPEC created a situation in which our ordinary view of fragmented interest groups no longer applied: that is, if the oil issue became so salient that politicians were obliged to respond to a populist sentiment with what was viewed as an effective solution to the shortage of liquid petroleum.

As expected, oil consumption has little to do with the SFC program. Because the coefficients are negative, although insignificant, on all the substantive votes except for those in 1979 and 1980, they suggest that the 1979 and 1980 votes departed from normal voting patterns. Otherwise, the pattern reflects the nonsupport expected because of either the pessimistic prospects for synfuels technology or the long-term nature of the benefits to this heterogeneous group.

Party and Ideology

An examination of the party and ACA (Americans for Constitutional Action) variables indicates that the synfuels program involved considerations that cannot be placed along the single-dimensional ideology scale defined by the ACA. The relation between ACA and synfuels support among Democrats is straightforward. For all votes except the Energy Security Act in 1979 and 1980, and the final vote on the SFC in 1985, conservative Democrats were more likely to support the program than liberals, with support increasing monotonically with ACA score. As with the other energy research programs considered in this book, conservatives support "hard" energy options. The exceptions bear special note. The 1979 bill passed with over 93 percent of the vote, including 95 percent of voting Democrats. In 1980 the coefficient on ACA for Democrats is negative and significant. Eighty-nine percent of the Democrats in Congress voted for the 1980 bill; the only exceptions were conservatives. Our interpretation of these votes (consistent with the results reported above for oil consumption) is that they fell out of the mold of "hard" energy option politics and were simply a response to the perceived energy crisis. The final vote also shows a breakdown of the conservative Democratic coalition.

The situation is more complicated for Republicans.[73] In general, the liberal wing of the Republican party supported the synfuels program—to a greater extent than did Democrats with the same ACA score—while support fell off among conservatives. Very conservative Democrats (a small set) were more likely to vote for the synfuels programs than were their Republican counterparts.

To investigate further ideological support among Republicans, a model was estimated that divides Republicans into two groups based on a cutoff value for ACA score. For each year a best cutoff was estimated using a Chi-squared criterion.[74] The results of these estimations for all but the 1979 and 1980 votes on the Energy Security Act are shown in table 10-11. The remaining variables were included in the estimations but are not reported since they did not change from table 10-10.

As table 10-11 shows, the ACA coefficients for Republicans with scores less than the cutoff are positive or approximately zero in all years; those for Republicans with ACA scores higher than the cutoff are negative and significant for all SFC votes and insignificant for the demonstration plants. The two Energy Security Act votes in 1979 and 1980 are omitted because no cutoff was discernible. As with Democrats, the synfuels program was supported by the liberal wing of the Republican party, indicating that a simple negative correlation exists with ACA score.

73. Using a Chi-square test, we reject the hypothesis that the coefficients for *ACA-D* and *ACA-R* are equivalent in the votes.

74. The cutoff maximizes the difference in (-2) times the likelihood ratios for the equations in table 10-10 and those including the additional two variables. The cutoffs vary from year to year because of differences in the votes chosen by the ACA for scoring purposes. In addition, the coefficients of the two *ACA-R* coefficients vary because the ACA scale obtains a maximum value at 100, so that a change in magnitude, but not in sign or significance, is not necessarily important. For the 1981 equation we used a local maximum at the cutoff value of 71. The global maximum for this equation is obtained at 88; however,

Table 10-11. *Results of the Analysis of Party and Conservatism of Support for the Synfuels Program*[a]

Vote	*Cutoff value (C)*	χ^2	Party	1 if ACA>C	ACA-R (ACA<C)	ACA-R (ACA≧C)
1975a	79	3.3	1.71	7.67	−0.007	−0.097
			(1.13)	(4.3)	(0.02)	(0.048)
1975b	60	5.1*	−4.04	6.82	0.11	−0.019
			(3.2)	(3.6)	(0.06)	(0.02)
1976	70	12.2**	0.58	9.23	0.029	−0.09
			(0.85)	(3.1)	(0.018)	(0.03)
1978	69	3.0	−0.49	0.72	0.014	−0.014
			(1.1)	(2.8)	(0.02)	(0.03)
1981	71	4.6*	−3.5	0.058	0.019	0.004
			(0.63)	(2.4)	(0.012)	(0.03)
1984a	75	7.4**	0.93	2.06	−0.067	−0.048
			(1.7)	(3.5)	(0.039)	(0.035)
1984b	75	3.8	−0.53	4.0	0.003	−0.038
			(1.34)	(2.6)	(0.02)	(0.025)
1985	75	2.6	−0.41	5.42	−0.009	−0.07
			(1.1)	(3.6)	(0.02)	(0.04)

*Significant at the 5 percent level.
**Significant at the 1 percent level.
a. The numbers in parentheses are standard errors. For a description of variables, see figure 10-1.

Ideological considerations regarding the synfuels program appear to have at least two dimensions. One dimension is conservatism, as measured by the ACA, which when applied to synfuels reflects an ideological tilt toward technocratic supply-enhancing solutions to energy shortages. This aspect of conservatism is shared by both political parties, and the party factor indicates a shift in favor of this rationale for synfuels investment among Republicans. The second dimension could be characterized as opposition to government interference in the market. It correlates with ACA conservatism among Republicans, but not among Democrats. Thus the SFC program, which is technocratic but interventionist, lost support among the most conservative Republicans.

This interpretation is borne out in the results for the demonstration votes. On pro-market grounds, we would expect conservative Republicans to be less opposed to a government demonstration program than to a federally subsidized quasi-private commercialization program. This result

we rejected this equation because it appeared to be dominated by the effects of the boundary constraint on the ACA scale.

is observed in all three votes on demonstration plants. Thus the regression results strongly support the hypothesis.

Committee Variables

The support of the oversight and appropriations committees for synthetic fuels programs is spotty. The Fossil and Synthetic Fuel Subcommittee of the House Committee on Energy and Commerce was opposed to the SFC program. The House appropriations subcommittee's lack of support for the loan guarantee program is not particularly important, since the proposals in 1975 and 1976 were not appropriations bills and had not been previously authorized. However, its lukewarm support for the SRC plants suggests the lack of an institutional support base within Congress for the synfuels program: the only votes in which its support is clear are the votes in 1984 to amend the appropriations bill. The House Science and Technology Committee coefficients show a similar ambivalent pattern.

Pooling

Selected pooling results are presented in table 10A-8. The votes on the loan guarantee program in 1975 and 1976 pool, but not with any other votes, including the oil shale vote in 1975. The two SRC votes and the 1985 SFC rescission vote pool. Otherwise the SFC votes do not pool with each other, with the demonstrations, or with the votes in 1984. The two 1984 votes do not pool, and we reject pooling as well for the coal and shale demonstration votes.[75]

These results confirm the lack of common interest among program components. The rationale for government-supported research for very speculative, long-term development projects is different from that for demonstrations, and both differ from the rationale for commercialization.

75. Lack of pooling between the two 1984 votes and between each and the 1985 vote indicates that these votes did involve other issues or sophisticated voting strategies. Different strategic maneuvering is possible, given expectations about the ultimate outcomes. For example, if a member favored the full appropriation, and calculated that a $10 billion reduction would fail against the bill, whereas a $5 billion reduction would pass, he or she would vote against the amendment to the amendment. In this analysis, this vote would be coded as an antisynfuels vote. A member opposed to the program having the same expectation would vote in favor of the amendment to the amendment; the vote would be similarly miscoded. The 1984 vote on the rule to allow amendments involves considerations apart from synfuels. Even a member opposed to the SFC might vote against the rule because it challenged the decision of the Rules Committee. We excluded the various amendments to the Energy Security Act in 1979 from this analysis because they too did not pool with other votes or present evidence on support during a different year.

Consequently, the lack of a common political base, reflected in particular in the ideology coefficients, reflects differences in acceptance of these rationales.

The absence of common support among districts that hosted different program components has several possible explanations. In addition to lack of a pork barrel logroll, one reason why districts with interests in research on advanced synfuels technologies may not have supported moving to the commercialization phases of the program is because the troubles at the R&D phase were more apparent where research and development were being performed and because commercialization and demonstration may have been perceived as limiting resources for research.

Differences in support for the program components are observed in resource-based distributive support (eastern coal supports the SRC plants), and in institutional support in Congress (the House appropriations subcommittee was consistent in its support of the demonstrations). Finally, the estimates show that the commercialization program was subject to different politics in 1979 and 1980 than in the earlier or later periods. Liberals in both parties strongly supported the bills, as did representatives from oil-consuming districts. About the only sour note to the 1979 consensus was the opposition of western coal regions with the richest coal deposits.[76]

Lack of a common support base is a good news/bad news phenomenon. Common support for synfuels would modify the expected gains from investing in a mixed strategy program. Different projects, instead of competing for resources, would collaborate. The absence of a logroll suggests that the mixed strategy approach was not undermined by pork barrel considerations. But if support groups differ among program stages, then there is further reason to expect a federal research effort to diverge from the hypothetical ideal. In particular, the program would lack the coherent view of research and development presented in chapter 3.

The final column in table 10-10 shows the results of the regression pooling the loan guarantee–SFC votes of 1975, 1976, 1979, 1980, and 1985. The dummy variables assume a value of one for the year denoted in the column and all years thereafter (that is, each dummy estimates an incremental change from the previous vote). These reflect a trend that is

76. We did not include a separate variable for oil shale, as all the relevant deposits are located in three congressional districts. These districts are in the states with large coal deposits, which may explain the positive coefficent for the high western coal variable for the oil shale demonstration vote. Concerns over rapid development of oil shale lands contribute to the negative coefficient of high W-coal in 1979.

also clear in the vote totals: support increased slightly from 1975 to 1976, increased dramatically in 1979, fell slightly in 1980, and dropped in 1985 to below the 1975 base. This trend does not follow the raw estimates of the cost of synfuels. While closer to the trend in oil prices, the change from 1979 to 1980 and the plunge in 1985 do not fully reflect imported oil prices. It is, however, possible to assume oil price scenarios and calculate program cost-benefit estimates (see table 10A-7) that closely resemble the trend.

Conclusion

For many years the U.S. government maintained a small research program on synthetic fuels from eastern coal. The program lacked political visibility, commercial pressures, and certainly significant governmental support. In a sense it was the archetypical government research and development program. Although the projects did not appear suited to eastern coal use, they would have been the most profitable had they worked out. For fairly small expenditures, the government tested very speculative research that was not of interest to private investors but was a long shot to benefit the declining eastern coal industry.

Despite poor results, the initial program became the basis for three major expansions motivated by energy shortages. The first was the shortage of natural gas. Faced with declining pipeline use and new pressures for cost-sharing projects with industry, the pipeline companies and the Office of Coal Research joined forces to investigate synthetic natural gas from eastern coal. The alliance was not a happy one, first because no project succeeded in making high-Btu gas from eastern coal and later because natural gas became plentiful; however, the coal research program managed to find new sponsors.

The second major expansion of the synfuels program followed the 1973 oil embargo. The Office of Coal Research received large budget increases. It responded to the energy concerns, as well as to the results of the previous program, by shifting to lower grade fuels that were considered to have a higher likelihood of near-term commercial availability. The shift to medium-Btu gas and fuel oil still represented a compromise between eastern coal interests, the existing institutional framework and support coalition for the OCR, and the new energy situation.

The 1979 oil crisis—the catalyst for the third expansion—created a temporary consensus in favor of synfuels development as well as numerous other energy policies. The resulting program headed west with the creation of the Synthetic Fuels Corporation, while simultaneously attempting to increase the pace of development at the Department of Energy, which had inherited the OCR program. The department again tried to shift investment to near-term options. Both the SFC and DOE programs, however, quickly fell apart because of administration opposition and changed circumstances regarding imported oil.

Although individual projects in the R&D program had the pork barrel character of some of the other programs discussed in this book, the program as a whole did not. We attribute this in part to luck. The program expanded so rapidly during the 1970s that unwise investments were easily accommodated, but it did not expand rapidly enough for the solvent-refined coal plants, the other big gasification demonstrations, or the SFC projects (with the exception of Great Plains) to become important for pork barrel reasons. While the potential for retrospective support may have encouraged the hasty decisions that were evident at the Office of Coal Research and the Department of Energy following the big budget increases, the failure to start massive projects probably saved the program from being a much larger fiasco than it was.

The analysis of congressional votes indicates that no significant logroll existed among districts on the DOE project. This result is in accord with the view that logroll coalitions need institutionalization to persist. In particular, the bills presented to the full Congress must include all pieces of the logroll. The voting agenda depends on the oversight and appropriation committees, which in this case were not committed to a multifaceted synfuels program. Thus, without institutional support, a mixed strategy program is not likely to exhibit the political rigidities of a single-project program that has the same cost and an equivalent number of districts with project pieces or subcontractors.

Our analysis of roll call votes demonstrates that ideological support for different segments of the program also differed. An R&D commercialization program faces different ideological opposition at different stages of its organization and development. In this case the opposition to commercialization programs among conservative Republicans, typically part of the coalition in favor of research expenditures on hard technologies, underscores the fragility of the original coalition upon which the Synthetic Fuels Corporation was founded. As a crisis response, it relied on support from liberal Democrats and free-market conservative Republicans—groups typically opposed to such programs. That support evaporated quickly, leaving no majority coalition in favor of the program.

Two views have been voiced regarding the crisis-response energy policy that resulted in the establishment and rapid demise of the SFC. According to one view, the fall in support for the SFC shows that the government makes decisions based on short-run circumstances and point estimates from very speculative projections. Thus the decision to begin several projects was a knee-jerk reaction to the 1973 and 1979 oil crises, and the subsequent decision to kill the program did not account for potential future embargoes and the "insurance policy" aspect of synfuels development. The second view is that the government acted efficiently in light of the best available information. Oil prices fell dramatically from 1980 to 1985. Meanwhile, projected synfuel costs were climbing, so that energy demand responses or other technologies were better suited to coping with future petroleum shortages.

It is difficult to argue with the second view. Certainly, the commercialization program cannot be justified on efficiency grounds, although this was as true in 1979 as in 1985. The technical and economic problems with synfuels are indisputable. A fraction of the cost increase is from substantial increases in coal prices, owing primarily to environmental and safety controls and secondarily to a response to the increased price of oil. Most of the increase, however, is in capital costs. Although construction costs increased during the 1970s at nearly twice the rate of inflation, these estimates surpass such considerations. Rather, they demonstrate their proponents' ubiquitous optimism about undeveloped processes. As a result, increasingly pessimistic cost estimates occurred simultaneously with the growth of the program. Thus in a convoluted way the cost increases were caused by the oil crises: the government's program intensified with oil price increases and resulted in information that led to upward revision in cost estimates.

Beyond its almost farcical economic basis, the program also demonstrates the difficulty of government support for long-term commercial development. Because many efficiency and environmental issues cannot be projected from pilot plants, commercialization of synthetic fuels required an expensive development program. Because part of the justification for the program was to develop a "backstop" technology that would hold down imported oil prices and protect against supply disruptions, the government would probably need to pay for at least part of a program were it desirable on efficiency grounds. But such a program would require stable coalitional support or, what might be worse, significant distributive benefits. That the synfuels program lacked both meant a disruptive history.

Table 10A-1. *Average Price of Imported Oil and Natural Gas, 1970–83*[a]

	Oil[b]		*Wellhead gas*[c]		
Year	*Current dollars*	*1972 dollars*	*Current dollars*	*1972 dollars*	*Delivered gas,*[c] *1972 dollars*
1970	2.16	2.36	0.171	0.187	0.70
1971	2.31	2.41	0.182	0.190	0.71
1972	2.47	2.47	0.186	0.186	0.73
1973	3.30	3.12	0.216	0.204	0.75
1974	10.81	9.39	0.304	0.264	0.83
1975	11.19	8.90	0.445	0.354	1.02
1976	12.15	9.18	0.580	0.438	1.21
1977	13.81	9.86	0.790	0.564	1.41
1978	14.12	9.39	0.905	0.602	1.45
1979	18.72	11.46	1.178	0.721	1.54
1980	30.60	17.15	1.588	0.890	1.76
1981	34.28	17.57	1.982	1.016	1.87
1982	31.48	15.22	2.457	1.188	2.15
1983	28.58	13.25	2.593	1.215	2.37

Sources: American Petroleum Institute, *Basic Petroleum Data Book, Petroleum Industry Statistics*, vol. 4, no. 3 (September 1984), sec. 4, table 14; and American Gas Association, *Gas Facts 1984: A Statistical Record of the Gas Utility Industry*, p. 112, table 9-5.

a. All deflation uses the GNP price deflator.

b. Dollars per barrel.

c. Dollars per thousand cubic feet.

Table 10A-2. *Synthetic Gas Price Estimates by Type of Process, 1969–83*[a]

Cents per million Btus (1972 dollars)

Year	*BIGAS (entrained bed SNG)*	*CO_2 Acceptor (lignite; fluid bed hydro-gasification, SNG)*	*HYGAS, steam oxygen (fluid bed hydro-gasification, SNG)*	*Lurgi (lignite; fixed bed, SNG)*	*Synthane (fluid bed SNG)*	*Medium-Btu gas*[b]	*SNG*
1969	66	44	65	...	...	...	...
1971	...	...	46–78	...	...	...	...
1972	79	...	...	...	...	...	...
1973a	99	68	103	113	...	...	...
1973b	...	...	...	123	...	...	...
1975a	129	...	...	...	...	...	...
1975b	...	...	130	149	...	...	...
1975c	...	...	...	273	...	...	...
1975d	...	...	...	...	...	...	223
1976	248	205	226	334	257	...	...
1977	...	...	...	336	...	...	...
1978a	279	265	219	333	382	...	...
1978b	...	...	...	...	...	333	...
1978c	507	...	403	...	...	...	...
1979a	...	...	...	...	...	295	357
1979b	...	...	...	...	...	216	274
1981	...	...	...	...	...	343	...
1983	...	...	...	...	...	256	...

Sources: For 1969: C. L. Tsaros and T. J. Joyce, "Synthetic Pipeline Gas Prices Calculated to Settle Within Range of 36.2–58.2 ¢/Mcf," *Oil and Gas Journal*, February 1969, pp. 80–83. For 1971: Dean Hale, "Large Scale High Btu Gas from Coal on Its Way," *Pipeline & Gas Journal*, March 1971, pp. 41–46. The range is for western coal at 8 cents per million Btus to eastern coal at 24 cents per million Btus. For 1972: Sidney Katell, "An Economic Evaluation of Pipeline Gas from Coal," *Gas*, August 1972, pp. 32–34. For 1973a: D. C. Mehta and B. L. Crynes, "How Coal-Gasification Common Base Costs Compare," *Oil and Gas Journal*, vol. 71 (February 5, 1973), pp. 68–71. For 1973b: Ted Wett, "SNG from Coal Involves Big Projects," *Oil and Gas Journal*, vol. 71 (June 25, 1973), pp. 131–34. For 1975a: "Bi-Gas," *Hydrocarbon Processing*, April 1975, p. 119. For 1975b: K. C. Vyas and W. W. Bodle, "Coal and Oil-Shale Conversion Looks Better," *Oil and Gas Journal*, vol. 73 (March 24, 1975), pp. 45–53. For 1975c: Ogden Hammond and Martin B. Zimmerman, "The Economics of Coal-Based Synthetic Gas," *Technology Review*, July/August 1975, pp. 43–51. Estimate for western Lurgi plant, excludes shipping cost. For 1975d: Synfuels Interagency Task Force to the President's Energy Resources Council, *Recommendations for a Synthetic Fuels Commercialization Program*, vol. 1, Overview Report (November 1975), stock no. 041-001-00112-3, p. 33. For 1976: Sidney Katell and Luella G. White, "Clean Fuels from Coal Are Expensive," *Hydrocarbon Processing*, vol. 55 (July 1976), pp. 85–90. For 1977: American Gas Associations "A Comparison of Coal Use for Gasification versus Electrification," mimeo, April 26, 1977. For 1978a: Robert Detman, "How Six Coal Gasification Processes Compare Economically," *Pipeline and Gas Journal*, February 1978, pp. 26–29. For 1978b: F. M. Barnett, J. J. Clawson, and K. C. Vyas, "Medium Btu Gas Fits Refiners Needs," *Hydrocarbon Processing*, vol. 57 (June 1978), pp. 131–33. For 1978c: Derived from Blazek, Baker, and Tison, *High-Btu Coal Gasification Processes*. See also notes to table 10-5. For 1979a: W. L. Nelson, "What Are Prices of Gas and Oil from Coal?" *Oil and Gas Journal*, vol. 77 (April 23, 1979), p. 67. For 1979b: "DOE Technology Matchup," in *Inside DOE*, May 18, 1979, quoting 1979 DOE cost estimates. For 1981: Robert B. Taylor and Robert A. Moon Jr., "Economic Comparison of Four Synthetic Fuels," *Oil and Gas Journal*, vol. 79 (June 29, 1981), pp. 120–24. For 1983: Reiner Kollrack, "Marketability of Chemically Cleaned Coals" (Washington, Engineering Societies Commission on Energy, 1983).

a. In most cases the authors give a range of values (based on uncertainty as well as a range for components like coal prices and financing). In the interest of space, this table presents an intermediate value. We apologize to the authors for imputing greater precision to their estimates than they had intended. All processes except CO_2 Acceptor and HYGAS use eastern coal.

b. Process not specified.

Table 10A-3. *Cost Estimates for Selected Gasification Processes, 1969, 1973, 1976, 1978*[a]
1972 dollars

	Capital cost		*Operating cost, dollars per million Btus*	*Coal cost*	
Year and process	*Dollars per million Btus*	*Millions of dollars*		*Dollars per million Btus*	*Dollars per ton*
1969					
HYGAS	0.35	166	0.03	0.26	4.92
CO_2 Acceptor	0.20	93	0.07	0.17	1.50
BIGAS	0.35	160	0.07	0.24	4.92
1973					
HYGAS	0.33	158	0.19	0.51	8.98
CO_2 Acceptor	0.23	107	0.14	0.33	3.15
BIGAS	0.34	170	0.24	0.42	8.98
1976					
HYGAS	0.98	351	0.60	0.68	10.33
CO_2 Acceptor	0.95	352	0.43	0.67	5.56
BIGAS	1.14	357	0.62	0.72	10.33
1978					
HYGAS	2.27	814	0.68	1.08	13.30
CO_2 Acceptor	1.72	636	n.a.	0.80	6.65
BIGAS	3.08	965	0.79	1.20	13.30

Sources: See notes for table 10A-2 on years applicable. The 1978 CO_2 Acceptor capital cost is from Detman, "How Six Coal Gasification Processes Compare"; other 1978 numbers are from Blazek, Baker, and Tison, *High Btu Coal Gasification Processes*. The 1978 publications do not give capital costs in dollars per million Btus; these figures are imputed using the proportions for the same process in 1976.

n.a. Not available.

a. Capital requirements per year in all estimates are about 20 percent of total investment.

Table 10A-4. *Cost Estimates for Liquefaction Processes, 1972–85*
Dollars per barrel of crude oil equivalent (1972 dollars)

Year	*Source*	*Direct*	*Indirect*	*Coal (both)*	*Shale*
1972	National Petroleum Council[a]	. . .	. . .	7.75–8.25	. . .
1973	Charles DiBona, President Nixon's energy adviser[a]	. . .	. . .	5.20–6.62	. . .
1973	Federal Power Commission[b]	. . .	. . .	6.62–9.46	. . .
1974	National Academy of Engineering[a]	. . .	. . .	7–8	. . .
1974	U.S Navy[a]	. . .	. . .	15	. . .
1975	Synfuels Interagency Task Force[c]	. . .	. . .	15–20	15
1975	[d]	11–13	20	. . .	9
1976	[e]	15–26	22	. . .	. . .
1978	[f]	14	. . .	. . .	. . .
1978	[g]	18	. . .	. . .	. . .
1979	[h]	18–23	21–24	. . .	13–16
1979	Department of Energy[i]	. . .	. . .	17–23	10–20
1980	Gas Research Institute[j]	. . .	. . .	22–26	15–22
1981	[k]	26	25	. . .	18
1983	[l]	31	28	. . .	. . .
1985	[m]	. . .	. . .	23–40	. . .

a. "As Oil Price Rises so Does Cost of Synthetic Crude," *Washington Post*, June 11, 1979, p. A-4.
b. Lawrence Lessing, "Capturing Clean Gas and Oil from Coal," *Fortune*, November 1973, p. 216.
c. Synfuels Interagency Task Force to the President's Energy Resources Council, *Recommendations for a Synthetic Fuels Commercialization Program*, vol. 1, overview report (GPO, November 1975), no. 041-001-00112-3, p. 33.
d. K. C. Vyas and W. W. Bodle, "Coal and Oil-Shale Conversion Looks Better," *Oil and Gas Journal*, vol. 73 (March 24, 1975), pp. 45–54.
e. J. B. O'Hara, "Coal Liquefaction," *Hydrocarbon Processing*, November 1976, pp. 221–26.
f. Gordon Friedlander, "Coal-Derived Liquids: A Continuing Study," *Electrical World*, vol. 189 (July 1, 1978), pp. 52–53.
g. E. W. Neben, "The Economics of Advanced Coal Liquefaction, *Chemical Engineering Progress*, vol. 74 (August 1978), pp. 43–48.
h. Cameron Engineers, "Overview of Synthetic Fuels Potential to 1990," reprinted as Appendix II in *Synthetic Fuels*, Report by the Subcommittee on Synthetic Fuels of the Senate Committee on the Budget (Government Printing Office, September 27, 1979), p. 177.
i. "DOE Technology Matchup," *Inside DOE*, May 18, 1979, p. 6.
j. "Synfuels Competitive Now, Study Argues," *Chemical and Engineering News*, August 11, 1980, pp. 5–6.
k. Robert B. Taylor and Robert A. Moon, Jr., "Economic Comparison of Four Synthetic Fuels," *Oil and Gas Journal*, June 19, 1981, pp. 120–24.
l. Reiner Kollrack, "Marketability of Chemically Cleaned Coals" (Washington, D.C.: Engineering Societies Commission on Energy, 1983), p. 12.
m. Testimony of Paul A. MacAvoy in *Corcoran, Reichl, and MacAvoy Nominations*, Hearing before the Senate Committee on Energy and Natural Resources, 99 Cong. 1 sess. (GPO, 1985), p. 20 (lower range of estimate); testimony of Robert Pyndyck in *Synthetic Fuels Policy*, Hearing before the Subcommittee on Fossil and Synthetic Fuels of the House Committee on Energy and Commerce, 99 Cong. 1 sess. (GPO, 1985), p. 20 (upper range of estimate).

Table 10A-5. *Cost Estimates for Selected Liquefaction Plants, 1975–76, 1978, 1981*[a]
1972 dollars

Year and process	Thousands of barrels of crude oil equivalent per day	Capital cost: Dollars per barrel	Capital cost: Millions of dollars	Operating cost, dollars per barrel	Coal cost: Dollars per barrel	Coal cost: Dollars per ton
1975–1976						
COED	28 + 830MWe[b]	14.90	1,246	8.52	3.03	(6.80)
Synthoil	50	7.81	(515)	4.48	2.82	(10.33)
Consol						
CSF	50	6.31	(479)	2.88	3.82	(7.66)
SRC-II	50	4.32	(328)	5.31	5.33	(6.80)
Shale	50	7.12	(421)	2.11	...	...
1978						
Synthoil	50	7.61	(502)	4.13	6.00	(15.11)
SRC-II	50	5.76	(585)	2.91	3.86	(5.67)
H-Coal	50	8.53	(866)	5.49	4.44	(5.67)
EDS	60	6.68	(1,192)	5.94	6.08	(15.11)
MIS shale	50	5.90	(585)	5.54	...	...
1981						
Methanol	50	12.62	(1,281)	1.90	10.08	(17.94)
EDS	66	8.61	(1,537)	6.93	9.98	(17.94)
Oil shale	50	12.93	(1,281)	5.21	...	...

Sources: See notes d, e, and k in table 10A-4; and Saman Majd, "A Financial Analysis of Selected Synthetic Fuel Technologies," MIT Energy Laboratory Working Paper, MIT-EL 79-004WP (January 1979).

a. In order to make the estimates more comparable across authors (although less so, unfortunately, across processes), the product adjustment factor used by Majd was omitted. Different product mixes produced by the different plants are corrected only up to an MMBtu basis for these estimates. Interpolations as described in the note to table 10A-3 are used here for 1975 and 1978 capital estimates.

b. The calculations for COED assume 55 percent of revenues come from sale of crude oil and 45 percent from electricity. All processes except methanol (indirect liquefaction) and oil shale are direct liquefaction processes. COED produces crude oil and gas for electricity; the remaining direct liquefaction processes produce mainly fuel oil. MIS shale is a modified-in-situ shale process, which was thought in the late 1970s to have environmental advantages and perhaps cost advantages over surface oil shale processing.

Table 10A-6. *Selected Statistics for Variables in Regression Analysis*[a]

Year and variable	*Number in cell/ number > 0*	*Mean*	*Standard deviation*
1976			
ACA-R (*R* only)	124	70.4	23.7
ACA-D (D only)	261	26.7	27.5
Sci &Tech	14		
Approp.	8		
E-coal (>0 only)	40	11.2	13.6
Low W-coal	15		
Med. W-coal	7		
High W-coal	6		
Oil res. (>0 only)	132	1.86	1.3
Oil con.		99.8	21.5
Liq. PP	10		
Gas PP	7	. . .	. . .
Fut SFC Prop	58	. . .	. . .
1984			
Fut SFC Prop	7	. . .	. . .
SFC Prop	18	. . .	. . .
Past SFC Prop	31	. . .	. . .

a. See table 10-9 for a description of the votes in 1976 and in 1984. See figure 10-1 for a description of the variables.

Table 10A-7. *Trends in Economic Factors, Selected Years, 1975–85*

Year	*OPEC (1972 dollars per barrel)*	*Synfuels liquefaction (1972 dollars per barrel)*	*Net benefits*		
			IATF[a] *(1975 dollars)*	*Shale*[b] *(1976 dollars)*	*SRC*[b] *(1976 dollars)*
1975	11.07	15.00	−1.63 billion	272 million	208 million
1976	10.18	15.00	. . .	. . .	. . .
1979	13.26	20.00	−3.5 billion	839 million	776 million
1980	18.99	22.00	. . .	. . .	. . .
1985	11.65	35.00	−9.0 billion	−455 million	−574 million
1985	. . .	. . .	. . .	−1.2 billion	−1.4 billion

a. Information program, expectation (1975); strong cartel, expensive synfuels cost (1979); weak cartel, expensive synfuels cost (1985). See Synfuels Interagency Task Force to the President's Energy Resources Council, *Recommendations for a Synthetic Fuels Commercialization Program*, vol. 1, Overview Report (Government Printing Office, November 1975), stock no. 041-001-00112-3.

b. 6 percent discounting case, 1975: oil scenario 2 (moderate annual increases), no cost overrun; 1979: oil scenario 1 (pessimistic), 20 percent cost overrun; 1985: oil scenario 2, 40 percent cost overrun; 1985 (second row): oil scenario 4 (constant real price), 40 percent cost overrun. See Majd, "Financial Analysis of Selected Synthetic Fuel Technologies."

Table 10A-8. *Selected Pooling Results*

Years[a]	χ^2	Degrees of freedom	Conclusion
1975a, 1976	14.37	14	Pool
1978, 1981	13.56	14	Pool
1981, 1985	19.7	19	Pool
1978, 1981, 1985	32.4	22	Pool
1975a, 1975b, 1976	23.0	15	Reject at 0.1
1979, 1980	36.4	9	Reject at 0.05
1984a, 1984b	78.6	18	Reject at 0.00
1975b, 1978	24.1	14	Reject at 0.05
1980, 1985	73.45	13	Reject at 0.00
1975a, 1976, 1985	70.27	22	Reject at 0.00
1975a, 1976, 1980	182.3	28	Reject at 0.00
1984a, 1985	34.6	16	Reject at 0.05
1984b, 1985	48.9	16	Reject at 0.05
1980, 1981	120.13	14	Reject at 0.00
1975a, 1976, 1978	50.6	28	Reject at 0.00

a. See table 10-9 for a description of the votes taken in these years.

Chapter 11

The Photovoltaics Commercialization Program

WILLIAM M. PEGRAM

PHOTOVOLTAIC (PV) devices convert sunlight into electricity. Federal support for PV technology was initially tied to the space program, where its first significant use was to provide power for the Vanguard satellite in 1958.[1] Because of the rising national concern in the early 1970s over the use, cost, and reliability of fossil fuel supplies, federal research was redirected toward terrestrial applications. From less than $1 million a year in fiscal 1972 and 1973, federal appropriations (excluding tax expenditures on credits) grew to $150 million in fiscal 1980, then shrunk to less than an annual average of $50 million in the Reagan years.

The PV program used an original management strategy, developing a kind of induced competition. It was a decentralized program, with parallel development of technical concepts and periodic competitions (tournaments) to determine which projects would be continued, deleted, or added. A single, simple cost-reduction parameter was used to provide interim and ultimate goals for the program as a whole and to serve as the prime criterion for tournament winners.

The program's structure seems sensible from both a practical and a theoretical perspective as a way of meeting defined goals. Indeed, according to some estimates from the mid-1980s, the interim cost reduction goals had been met, and with an expenditure of only about half the original amount deemed necessary. Though still far from a commercial success, the number of photovoltaic modules manufactured or imported into the United States

I would like to thank Jeffrey Banks, Jeffrey A. Dubin, D. Roderick Kiewiet, and R. Douglas Rivers for helpful comments.

1. Solar Energy Research Institute (SERI), *Basic Photovoltaic Principles and Methods* (Washington: National Technical Information Service, 1982).

grew dramatically in the 1980s (table 11-1). Nevertheless, the program was greatly contracted in scope long before it was completed.

This chapter explores the history of photovoltaic development, concentrating on political and economic features of the R&D program.[2] Solar power advocates were the core support group for the PV program. But as the program evolved to fit into overall energy policy perceptions in the late 1970s, the solar coalition split, and PV eventually lost majority backing. The changed role of PV in energy policy, the changed short-term outlook for energy, the innovative management structure of the program, and the outcome of the program itself finally caused the federal effort to be sharply reduced.

Basics of PV Technology

To understand the PV program requires a rudimentary knowledge of how a PV system works.[3] At the heart of the system is the solar cell, composed of material that releases electrons when exposed to light. Although many different materials have this property, which ones will be used in commercial PV systems depends on four criteria: the cost of the basic material, the cost of manufacturing solar cells from it, the efficiency with which the material converts sunlight to electricity, and the durability of the cells in the field.

Silicon meets two of these criteria: because it is the principal component of ubiquitous materials such as sand and quartz, its material costs are very low; and it is quite durable. But silicon solar cells are expensive to manufacture and are not as efficient as some other materials in converting sunlight to electricity. Manufacturing costs are high for two reasons. First, natural silica contains impurities that must be removed to make solar cells, and the purification process is expensive and wastes most of the raw material. Second, purified silicon is polycrystalline in form, which is undesirable because it makes cells of low efficiency. To make more efficient silicon cells requires either inventing ways to make better use of forms of silicon other than single crystal, or "regrowing" the silicon in a single crystal form and then slicing it into thin wafers about 0.5 millimeters thick. In the mid-1970s known processes for making solar cells wasted about 75 percent of the purified silicon.

2. For a more detailed analysis of the program, see William M. Pegram, *Sunshine for Oil*, forthcoming.
3. SERI, *Basic Photovoltaic Principles and Methods* (Van Nostrand Reinhold, 1984).

Table 11-1. *Photovoltaic Modules by End Use, 1982–88*

End use	Shipments, peak kw 1985	1986	1987	1988
Water pumping	545	591	438	655
Transportation	370	419	322	435
Communication	1,292	1,375	1,319	2,024
Consumer goods	244	294	793	1,747
Military	112	101	W[a]	122
Residential[b]	1,800	2,029	1,745	2,139
Industrial/commercial	826	895	1,413	1,584
Utility	518	553	534	699
Other	63	76	W[a]	272
Total	5,769	6,333	6,856	9,676

	1984		1983		1982	
End use	*Shipment (peak kw)*	*Percent of total*	*Shipment (peak kw)*	*Percent of total*	*Shipment (peak kw)*	*Percent of total*
Specialty	470	4.7	242	1.9	88	1.3
Stand-alone[c]	3,475	35.1	3,334	26.4	3,265	47.3
Residential	492	5.0	160	1.3	51	0.7
Intermediate	204	2.1	93	0.7	454	6.6
Central power	5,271	53.2	8,791	69.6	3,040	44.1
Total	9,912	100.0	12,620	100.0	6,897	100.0

Sources: For 1987–88: Energy Information Administration, *Solar Collector Manufacturing Activity, 1988*, DOE/EIA-0174(88) (Department of Energy, November 1989), p. 15; for 1985–86: EIA, *Solar Collector Manufacturing Activity, 1986*, DOE/EIA-0174(86) (October 1987), p. 24; and for 1982–84: EIA, *Solar Collector Manufacturing Activity, 1984*, DOE/EIA-0174(84) (August 1985), pp. 21, 26. Figures are rounded.

a. Withheld to avoid disclosure of individual company data.

b. Includes grid and non-grid-connected applications.

c. Denotes not grid-connected. Residential and intermediate categories include only grid-connected.

A PV system places the solar cells on a surface that exposes them to sunlight, protects them against deterioration, and collects the electricity they generate. The combination of solar cells and the surface on which they are placed is called an array or module.[4] Because sunlight is diffuse, generating significant amounts of electricity requires using a large number of solar cells. Hence arrays are large, and a big part of the cost of PV is the material on which the cells are placed.

The federal PV research program attacked three broad categories of research problems.[5] First, several activities were directed at lowering the

4. The preferred terminology has changed over time. In current usage, cells are connected together on a single plate to form *modules*, which in turn are connected to form *arrays*.

5. Energy Research and Development Administration (ERDA), *Photovoltaic Conversion Program: Summary Report*, ERDA 76-161 (November 1976), pp. 2–3.

manufacturing cost of silicon solar cells by reducing the cost of solar grade silicon from $65 to $10 per kilogram, reducing the waste in manufacturing, and producing cells that could be packed more tightly onto collectors, thereby reducing the cost of supporting a given number of solar cells. Second, several projects were directed at reducing the costs or improving the performance of modules, such as by developing low-cost ways to encapsulate the solar cells to produce a longer system life or by automating the manufacture of modules. Third, the program sought ways to produce more electricity from a given cell by developing tracking systems, concentrators, and more efficient cells.

The last category of projects attempted to produce more electricity from a given amount of sunlight. Sunlight contains a broad spectrum of light waves, and the ideal PV material would absorb a high fraction of the sunlight falling upon it and would produce maximal electricity at frequencies that correspond to most of the light in sunshine. In both respects, silicon is a good, but not ideal, material. In the early 1970s relatively little research had been done to construct solar cells from more promising materials, so that much less was known about whether their theoretical superiority could be effectively used in practice. A primary aim of the PV program was to resolve this issue.

The simplest form of PV module is a flat surface that is permanently placed in a fixed position. To maximize the value of the power that is generated from a PV module, solar cells must be exposed to the most sunlight at the times when electricity is most valuable. To increase the electricity produced by a cell, the module can be placed on a tracking device that keeps the cell facing the sun during all daylight hours throughout the year. Another way to make good use of a cell is to use lenses or mirrors to concentrate the sunlight that falls on a large surface area onto a single cell. All concentrators have three problems: cooling, tracking, and varying sunlight intensity. The efficiency with which a solar cell converts sunlight to electricity varies with the temperature of the solar cell. Flat-plate collectors operate at relatively efficient temperatures, but concentrators, because they concentrate heat as well as light, can cause the temperature of the solar cells to be too high for efficient conversion. To solve this problem requires devising a means for cooling the solar cells. Tracking is useful for flat-plate collectors but is essential for concentrator systems. Lenses or mirrors are expensive and must point at the sun for the system to be economic. Finally, flat-plate collectors still function with clouds, but concentrators are much less efficient when exposed to diffuse light.

History of the Program

Although photovoltaic cells were used extensively in the space program, federal support for terrestrial applications of solar energy (including photovoltaics) was only about $100,000 a year between 1950 and 1970.[6] In the early 1970s federal support for terrestrial solar technology came primarily from the National Science Foundation (NSF). The NSF program enjoyed relatively small budgets, but it provided the analytical and informational foundations for later efforts (table 11-2).

During the NSF period, five key studies assessed the long-term prospects for solar energy and photovoltaics.[7] Four of these studies estimated the future contribution of solar energy to the national energy supply, including separate estimates for each technology (table 11-3).

The early studies foresaw a significant penetration of PV by the year 2000, even though hundredfold cost reductions were necessary to make PV economic. These studies concluded that such cost reductions were achievable given sufficient time and R&D (table 11-4).

The early reports attempted to justify federal R&D by rudimentary cost-benefit calculations. The report by the NSF and NASA in 1972 showed annual savings in fossil fuel of $750 million in 2000 and $16 billion in 2020,

6. Estimate of Dr. John Teem, then assistant administrator for solar, geothermal, and advanced energy systems, ERDA, in congressional testimony. See *Energy Research and Development and Small Business*, Hearings before the Senate Select Committee on Small Business, 94 Cong. 1 sess. (Government Printing Office, 1975), pt. 1, pp. 169–245.

7. The studies are the following: (1) NSF/NASA Solar Energy Panel, "An Assessment of Solar Energy as a National Energy Resource," December 1972, reprinted in *Solar Energy Research and Development*, Hearings before the Joint Committee on Atomic Energy, 93 Cong. 2 sess. (GPO, 1974), pp. 676–759; (2) "Report of Subpanel IX to the Atomic Energy Commission Chairman in Support of Her Development of a Comprehensive Federal Energy Research and Development Program," reprinted in *Solar Energy*, Hearings, pp. 337–418; (3) National Science Foundation, *Workshop Proceedings, Photovoltaic Conversion of Solar Energy for Terrestrial Applications*, vols. 1 and 2, NSF-RA-N-74-013 (Washington, 1973); Jet Propulsion Laboratory, *Assessment of the Technology Required to Develop Photovoltaic Power Systems for Large-Scale National Energy Applications*, JPL Special Publication 43-11 (Pasadena, Calif., 1974); Jet Propulsion Laboratory, *Executive Report of Workshop Conference on Photovoltaic Conversion of Solar Energy for Terrestrial Applications* (Washington: National Science Foundation, 1974); (4) Mitre Corporation, *Systems Analysis of Solar Energy Programs*, MTR-6513 (December 1973); Mitre Corporation, "Recommendations to RANN/NSF Solar Energy R&T Program," M-74-21 (February 1974), reprinted in *Solar Energy*, Hearings (1974), pp. 419–501; and (5) Federal Energy Administration, *Final Solar Energy Task Force Report: Project Independence Blueprint*, prepared by the Interagency Task Force on Solar Energy under the direction of the National Science Foundation (1974).

Table 11-2. *Federal Government Expenditures on Terrestrial Photovoltaics, Fiscal Years 1971–91*[a]
Dollars in millions

Year	*Current dollars*	*Constant fiscal 1982 dollars*
1971	Small or none	Small or none
1972	0.3	0.7
1973	0.8	1.6
1974	2.4	4.6
1975	8.0	13.9
1976	21.6	34.8
1977	64.0	95.5
1978	76.5	106.7
1979	118.8	152.5
1980	150.0	177.0
1981	139.2	149.5
1982	74.0	74.0
1983	58.0	55.7
1984	50.4	46.6
1985	57.0	51.1
1986	40.7	35.5
1987	40.4	34.2
1988	35.0	28.8
1989	35.5	28.0
1990	34.7	26.3
1991	46.9	34.1
Total	1,054.2	1,151.2

Sources: For 1971–75: Lloyd O. Herwig, "Solar Research and Technology," in *Energy Research and Development and Small Business*, Hearings before the Senate Select Committee on Small Business, 94 Cong. 1 sess. (GPO, 1975), p. 1251 (actual figures for 1971–74; estimated for 1975); and for 1976–91, see table 11-7. Figures are rounded.

a. Figures include the effects of supplementals and rescissions. The 1976–91 figures represent appropriations. The 1971–75 figures probably represent costs (approximating outlays). Appropriations figures would be somewhat larger because program is growing at this time. Figures do not reflect spending in the transition quarter (July 1, 1976–September 30, 1976). For 1972–89 constant dollars computed by constructing fiscal year deflator as average of GNP implicit price deflator for quarters in fiscal year (found in *Survey of Current Business*). Fiscal 1990 and 1991 inflation assumed at fiscal 1989 rate (4.2 percent).

compared with total federal R&D expenditures of $780 million. The Solar Energy Task Force Report, part of the *Project Independence Blueprint* in 1974, upped the benefits from displaced oil use to $4 billion a year by the year 2000.[8]

Although differing in details, the early studies reveal much agreement

8. Federal Energy Administration, *Final Solar Energy Task Force Report*, p. VII-10. The tables on pp. VII-4 and VII-5 estimate the capital costs of the photoelectric energy power system (PEPS) in service by the year 2000 to be between $80 billion and $440 billion.

Table 11-3. *Contribution of Photovoltaics*[a]
10^{15} Btus per year quadrillion Btus, or "quads"

Study	*1980*	*1985*	*1990*	*2000*	*2020*
NSF/NASA (1972)	n.a.	n.a.	n.a.	1.5	26.5
Subpanel IX (1973)	0	0.07	0.7	7.0	n.a.
MITRE (1973)					
Minimum	0	0.01	n.a.	5.3	17.8
Accelerated	0	0.04	n.a.	17.8	59.4
Solar Energy Task Force (1974)					
Business as usual	Neg.	0.10	0.2	5.0	n.a.
Accelerated	Neg.	0.03	1.0	23.1	n.a.
ERDA (1975)	n.a.	0.01	n.a.	2.1	5.6
National Energy Plan (1979)					
Base	n.a.	n.a.	n.a.	0.6	n.a.
Maximum practical	n.a.	n.a.	n.a.	1.7	n.a.
Technical limit	n.a.	n.a.	n.a.	3.0	n.a.
DOE (1980)					
Low case	n.a.	0	Small	n.a.	n.a.
Best estimate	n.a.	0	0.02	n.a.	n.a.
High case	n.a.	Small	0.04	n.a.	n.a.

Sources: NSF/NASA: NSF/NASA Solar Energy Panel, "An Assessment of Solar Energy as a National Resource," December 1972, reprinted in *Solar Energy Research and Development*, Hearings before the Joint Committee on Atomic Energy, 93 Cong. 2 sess. (GPO, 1974), pp. 676–759; Subpanel IX: *Report of Subpanel IX to the Atomic Energy Commission Chairman in Support of Her Development of a Comprehensive Federal Energy Research and Development Program*, reprinted in *Solar Energy Research and Development*, Hearings, p. 411; MITRE: Mitre Corporation, *Systems Analysis of Solar Energy Programs*, MTR-6513 (December 1973), pp. 178, 181; Solar Energy Task Force: Federal Energy Administration, *Final Solar Energy Task Force Report: Project Independence Blueprint*, prepared by the Interagency Task Force on Solar Energy under the direction of the National Science Foundation (1974), p. I-7; ERDA: Energy Research and Development Administration, *Definition Report: National Solar Energy Research, Development and Demonstration Program*, ERDA-49 (June 1974), reprinted in *Energy Research and Development and Small Business*, Hearings before the Senate Select Committee on Small Business, 94 Cong. 1 sess. (GPO, 1975), pt. 2, p. 4976; National Energy Plan: Department of Energy, *National Energy Plan II* (May 1979), p. VI-4; and DOE: Department of Energy, *Reducing U.S. Oil Vulnerability: Energy Policy for the 1980s*, An Analytical Report to the Secretary of Energy, prepared by the Assistant Secretary for Policy and Evaluation, DOE/PE-0021 (November 1980), p. IV-G-4.

n.a. Not available.

a. The amount of conventional energy that can be displaced by photovoltaic energy, not the amount of energy actually delivered by solar systems. Because the efficiency of conversion of fuel to electricity is approximately 0.30, the displacement figures shown above are 3.3 times as large as the corresponding PV production figures.

about the prospects for PV. One consistent theme is that commercializing the technology would take more than a decade. Little commercial use was expected by 1985, but significant use could occur by 2000. To achieve these results, the R&D program was expected to cost $300 million to $800 million. The early studies also recommended a broad R&D strategy that would explore several basic technical alternatives and that would seek large cost reductions in all components of a commercial PV system.

Table 11-4. *Estimates of Federal Photovoltaic Program Budget Requirements*
Millions of dollars[a]

Source	*Total amount required*	*Number of years*
Office of Science and Technology (1972)	322	Not specified
NSF/NASA (1972)	780	11
Cherry Hill (1973)	560[b]	11
Subpanel IX (1973)	248	First 5 years
	373	First 10 years
	378	First 25 years
MITRE (1973)[c]		
Minimum viable	39.6	First 7 years
Accelerated	72.9	First 7 years
Solar Energy Task Force (1974)	No budget given	
NSF Testimony (1974)	815[d]	10
NSF Report (1974)	816	10

Sources: Office of Science and Technology: "Data for Use in the Assessment of Energy Technologies," AET-8 (April 1972), cited in American Institute of Aeronautics and Astronautics, *Solar Energy for Earth: An AIAA Assessment* (April 1975), p. 57; NSF/NASA (1972), p. 65 (see table 11-3); Cherry Hill: Jet Propulsion Laboratory, *Executive Report of Workshop Conference on Photovoltaic Conversion of Solar Energy for Terrestrial Applications* (Washington: National Science Foundation, 1974), pp. 9, 13, 19, 22, 23, 26; Subpanel IX (1973), fig. 24 (see table 11-3); MITRE (1973), pp. 439–41 (see table 11-3); Solar Energy Task Force (see table 11-3); NSF Testimony: *Solar Photovoltaic Energy*, Hearings before the Subcommittee on Energy of the House Committee on Science and Astronautics, 93 Cong. 2 sess. (GPO, 1974), p. 131; and NSF Report: National Solar Energy Programs, in *ERDA Authorization—Part IV 1976 and Transition Period, Solar and Physical Research*, Hearings before the Subcommittee on Energy Research, Development, and Demonstration of the House Committee on Science and Technology, 94 Cong. 1 sess. (GPO, 1975), p. VII-25.

a. The reports do not indicate whether figures are in constant or current dollars, or whether these are discounted present values.

b. This estimate does not include estimates for some categories for the final six years of the proposed program. If these were included, the expenditure requirement would probably be about $30 million more.

c. The MITRE report says that this is the amount to conduct all the research tasks for PV. It is unclear what would be achieved by expenditure of the amounts shown and what support beyond the initial seven years would be required—especially since elsewhere in the report MITRE predicts that PV will cost from one to four times as much as conventional energy in the year 2000.

d. This is the funding required from public and private sources to establish a market price of about 50 cents per peak watt. It was expected that federal funding would dominate in the 1970s and private funding in the 1980s, if progress continued. This figure appears to conflict with the $816 million figure given in the NSF Report in the same year, since the latter only estimated federal funding requirements.

The NSF R&D Program

The R&D program developed by the NSF relied on an unusual management strategy. The program was characterized by decentralization, coordination, and competition.

DECENTRALIZATION. Federal R&D programs are typically organized with one, two, or three levels of decisionmaking. An example of a single-level structure is government in-house research. A two-level approach is the standard government-contractor form, and the three-level approach is the government-contractor-subcontractor form. PV used a three- and four-level form throughout its history. In this structure, the primary management of the program took place in several level B lead centers and level C field centers rather than from level A headquarters in Washington.[9] The lead and field centers, in turn, awarded contracts to level D private contractors.

The headquarters of the PV program resided with the Research Applied to National Needs program in the National Science Foundation until January 1975, when responsibility was shifted to the newly created Energy Research and Development Administration (ERDA).[10] Responsibility for all ERDA programs, including photovoltaics, was shifted to the Department of Energy in October 1977 by the Department of Energy Organization Act.[11]

The number and responsibilities of lead centers changed over time. At the height of the program in 1980, the two lead centers were the Solar Energy Research Institute (SERI) in Golden, Colorado (for advanced research and development), and the Jet Propulsion Laboratory (JPL) in Pasadena, California (for technology development and applications and the federal PV utilization program). The JPL had management responsibility over seven level C field centers (table 11-5).[12] In addition, the Department of Energy's Albuquerque and Oak Ridge Operations Office and the Oak Ridge National Laboratory had management responsibility for several PV demonstration projects.[13]

COORDINATION. Private sector research is usually characterized by duplication among firms. By contrast, government R&D typically assigns each contractor or laboratory a specific part of the project to avoid overlap.

9. Levels A, B, and C terminology does not appear in program documents, but is common in NASA.

10. Responsibility for photovoltaics was transferred to ERDA through three acts passed in 1974. ERDA was established by the Energy Reorganization Act of 1974, which transferred NSF solar heating and cooling and geothermal power development programs to ERDA. The Solar Energy Research, Development, and Demonstration Act of 1974 authorized a federal research, development, and demonstration program to commercialize solar energy and provided for these functions to be carried out by ERDA. These responsibilities were amplified in the Federal Nonnuclear Energy Research and Development Act of 1974.

11. See Department of Energy, *Photovoltaic Energy Systems Program Summary*, DOE/CE-0012 (January 1981), p. 2-1.

12. Department of Energy, *Photovoltaic Energy*, p. 2-4.

13. Department of Energy, *Photovoltaic Energy Systems Program Summary, Fiscal Year 1983*, prepared by the Solar Energy Research Institute and the Jet Propulsion Laboratory for the Assistant Secretary, Conservation and Renewable Energy, DOE/CE-0033/2 (January 1984), pp. 319–38.

Table 11-5. *Field Centers Directed by the Jet Propulsion Laboratory as of January 1981*

Field center	*Responsibility*
Jet Propulsion Laboratory, Pasadena, California	Low-cost solar array collectors
Sandia National Laboratories, Albuquerque, New Mexico	Systems design Concentrator collectors Subsystem development Intermediate load center applications
MIT Lincoln Laboratory, Lexington, Massachusetts	Residential applications
Aerospace Corporation, El Segundo, California	Central station
MIT Energy Laboratory, Cambridge, Massachusetts	Mission and policy analysis
Brookhaven National Laboratory, Upton, New York	Environmental health and safety requirements
NASA Lewis Research Center, Cleveland, Ohio	Remote stand-alone applications International applications

For example, in standard defense R&D, one contractor may design the engine, another the radar, and so on. Because successful completion of all parts is required for success, the program is only as strong as its weakest link.

PV used the standard approach of assigning different responsibilities to each management organization. The JPL, SERI, and Sandia National Laboratories (Sandia) in Albuquerque were each assigned responsibility for one of the three major technical alternatives: flat-plate single-crystal silicon (JPL), other materials (SERI), and concentrators (Sandia). Other field centers were assigned different responsibilities. In some cases their responsibilities were complementary. For example, Brookhaven's work on environmental effects complemented JPL work on reducing the cost of flat-plate single-crystal arrays. In other cases, the tasks assigned to different organizations were substitutes. For example, although exotic materials, concentrators, and flat-plate single-crystal silicon were different technically, they sought to serve the same economic function. In either the complementary or competitive case, the element of coordination was that no other organization could undertake technically overlapping work.

COMPETITION. The distinctive aspect of the PV program was that competition existed. Competition was constrained by the assignment of specific responsibilities to a single organization. Because flat-plate single-crystal silicon, concentrators, and exotic materials were substitutes, the organizations pursuing them were competitors.

Competition was further enhanced by a contracting strategy that supported parallel approaches to a particular task, with only the most successful receiving follow-on contracts. In flat-plate single-crystal silicon arrays, several alternative methods to obtain the silicon material and to make the cells were supported. For exotic materials, polycrystalline silicon, amorphous silicon, and a variety of other materials and cell types were investigated. The concentrator program pursued three primary alternatives: parabolic trough, linear point-focus Fresnel lens, and point-focus Fresnel lens.[14]

The program provided parallel support for multiple concepts during the initial R&D phases. As each concept progressed, the menu of alternatives was to be gradually reduced.[15] This approach offered several advantages. First, it increased the probability of success. Although the outcome of any one approach was uncertain, if the probabilities of success in each alternative were independent, the chance of a favorable outcome from at least one approach was good. Second, by dropping projects that made slow progress, the program created a competitive spur to achieve good results. Third, decomposing the effort into numerous specific tasks allowed small firms, groups, or individuals to participate by focusing research on a single aspect of the problem.[16] Thus the technology was amenable to a "fragmentation" strategy, which spread the distributive benefits among a large number of firms and organizations.

The political downside of the strategy is that, as with the satellite program, losers are created. However, the program avoided controversy on these grounds, for four reasons. First, because the program never did approach the commercialization phase when it would have focused on at most a few technical options, the program was sufficiently fragmented to

14. Paul D. Maycock and Edward N. Stirewalt, *A Guide to the Photovoltaic Revolution* (Rodale Press, 1985), p. 52.

15. As early as October 1973, the same month as the Cherry Hill conference on photovoltaics just discussed, Lloyd Herwig, director of Advanced Solar Energy Research and Technology, Research Applied to National Needs, National Science Foundation, described the NSF R&D strategy as a "phased program planning approach." Lloyd O. Herwig, "Solar Research and Technology," presentation to the American Physical Society, February 7, 1974, reprinted in *Energy Research and Development and Small Business*, Senate Hearings, pt. 1A, pp. 1228–53.

16. Jeffrey L. Smith, "Photovoltaics," *Science*, vol. 212 (June 26, 1981), pp. 1472–78.

allow at all times for many "winners." Second, a commercial competitive advantage was anticipated to be rather far off in the future, so that losers lost only a contract, not a position in an important market. Third, while the strategy was designed to eliminate competitors, the distribution and magnitude of expenditures among states and congressional districts remained fairly constant. The year-to-year correlations for contract obligations always exceeded 0.75 for states and 0.60 for districts. Fourth, because the program was so fragmented, most contracts were very small, so that canceling one was not likely to cross the threshold of political salience in the contractor's home district.

PV Program Expansion: The DOE Era

After ERDA's programs were transferred to the Department of Energy in 1977, the PV program reached its high-water mark in both budget and ambition. The Solar Photovoltaic Energy Research, Development, and Demonstration Act of 1978 set a production goal of 2 million peak kilowatts by 1988 and contemplated expenditures of $1.5 billion over ten years. As in other energy research programs at that time (see, for example, chapter 10), near-term commercialization received greater emphasis. The program underwent two dramatic changes: an effort to stimulate demand and a shift from distributed to central station applications. In addition, the expansion increased both budget and off-budget (tax credits) expenditures. Three significant demand-side programs were pursued: government procurement, tax credits, and restrictions on utility fuel use (see figure 11-1).

PROCUREMENT. The rationale for government purchases is to achieve economies of scale and to speed the learning curve. Government purchases, it was hoped, would enable the industry to invest in more automated, lower-cost production techniques. In addition, solar cell technologies were expected to exhibit a "learning curve" like that in related industries. Semiconductors, in particular, operate on the same principle as photovoltaic cells, and their manufacture has exhibited sharply declining costs because of learning. Government purchases of PV were expected to speed these cost reductions and enable a rapid transition across the large gap between actual and commercially acceptable costs.

Opponents of procurement made two main arguments. First, creating a large surge in demand ran the risk of generating a large expansion of capacity in inefficient technology. Once in place, this capacity would retard the development and introduction of new technology. Second, the analogy to semiconductors was imperfect. Cost reductions in semiconductors had

Figure 11-1. *Chronological Summary of Demand Stimulus Measures Applying to Photovoltaics*

September 1977: House approved Tsongas amendment authorizing $28 million for federal purchase of PV systems.

October 1977: House approved Tsongas amendment appropriating $12.2 million for federal purchase of PV systems.

February 1978: The Department of Energy Act of 1978 authorized $13 million for purchase of cost-effective PV for federal facilities.

November 1978:

The National Energy Conservation Policy Act authorized $98 million over fiscal 1979–81 for federal PV purchases.

The Energy Tax Act of 1978 established a 10 percent business credit, which applied to PV; it established a residential tax credit which did not.

The Powerplant and Industrial Fuel Use Act of 1978 limited oil and gas use in electric utilities.

The Public Utility Regulatory Policies Act of 1978 required utilities to buy power from cogenerators and small producers (such as PV) at avoided cost.

April 1980: The Crude Oil Windfall Profit Tax of 1980 was enacted. It raised the residential tax credit to 40 percent of the first $10,000 and applied the credit to PV, and it raised the business tax credit to 15 percent and extended the credit for three years until the end of 1985.

December 1985: Tax credits expired.

October 1986: The Tax Reform Act of 1986 restored business credits at 15 percent in 1986, 12 percent in 1987, and 10 percent in 1988.

November 1988: The Technical and Miscellaneous Revenue Act of 1988 extended 10 percent credit through 1989.

December 1989: The Omnibus Budget Reconciliation Act of 1989 extended 10 percent credit until September 30, 1990.

November 1990: The Omnibus Budget Reconciliation Act of 1990 extended 10 percent credit throughout 1991.

resulted largely from miniaturization. This was not possible in photovoltaics because the output of a photovoltaic cell of given efficiency was proportional to its surface area. The theoretical maximum improvement in efficiency for PV was about a factor of three, rather than the orders of magnitude experienced in semiconductors.

The procurement strategy was implemented by constructing demonstration facilities (table 11-6) that had three features. First, most of the projects were uneconomic. Only the remote projects (Mt. Laguna Air Force Base in California, Natural Bridges National Monument in Utah, and the Papago Indian village in Arizona) could possibly be justified economically. Second,

Table 11-6. *Major PV Projects Receiving Federal Funds*

Contractor	*Location*	*Application*	*Size (peak kw)*
PVUSA	Davis, Calif.	Utility	n.a.
Sacramento Municipal Utility District	Sacramento, Calif.	Utility	1,000
Georgetown University International Center	Washington, D.C.	University building	300
Mississippi County Community College	Blytheville, Ariz.	College	250
Arizona Public Service	Phoenix, Ariz.	Airport	225
Science Applications	Oklahoma City, Okla.	Science and art center	150
Lea County Electric Cooperative	Lovington, N. Mex.	Shopping center	100
Solar Power Corp.	Beverly, Mass.	High school	100
Natural Bridges National Monument	Utah	National Monument community	100
Acurex	Kauai, Hawaii	Hospital	60
Mt. Laguna Air Force Base	Mt. Laguna, Calif.	Radar station	60
BDM	Albuquerque, N. Mex.	Office building	47
E-Systems	Dallas, Tex.	Airport	27
	Mead, Nebr.	Agricultural test facility	25
AM radio station	Bryan, Ohio	AM radio station	25
New Mexico State	El Paso, Tex.	Computer at power station	20
Papago Indian Village	Schuchuli, Ariz.	Indian village	3.5
MIT Energy Laboratory	Concord, Mass.	Northeast Residential Experimental Station	n.a.
Florida Solar Energy Center	Cape Canaveral, Fla.	Southeast Residential Experimental Station	n.a.
New Mexico Solar Energy Institute	Las Cruces, N. Mex.	Southwest Residential Experimental Station	n.a.

Source: Paul D. Maycock and Edward N. Stirewalt, *A Guide to the Photovoltaic Revolution* (Rodale Press, 1985), pp. 212–13, except for the PVUSA project. List does not include Northwest Mississippi Junior College, a $6.6 million project never completed because of fraud.

n.a. Not available.

only two central station utility projects received federal grants. All other projects were for intermediate-load and residential applications. Third, the projects were geographically dispersed, rather than focused on a small number of producers to help them achieve scale economies. Approximately $120 million was spent on federal purchases of PV systems, or 13 percent of total appropriations through 1988.[17]

TAX CREDITS. The federal government subsidized the private procurement of photovoltaic systems through tax credits. These subsidies originated in the Energy Tax Act of 1978. The Crude Oil Windfall Profit Tax Act of 1980 expanded the credits and extended the life of the business credits. The credits expired at the end of 1985, but the business energy credits were restored by the 1986 tax bill, and have been repeatedly extended by subsequent legislation.[18] Furthermore, owners of solar systems can depreciate them over five years, and some states have adopted additional tax incentives for PV. These tax provisions, at a sizable cost to the Treasury,[19] have been instrumental in the construction of nearly 20 megawatts of central station PV.[20]

FUEL USE RESTRICTIONS. Additional demand-side stimulus was given to photovoltaics through restrictions on fuel use by utilities. These were of three types. First, environmental policy stimulated all solar technologies because solar systems do not emit pollutants. Second, direct restrictions on the use of petroleum and natural gas by utilities were adopted in the Powerplant and Industrial Fuel Use Act of 1978. This encouraged the use of alternative generation sources. Third, utilities were required to purchase power from cogenerators or small generators at "avoided cost" by section 210 of the Public Utility Regulatory Policies Act of 1978. The methods of estimating "avoided cost" were generous to alternative power sources and stimulated investment in them.

System Configuration Issues

The most important system configuration issues were the role of storage and the choice between on-site (distributed) versus central station applications. The storage issue was resolved in the late 1970s, and on-site versus

17. Personal communication, Andrew Krantz, Sandia National Laboratories.

18. Extensions were provided by the 1988 tax bill and the 1989 and 1990 deficit reduction measures. The credits are now scheduled to expire at the end of 1991.

19. A precise estimate is not available for PV, since the Treasury's estimates in the annual budget submission on tax expenditures apply to larger aggregates.

20. Maycock and Stirewalt, *Guide to the Photovoltaic Revolution*, p. 198.

central station in 1983. The later date for the second issue reflects its greater complexity as well as its more contentious politics.

STORAGE. Most early PV studies assumed that storage was necessary. The MITRE study, the Solar Task Force, and a 1979 Resources for the Future energy study found that energy storage was not only economical but "critical to making solar energy economic on a large scale."[21]

Subsequent analysis of the interaction of PV with electric grids and of the costs of storage reversed this view. The new consensus is that the cost of PV without storage should be compared with the value of the electricity generated. A utility system with PV as a very high percentage of its generation would require a reliable source of power for the remainder, but this need not be provided by storage. Thus including the cost of storage in PV costs is no more appropriate than including it in the costs of other generation options.[22]

ON-SITE VERSUS CENTRAL STATION. The PV program initially was not based on an assessment of the relative economics of distributed versus central station applications. In 1978 residences were identified as the first major market in which PV would be economic. In 1983 the program reversed itself and declared central stations as the first market that would become economic.

The controversy surrounding the most attractive applications for PV touched on several different issues. At the core are the inherent technical and economic differences between distributed and central power. Distributed systems reduce the need for transmission and distribution of electricity, although they do not eliminate it because distributed users will still buy some power from the central system. Distributed systems also avoid the need to assemble tracts of land for large generation installations, and the transactions costs connected with measuring and billing for power. Finally, such systems offer better opportunities for using the heat created by collecting sunlight, such as by heating water.

The advantage of central systems is potential scale economies. Maintenance and repair are likely to be less costly in centralized operations, for they can economically employ full-time on-site personnel for this purpose. Furthermore, the economic value of photovoltaics must be assessed within

21. Hans H. Landsberg, ed., *Energy, The Next Twenty Years*, report by a Study Group sponsored by the Ford Foundation and administered by Resources for the Future (Ballinger, 1979), p. 493.

22. A good introduction to this issue is provided in William D. Metz, "Energy Storage and Solar Power: An Exaggerated Problem," *Science*, vol. 200 (June 30, 1978), pp. 1471–73.

the context of the entire mix of generation facilities connected to a grid, including the cumulative effect of all PV systems. Centralization facilitates the management of all facilities in a coordinated grid.[23] Finally, central systems can be located where PV is most efficient. For example, inland desert locations may have enough greater insolation than coastal areas to generate additional power that will more than compensate for the cost and power losses of transmission facilities.

Besides these efficiency issues, the debate also included financial issues affecting private but not social costs. Among these were differential tax treatments of residences, other buildings, and utilities (investor owned and municipal), and the cost of financial capital to various potential investors in solar technology. Beyond the financial issues were more fundamental political questions. Environmental and conservation groups had become enamored with the "small is beautiful" concept and the idea that individual independence from large, powerful organizations was a valued end in itself.[24] Their opponents were electric utilities. Distributed technologies brought a competitive threat to their monopoly.[25]

The program's pronouncements in 1983 that the utility market would be the first to be economic imply a realignment of supporters from ideological proponents of decentralized energy systems to electric utilities, a hypothesis considered below. In addition, the reorientation had implications for how PV spending was allocated: the research effort on flat-plate technology would be similar under both programs, but more money would

23. It must be stressed that this is a management issue. The value of PV electricity is independent of whether it is generated at distributed facilities or centralized ones, a point that was often missing from early analyses of the distributed versus centralized issue.

24. The case for dispersed PV systems was put strongly by Barry Commoner, *The Politics of Energy* (Knopf, 1979), p. 44: "However, unlike conventional energy sources, *there is no economy of scale in the acquisition of solar energy* . . . a large, centralized solar plant produces energy no more cheaply than a small one (apart from minor savings in maintenance costs). Thus, with transmission costs eliminated or at least greatly reduced, the small plant is by far the most efficient way to deliver the energy." Commoner's position was attacked in Samuel McCracken, "Solar Energy: A False Hope," *Commentary*, vol. 68 (November 1979), pp. 61–67. The controversy is reviewed in Tracy Kidder, "The Future of the Photovoltaic Cell," *Atlantic Monthly*, June 1980, pp. 68–76. Despite the emphasis in the mid-1980s on central stations, some populist advocates of the distributed systems remain. See Maycock and Stirewalt, *Guide to the Photovoltaic Revolution*.

25. Distributed PV technology was not consistent with the natural monopoly rationale for public utilities, which is that electrical energy is most efficiently generated in large facilities serving thousands of users. Roger G. Noll, "Maintaining Competition in Solar Energy Technology," in Federal Trade Commission, *The Solar Market: Proceedings of the Symposium on Competition in the Solar Energy Industry* (Washington, 1978).

be spent on concentrators and less on residential and institutional issues, such as building codes. Less attention would be given to issues relating to the interconnection of dispersed systems to the grid. Participation of utilities in the PV program would thereby increase.

In the 1980s the Reagan administration sought to eliminate demonstrations, in keeping with its general philosophy about government support of R&D. This effort was not completely successful. In 1986 Congress earmarked $2 million for the Austin, Texas, residential experimental station and $1 million for the Massachusetts Photovoltaic Center.[26] Even more was spent on central station projects. Millions of federal dollars were spent on the Sacramento Municipal Utility District (SMUD) project and the Georgetown University National Exemplar. Although the latter is an on-site use, its large scale and long horizon distinguish it from residential applications. Finally, in 1986 the House Appropriations Committee directed that $7.5 million of the $35.6 million PV budget be given to the utility scale applications (PVUSA) project proposed by Pacific Gas and Electric.

Budgetary History

Besides reflecting the boom-bust cycle just discussed, budgetary statistics show several interesting trends. First, the annual budgets for photovoltaics (table 11-7) and all solar energy programs (table 11-8) exhibit strong similarities.[27] The photovoltaic share of the solar budget rose steadily, from 19 percent in 1976 to a maximum of 39 percent in 1989. Despite this rise, the budgetary trend for PV parallels the entire solar program appropriation (table 11-9).

The budgetary history of solar energy shows that the program was driven

26. *Making Appropriations for Energy and Water Development for the Fiscal Year Ending September 30, 1986, and for Other Purposes*, H. Rept. 99-307, 99 Cong. 1 sess. (GPO, 1985), p. 40.

27. Whatever their importance in the congressional process, authorizations have played a rather minor role in photovoltaics, especially in recent years. Congress failed to pass the relevant energy authorization bill for fiscal 1979–81 and 1985, and for 1983–84 it set authorizations in terms of a formula based on the previous year's appropriation plus 10 percent. Such a rule is not likely to be binding during program contraction. Furthermore, this rule appears to apply at the level of the appropriation account; that is, "operating expenses—energy supply, research and development activities," and photovoltaics is but a small part of this account. See *Department of Energy Authorization for Fiscal Year 1981—Civilian Applications*, S. Rept. 96-687, 96 Cong. 2 sess. (GPO, 1980), p. 9; and *Authorizations for Department of Energy Civilian Programs, Fiscal Year 1982*, S. Rept. 97-81 (GPO, 1981), p. 6.

Table 11-7. *Photovoltaics Appropriations, Fiscal Years 1976–91*[a]
Millions of current dollars

Year	*Administration request*	*House committee*	*House*	*Senate committee*	*Senate*	*Final*
1976[b]	10.0	12.0[c]	29.5	10.7[c]	n.a.[d]	21.6[c]
1977	32.8	34.7	68.8	49.6	n.a.[e]	64.0
1978	57.5	57.5	57.5	57.5	57.5	57.5
Supplemental	57.5	57.5	76.5	76.5	76.5	76.5
1979	76.1	125.3	125.3	106.1	106.1	118.8
1980	130.0	157.0	157.0	152.0	152.0	157.0
Rescission[f]	n.a.	n.a.	n.a.	n.a.	n.a.	150.0
1981	160.2	149.0	160.2	152.2	156.0[g]	160.2
Rescission	139.2	139.2	139.2	139.2	139.2	139.2
1982	62.9	84.0	84.0	78.0	78.0	74.0[h]
1983	27.0	47.0	n.a.[i]	58.0	n.a.[i]	58.0
1984	32.7	46.7	46.7	55.7	55.7	50.4[j]
1985	47.5	57.0	57.0	56.5	56.5	57.0
1986	44.8	45.8	45.8	49.0	49.0	40.7[k]
1987	20.6	35.6	35.6	40.6	40.6	40.4
1988	20.4	35.0	35.0	39.0	39.0	35.0
1989	24.2	24.2	24.2[l]	37.0	37.0	35.5
1990	25.1	36.5	36.5	35.5	35.5	34.7
1991	43.6	46.9	46.9	46.9	46.9	46.9

Sources: Numbers primarily from the report of the Appropriations Committee when the bill is reported, supplemented by the *Congressional Record* for floor amendments and the Department of Energy Budget Request documents. The revised budget request of the incoming president is used in fiscal 1978 and 1982. The revised budget request in 1979 of $106.1 million is not used, since there was no change in administration.

n.a. Not available.

a. Includes operating expenses, capital equipment, and construction.

b. All 1976 appropriation figures represent costs (approximating outlays) except for the final figure (obligations) and the House floor figure (unknown whether it is costs or obligations). The 1976 appropriation figures also do not include $8.2 million for solar research in the 1976 NSF budget. See *Congressional Record*, December 5, 1975, p. S38926.

c. These figures do not include capital equipment funds, which were not allocated among the solar energy accounts. Including these would probably increase the PV amounts by less than $1 million.

d. Glenn amendment, to increase solar appropriation by $30 million, did not specify the increase for photovoltaics.

e. Hart amendment, to increase solar appropriations by $16.4 million, did not specify the increase for photovoltaics.

f. Precise PV totals in rescission bill not clear from House and Senate reports.

g. Description of Tsongas amendment in *Congressional Record* appears to leave PV funding unchanged, with only the Dole amendment increasing the PV total.

h. Although the conference total was $78.0 million, a general reduction taken by the administration reduced this to $74.0 million. See *Department of Energy FY 1983 Congressional Budget Request*, vol. 2, p. 24.

i. Neither House nor Senate took action on the 1983 appropriation bill, so funds were provided through the 1983 continuing appropriations resolution which did not specify amounts for PV.

j. Although the conference total was $51.2 million, a general reduction taken by the Administration reduced the PV total to $50.4.

k. Conference total of $49.0 million reduced by $6.5 million by "management initiatives" and $1.1 million by Gramm-Rudman-Hollings reductions. See *Department of Energy Fiscal 1987 Congressional Budget Request*, vol. 2, p. 25.

l. The $20 million increase in solar from the Brown amendment is not allotted among the solar accounts.

Table 11-8. *Solar Program Appropriations for Fiscal Years 1976–91*[a]

Year	*Administration request*	*House committee*	*House*	*Senate committee*	*Senate*	*Final*
1976	57.1	75.0	84.7	71.7	107.1	114.7
1977	160.0	213.7	304.8	261.9	278.3	290.4
1978	320.0	368.2	368.2	358.5	364.0	368.3
Supplemental	368.3	368.3	392.3	392.3	392.3	389.3
1979	372.1	526.3	526.3	450.1	455.1	485.3
1980	563.8	561.0	561.0	569.9	569.9	577.7
Rescission	. . .	549.2	549.2	517.2	553.7	552.7
1981	613.2	553.1	602.1	564.9	574.6	596.4
Rescission	505.6	500.0	500.0	503.6	503.6	500.0
1982	193.2	304.4	304.4	253.4	253.4	268.2
1983	72.2	180.4	n.a.	187.9	n.a.	201.9
1984	86.4	180.0	180.0	176.0	176.0	181.7
1985	163.6	174.5	174.5	183.0	183.0	179.4
1986	148.0	147.0	147.0	161.7	161.7	144.6
1987	72.3	113.4	113.4	125.8	123.5	123.5
1988	71.2	101.2	101.2	105.1	105.1	96.9
1989	80.4	80.9	100.9	98.8	92.2	92.2
1990	71.2	95.0	95.0	94.6	94.6	92.4
1991	115.9	130.4	130.4	129.7	129.7	131.1

Sources: See table 11-7. As in table 11-7, the 1976 figures represent a mixture of costs and obligations. The original budget request for 1976 and 1979 is shown, since no change in administration took place. In 1978 and 1982, when there was a change in administration, the request of the incoming president is shown. For 1986 the conference figure of $157.2 million was reduced $6.5 million by management initiative (solely in PV), $5.9 million by Gramm-Rudman-Hollings cuts, and $0.2 million by other adjustments.

n.a. Not available.

a. Includes operating expenses, capital equipment, and construction.

Table 11-9. *Percentage Changes in Appropriations for PV and Solar Programs, Fiscal Years 1976–91*

Appropriation	*Photovoltaics*	*Solar*
Average annual increase, 1976–80	74.4	56.4
1981 rescission over previous 1981 level	−13.1	−16.2
1982 over 1981 rescission	−46.8	−46.4
1983 over 1982	−21.6	−24.3
1984 over 1983	−13.1	−10.0
1986 over 1984	−19.2	−20.4
1989 over 1986	−12.8	−36.2
1991 over 1989	32.1	42.2

Sources: See tables 11-7 and 11-8.

Table 11-10. *Percentage Change of Final Appropriation over Administration Request for PV and Solar Program, Fiscal Years 1976–91*[a]

Year	*PV*	*Solar*	*Year*	*PV*	*Solar*
1976	116.0	100.9	1985	20.0	9.7
1977	95.1	81.5	1986	−9.2	−2.3
1978	0.0	15.1	1987	97.1	70.8
1979	56.1	30.4	1988	71.6	36.1
1980	20.8	2.5	1989	46.7	14.7
1981	0.0	−2.7	1990	38.2	29.8
1982	17.6	38.8	1991	7.6	13.1
1983	114.8	179.6			
1984	54.1	110.3	Average change	46.7	45.5

Sources: See tables 11-7 and 11-8.
a. In 1986 the final figures for PV and solar are below the budget request because of Gramm-Rudman-Hollings reductions.

primarily by congressional enthusiasm. In all but two years the amount appropriated by Congress equaled or exceeded the president's request. For PV, appropriations equaled or exceeded the administration's request in every year. The average increase in appropriations over the president's proposed budget was approximately 46 percent for both photovoltaics and all of solar energy (table 11-10).

Congressional enthusiasm for solar energy was not confined to its oversight committees. Neither the House nor the Senate ever voted on an amendment to reduce funds below the amounts proposed by their appropriations and authorization committees. During the period of growth in the 1970s, proposals to increase funds above committee proposals were often offered and passed. Moreover, with few exceptions, proposals to increase spending on PV were packaged with increases in all or most other solar energy activities. During the harder times of the 1980s, appropriations committee proposals were accepted without amendment and were always higher than the administration had requested.[28]

Finally, the budget patterns suggest a standard of assessment for the program. If one assumes that the program managers, executive branch budget officials, and members of the oversight committees usually had more expertise about the program than did Congress in general, these figures imply that the program grew too fast in the 1970s. If so, the

28. The exception is fiscal 1989, when an amendment to increase solar funding by $20 million over the committee level was adopted.

incremental dollars were spent on activities with fairly low productivity. Moreover, because it was a long-term program, one would expect that its chance for success would have improved if the boom and bust pattern of the decade after 1975 had been replaced by a smoother path of expenditures.

Economic Assessments of Photovoltaics

The economic assessment of photovoltaics changed over the life of the program. Throughout the 1970s the assessments were generally optimistic, whereas in the 1980s they became more pessimistic. To understand the factors underlying the shifting "bottom line" evaluation, this section interprets, first, the information provided to decisionmakers at different times, and second, the information that could have been used but was not. The distinction arises because the early goals were based on a simple statistic measuring module costs, which is only one factor in assessing the potential contribution of PV to domestic energy supply.

Module and System Costs

Module cost refers to the costs of a PV module in relation to its electricity output under ideal insolation and weather conditions. Module costs are distinguished from system costs, which also include "balance-of-system" costs (for example, costs of land, support structure, and power conditioning, as well as batteries for systems not connected to power grids). Balance-of-system costs have been estimated as about 50 percent of total costs.

The PV program used two approaches to set goals for the price of modules. The first was to choose a module price goal based on an assessment of technical obstacles and possibilities and then to derive the energy cost resulting from it. The second was to choose an energy cost based on competitive requirements in the market and then to derive the module price necessary to achieve it. The program used the first approach from 1973 to 1979 and the second after 1983. From 1979 to 1983 the program used both. The first approach was used to set the original goals for single-crystal silicon cells of 50 cents per peak watt by 1985 and 10 cents per peak watt by the year 2000, and for more exotic technologies of 20 cents per peak watt in 1985 and five cents per peak watt in 2000.[29]

29. Jet Propulsion Laboratory, *Executive Report of Workshop Conference on Photovoltaic Conversion*, pp. 1–4.

The original goals were altered only in minor ways as the scale-up of the federal program proceeded. The JPL plan for the low-cost silicon solar array program expressed its primary goal as follows: "to develop the technological and industrial capacity for producing more than 500MW of single crystal silicon solar photovoltaic arrays per year at a cost of less than $0.50 per peak watt by 1984."[30] The Solar Energy Task Force Report of Project Independence, prepared under the direction of the NSF, proposed goals of 50 cents per peak watt in 1985 and 10 cents in 1995.[31] ERDA later pushed back the year of attainment of the 50-cent goal to 1986.[32] The first national photovoltaics program plan separated the module price goals into near term (1982), midterm (1986), and far term (1990). The goal of 50 cents (1975 dollars) per peak watt in 1986 was retained, and goals of $2 per peak watt in 1982 and 10 to 30 cents per peak watt in 1990 were added.[33]

Meanwhile, Congress entered the goal-setting process. Section 2 of the Solar Photovoltaic Energy Research, Development, and Demonstration Act of 1978 stated that the purpose of the program was "to reduce the average cost of installed solar photovoltaic energy systems to $1 per peak watt by fiscal year 1988." This formulation shifted the focus of the goals from modules to systems, requiring reconciliation in subsequent program plans. Instead of a single goal for a particular year as provided in the act, the program documents continued to provide different goals for different years (1982, 1986, and 1990) as in the 1978 program plan. However, consistent with the act's emphasis on system prices, goals were now stated for both arrays and the entire system. The array goals (all in 1980 dollars)[34] were $2.80 per peak watt in 1982, $0.70 in 1986, and $0.15 to $0.40 in

30. Jet Propulsion Laboratory, *Proposed Program Development Plan for the National Science Foundation: Low Cost Silicon Solar Array Program*, 1200-172 (Pasadena, Calif., 1974), p. 1-1.

31. Federal Energy Administration, *Final Solar Energy Task Force Report*, p. VII-15.

32. ERDA, *Photovoltaic Conversion*, p. 1. The reason for the change in the year of the goal from the Cherry Hill conference and Solar Energy Task Force (1973–74), to the JPL plan (1974), and finally to the ERDA Program Summary (1976), is not apparent. It may reflect the distinction between the development of a manufacturing capability in 1984 and actual selling prices in 1986, but that is speculation.

33. Department of Energy, Assistant Secretary for Energy Technology, Division of Solar Technology, *National Photovoltaic Systems Program Plan, February 3, 1978*, DOE/ET-0035(78) (1978), pp. 5, 7.

34. During 1979, program goals were recalculated in 1980 dollars, which raised the module goal from 50 cents to 70 cents per peak watt. Department of Energy, Assistant Secretary for Conservation and Solar Energy, *Federal Policies to Promote the Widespread Utilization of Photovoltaic Systems*, vol. 1, DOE/CS-0114/1 (February 1980), p. ES-12; vol. 2, DOE/CS-0114/2 (March 1980), p. 12-5.

1990. The system goals were $6–$13 per peak watt in 1982, $1.60–$2.60 in 1986, and $1.10 to $1.30 in 1990.[35]

For the first time the goals for a particular year were keyed to specific applications: remote in 1982, distributed (grid-connected) in 1986, and central station in 1990. In addition, the goals were designed to satisfy two constraints: the price necessary to compete with conventional sources and the extent of PV cost reduction believed possible.[36] Thus the program operated under the old NSF module price goal and a system price goal, derived partly from considerations of competitiveness. The 1979–83 period thus represents a transition from the technological approach (what is possible to achieve) to the economic approach (competitiveness).

By 1983 the program became driven entirely by cost competitiveness. The 1983 five-year research plan abandoned "costs per peak watt".[37] Instead a late 1990s' energy cost goal was chosen—15 cents per kilowatt hour (in 1982 dollars), based on work by Roger Taylor of the Electric Power Research Institute (EPRI).[38] From this energy cost goal, the plan derived goals for

35. Program documents in 1979–80 contain different versions of these goals. Some omit a 1982 goal—for example, Department of Energy, "National Photovoltaics Program Multi-Year Program Plan" (draft), September 30, 1980, p. 4-4; Department of Energy, *Federal Policies*, vol. 1, p. ES-12; and Jeffrey Smith, *Photovoltaics as a Terrestrial Energy Source*, prepared for Department of Energy, DOE/ET-20356-6 (October 1980), vol. 1, pp. 26–27. These three documents also state the 1986 goal as a point estimate (1.60). Finally, the 1990 central station goal appears to have soon been changed to $1.10–1.80. The figure $1.30 is used in Smith, *Photovoltaics*, vol. 1, p. 27, citing Department of Energy, "National Photovoltaics Program Multi-Year Program Plan" (draft), June 6, 1979, p. 2-3; $1.30 is also used in Department of Energy, "National Photovoltaics Program Multi-Year Program Plan" (draft), September 30, 1980, p. 4-4. The figure $1.80 is used in Department of Energy, Assistant Secretary for Conservation and Solar Energy, *Photovoltaic Energy Systems Program Summary*, DOE/CS-0146 (January 1980), p. 9; Department of Energy, *National Photovoltaics Program: Electrical Power from Solar Cells*, prepared by the Jet Propulsion Laboratory (September 1980), p. 3; Department of Energy, "National Photovoltaics Program Systems Development Plan" (draft), December 1980, p. 1-5; and Department of Energy, *Photovoltaic Energy Systems Program Summary*, DOE/CE-0012 (January 1981), p. 2–8.

36. Jeffrey L. Smith, *Federal Policies to Promote the Widespread Utilization of Photovoltaic Systems, Supplement: Review and Critique*, prepared by the Jet Propulsion Laboratory for the Department of Energy, DOE/JPL-1012-45, Pasadena, Calif. (April 1980), pp. 104–05.

37. Department of Energy, Photovoltaic Energy Technology Division, Office of Solar Electric Technologies, *National Photovoltaics Program Five Year Research Plan 1984–1988* (May 1983), pp. 6, 28–29.

38. Roger W. Taylor, "Utility Requirements for Photovoltaic Power," presented at 161st Meeting of the Electrochemical Society, Montreal, Canada, May 10, 1982; Roger W. Taylor,

area-related module costs and conversion efficiency. Modules achieving 13 percent conversion efficiency were to cost no more than $40 per square meter, and modules achieving 17 percent efficiency were to cost no more than $75 per square meter.[39]

A straightforward calculation based on assumptions in the EPRI report shows that these goals were more ambitious module price goals than the 50 cents per peak watt that had been used from 1973 to 1983, and are equivalent to a cost of 22.4 and 32.2 cents per peak watt (in 1975 dollars) for 13 percent and 17 percent efficient modules, respectively.[40] This discrepancy was noted in the draft of the five-year plan prepared by the JPL, although the comparison was omitted from the final report. The JPL draft report states that the new goals were more strict than the old goals because "they are aimed at long-term economic viability and at achieving significant levels of energy production. The change in goals, of course, reflects the change in the program's objective, that is, to conduct long-term, high-risk research with the potential for high payoff."[41]

The five-year research plan was updated in 1987 in several minor respects.[42] First, the energy cost goal was reformulated from a goal levelized in current dollars to one levelized in constant dollars.[43] Second, the goal

---

"Photovoltaic System Requirements: Central and Distributed Applications," in *EPRI Conference Proceedings: Solar and Wind Power—1982 Status and Outlook*, EPRI AP-2884 (Palo Alto, Calif.: Electric Power Research Institute, 1983); and EPRI, *Photovoltaic Systems Assessment: An Integrated Perspective*, Special Report prepared by Roger W. Taylor, AP-3176-SR (September 1983).

39. Department of Energy, *National Photovoltaics Program Five Year Research Plan, 1984–1988*, pp. 28–32.

40. The module area required to generate a kilowatt of AC power is given in the five-year plan by:

$$A = 1/(average\ peak\ insolation \times system\ efficiency),$$

where system efficiency equals the product of balance-of-system efficiency and module efficiency. The five-year plan assumes that average peak insolation is 1 kw/m^2 and balance of system efficiency is 0.81. With these assumptions, A equals 9.50 m^2/kw for 13 percent modules and 7.26 m^2/kw for 17 percent modules. Multiplying by the area-related module cost and converting to 1975 dollars gives a cost of 22.4 cents and 32.2 cents per peak watt for 13 percent and 17 percent efficient modules, respectively.

41. Department of Energy, "National Photovoltaics Program Five Year Research Plan, 1984–1988" (draft), March 1983, p. 39.

42. Department of Energy, *Five Year Research Plan 1987–1991 Photovoltaics: USA's Energy Opportunity*, DOE/CH100093-7 (May 1987).

43. Conventional, nonrenewable energy prices were expected to rise over time. For any such price path, one can determine two other price paths, each where the energy price remains constant over the time frame of the analysis and whose net present value is equal to the net present value of the conventional energy price path. For one of these paths, the

was tightened from 6.5 cents per kilowatt hour to 6.0 cents in levelized constant dollars. Third, the time of accomplishment was delayed from late 1990s to the year 2000. Fourth, the module efficiency goals were made more stringent.

Probability of Attaining Goals

Congress received many assurances during the initial years of the program that 50 cents per peak watt was achievable.[44] In 1978 the JPL published a detailed set of cost targets for the components of a manufacturing process that could meet the 50 cents per peak watt goal for flat-plate silicon photovoltaic modules in 1986.[45] In 1979 the American Physical Society Study Group on Photovoltaics and the General Accounting Office released guardedly optimistic assessments.[46]

By 1982 the prospects for achieving the 1986 goals had dimmed. The General Accounting Office assessed the effects of fiscal 1982 budget reductions on achieving the PV cost and production goals and concluded that "the Act's goals had little likelihood of being achieved prior to the fiscal year 1982 budget reductions; however, due to these reductions, the goals now have even less likelihood of being reached."[47] In 1983 the Department of Energy adopted even more ambitious cost targets, making success even less likely.

The flat-plate solar array (FSA) project was terminated in 1986. After-

---

path is level in current dollars; for the other path, the path is level in constant, inflation-adjusted dollars. A price path of 15 cents per kilowatt hour, levelized in current dollars, is equal to 6.5 cents, levelized in constant dollars, for the inflation rate, discount rate, and plant life used in the analysis.

44. See, for example, John V. Goldsmith, testimony in *Solar Photovoltaic Energy*, Hearings before the Subcommittee on Energy of the House Committee on Science and Astronautics, 93 Cong. 2 sess. (GPO, 1974), p. 75; and *ERDA Authorization Fiscal Year 1977: Part II—Solar Energy*, Hearings before the Subcommittee on Energy Research, Development, and Demonstration of the House Committee on Science and Technology, 94 Cong. 2 sess. (GPO, 1976), p. 269.

45. Robert W. Aster, *Economic Analysis of a Candidate 50¢/Wpk Flat-Plate Photovoltaic Manufacturing Technology*, prepared by the Jet Propulsion Laboratory for the Department of Energy, DOE/JPL-1012-78/17 (Pasadena, Calif., December 1978).

46. American Physical Society, *Principal Conclusions of the American Physical Society Study Group on Solar Photovoltaic Energy Conversion* (New York, 1979), p. 6; and General Accounting Office, letter to the Honorable James R. Schlesinger, Secretary of Energy, EMD-79-40 (30712), April 19, 1979, p. 4.

47. General Accounting Office, *Probable Impacts of Budget Reductions on the Development and Use of Photovoltaic Energy Systems*, Report to the Chairman, Subcommittee on Energy Conservation and Power of the House Committee on Energy and Commerce, EMD-82-60, B-206660 (1982), p. 2.

Table 11-11. *Module Prices in Block Buys, Selected Years, 1975–83*
1980 dollars

Year	*Price per peak watt*
1975	29.40
1976	19.18
1977	14.56
1983	4.24

Sources: Prices for 1975–77 are for Jet Propulsion Laboratory block buys 1–3, given in Marie Slonski and Bob Easter, "Cost Effectiveness Aspects of FPUP Strategy," JPL memorandum, September 12, 1979; and for 1983, for 1 mw Sacramento Utility District (SMUD) purchase, given in W. T. Callaghan, P. K. Henry, and P. A. McGuire, *Cherry Hill Revisited: Economic Payoff for Technical Achievements in Flat-Plate Photovoltaics, 1975–1985*, Proceedings of the 18th IEEE Photovoltaics Specialists Conference (Pasadena, Calif.: Jet Propulsion Laboratory).

ward, several reports stated that the FSA project essentially achieved the original goal. For example, the manager of the FSA project claimed that these goals were achieved with approximately 50.2 percent of the budget originally recommended.[48] These claims are speculative because they are based on projections using SAMICS, an elaborate costing model developed by the JPL for estimating manufacturing costs in a full-scale, state-of-the-art process, not on the presence of any manufacturer willing to sell at a price equal to the goal. The lowest price available for a large quantity of modules was $4.95 per peak watt in 1983 (in 1982 dollars).[49]

Another indicator of the success of the program is the price trend for "block buys" of PV modules during the procurement phase of the program. The decline in average prices for each block buy (table 11-11) provides evidence that costs were falling, although the relationship between price and cost is imperfect. A relatively small number of modules were purchased, so that prices would have been lower for larger orders if there were scale economies and a learning curve. But if firms selling modules conclude that the prices in these buys would affect future sales, they might submit "buying in" bids below manufacturing costs. In any case, the evidence about prices and costs indicates that while the program did not achieve its goals, it was nonetheless highly successful. Although the 1983 prices were

48. W. T. Callaghan, P. K. Henry, and P. A. McGuire, *Cherry Hill Revisited: Economic Payoff for Technical Achievements in Flat-Plate Photovoltaics 1975–1985*, Proceedings of the 18th IEEE Photovoltaics Specialists Conference (Pasadena, Calif.: Jet Propulsion Laboratory), p. 3. Others argue that flat-plate solar array funding was only 44.1 percent of funding recommended in the 1978 National Photovoltaics Program Plan, and only 39.4 percent of the funding recommended in the original plan for the project in 1974. See Paul D. Maycock, "The Jet Propulsion Laboratory Low Cost Solar Array Project, 1974–1986," presentation at Flat-Plate Solar Array Project Integration Meeting, Pasadena, Calif., April 29, 1986, table 3.

49. Callaghan, Henry, and McGuire, *Cherry Hill Revisited*, p. 4.

about three times higher than the original goal, they represent a reduction in PV costs of approximately 90 percent in about a decade, reflecting an annual rate of technical progress of about 20 percent. This performance is indeed impressive, especially in comparison with other federal energy R&D projects.

The Economic Benefits of PV

Until the 1980s official forecasts of the contribution of photovoltaics to national energy supply were relatively optimistic for the year 2000 and beyond (table 11-3). These conclusions assumed that if the module cost targets were attained, PV would prove competitive with other sources of electricity. But the justification for a long-term R&D program costing hundreds of millions of dollars was that it would provide net economic benefits. To do so, PV must supply electricity at a net cost savings for a large enough number of users to justify the R&D. Generally speaking, the early assessments of the program significantly overstated these net economic benefits.

The first source of error is that between 1974 and 1975 the bottom fell out of the forecasts of the total contribution of PV to national energy supply. In 1973 and 1974 the low, business-as-usual estimates of PV displacement were approximately 5.0 quads in the year 2000; by 1975 ERDA was projecting 2.1 quads with a fully functioning PV program. Subsequent estimates have successively lowered the PV contribution in 2000. An important element in the revisions was the growing realization that energy demand was not going to grow according to its historical trend rates and that, consequently, the demand for new electrical generation capacity was going to be much weaker than had been expected in the early 1970s.

The early analyses also overstated the value of PV penetration. One source of error was the assumption concerning the type of fuel displaced. Both the 1974 Solar Energy Task Force and 1975 ERDA testimony assumed that 100 percent of the displaced fuel would be oil;[50] however, the 1979 American Physical Society Study Group concluded that very little of the displaced fuel would be oil, a conclusion that was challenged by the JPL.[51]

50. Federal Energy Administration, *Final Solar Energy Task Force Report*, p. VII-10; and *Energy Research and Development*, Senate Hearings, pp. 207–08.

51. American Physical Society, *Principal Conclusions*, pp. 45–47; and Jet Propulsion Laboratory, *A Review of the American Physical Society Report on Solar Photovoltaic Energy Conversion*, Photovoltaics Program Technology Development and Applications Lead Center Report 5230-2 (Pasadena, Calif., February 1980), pp. 13–14.

The issue of which technologies PV will displace is complex and requires an accurate assessment of all other new generation options that could compete in an optimized generation mix.[52] If 100 percent was too high a percentage for oil displacement, that assumption would have overstated PV value in that energy policies during the 1970s were usually evaluated in terms of their ability to reduce U.S. dependence on imported petroleum.

There were at least three additional overstatements of the value of the PV program. First, the dollar value of oil savings was compared with the R&D costs of the PV program rather than with the total cost of the PV systems.[53] Second, the benefits from PV use (for example, fuel savings) were sometimes not discounted. The high ratio of capital costs to operating costs and the lifetime of twenty to thirty years assumed for PV systems make the failure to discount especially serious. Third, the benefits of PV were sometimes confused with the benefits of a federal government PV research and development program. The benefits of PV use should be evaluated as suggested earlier. The benefits of a federal R&D program, by contrast, depend on realizing the benefits of PV sooner than would otherwise occur.[54]

The principal conclusions to be drawn from studying the wealth of analysis of PV's prospects are as follows. In the initial phase of the program, the economic benefits of the program were consistently overstated. Note, however, that the choice of module costs as a goal is consistent with the growth of the program out of the NSF and NASA programs, as well as with the goals of the more extreme proponents of decentralized power systems. Module costs are the most relevant feature of PV for both space applications and for true stand-alone (that is, not grid-connected) residential use. The lack of discussion of system costs, for example, is consistent with a pure

52. Taylor, "Photovoltaic Systems Assessment," p. 4-25.

53. This is done implicitly in NSF/NASA, "An Assessment of Solar Energy," tables 3, 4; and explicitly in Federal Energy Administration, *Final Solar Energy Task Force Report,* p. VII-10; and *Energy Research and Development*, Senate Hearings, pp. 207–08. In addition, the benefits of various technologies were sometimes evaluated by comparing the energy production of the technology with the federal funds devoted to the technology. The problem here is that the extent of displacement may not be perfectly correlated with cost savings. See, for example, the response of the Policy and Evaluation group in the Department of Energy to Senator Johnston, *Inside DOE*, May 18, 1979, or the suggested approach in Richard Schmalensee, "Appropriate Government Policy toward Commercialization of New Energy Supply Technologies," *Energy Journal*, vol. 1 (April 1980), p. 18.

54. This point is made clear in T. W. Hamilton, *Potential Benefits of Photovoltaic System Development—Methodology and Results*, Report prepared by Jet Propulsion Laboratory, 5210-16 (Pasadena, Calif., 1981), in which the proper approach is developed.

research mind-set. Balance-of-system costs were viewed as dictated largely by the mature technology of the construction industry, so that R&D was unlikely to reduce them significantly.

Only when the goal became efficient grid-connected use of PV did the complications of avoided capacity and avoided energy costs become relevant. The optimistic biases introduced by the early goals were not necessarily naked attempts to make the technology look better than it was. Instead, they may have been by-products of the program's evolution.

The Politics of the Photovoltaics Program

Although the demise of solar programs began with the election of Ronald Reagan, the question remains whether Reagan's opposition to the program reflected fundamental ideological and political differences with his predecessor, a simultaneous change in the long-term prospects of the solar energy program, or differences between Republicans and Democrats over the constituencies they seek to help with targeted federal expenditures.

Party and Ideology

The policy differences between Jimmy Carter and Ronald Reagan on the issue of solar energy are abundantly clear. President Carter was an advocate of a broad-based, research-oriented energy program, and specifically of solar energy. On Sun Day, May 3, 1978, President Carter instituted the domestic policy review of solar energy. On June 20, 1979, he proposed a national goal of meeting 20 percent of our energy needs with solar and renewable resources by the end of this century and outlined a comprehensive program enlisting a number of government agencies to accelerate the use of solar energy.[55] In contrast, President Reagan, soon after taking office, proposed shifting "the Department of Energy's solar activities away from costly near-term development, demonstration, and commercialization efforts and into longer-range research and development projects that are too risky for private firms to undertake." Accompanying this redirection was a proposed reduction of Department of Energy solar spending by more than 60 percent in 1982, with cumulative savings of nearly $1.9 billion by the end of 1986.[56]

55. *Solar Energy: Message from the President of the United States Transmitting Proposals for a National Solar Energy Strategy*, H. Doc. 96-154 (GPO, June 1979).

56. *Program for Economic Recovery: Message from the President of the United States*, H. Doc. 97-21, 96 Cong. 1 sess. (GPO, February 1981), pp. 4–16.

A similar, though less dramatic, shift in the composition of Congress took place in 1981. Republican strength in the House and Senate was much greater in 1981 than in the late 1970s. In the House, Republican strength in 1981 was approximately equal to that in 1973–74, about 45 percent of the membership. In all intervening years, Republicans had controlled fewer than 40 percent of the seats. In the Senate, Republicans took control in 1981 for the first time in thirty years. As in the House, the previous peak in Republican strength was in the early 1970s. Thus rapid growth in solar energy spending took place at the peak of Democratic congressional influence and during a four-year period of Democratic control of the presidency.

The politics of solar energy must also be placed in the larger context of energy policy. Until the early 1970s energy R&D policy consisted primarily of support for nuclear power, which was then still relatively uncontroversial. By the mid-1970s nuclear power had become a point of controversy, with liberal members of Congress more likely to oppose nuclear power and conservatives likely to advocate it. Exotic energy technologies, like solar energy, arose in a milieu of nuclear controversy, yet also of great national concern about future energy resources. Such technologies became the liberal alternative to nuclear power for dealing with the energy crisis. Indeed, solar advocates often argued for more funds to balance the funds received by the breeder (chapter 9). The fall in oil prices that began shortly after Reagan took office was accompanied by a drop in the salience of energy policy and a decline in all energy research programs in the federal government.

Distributive Benefits

In the photovoltaics program, short-run program benefits, represented by program expenditures, were fairly small for each congressional district. However, the long-run benefits also had distributive aspects, for not all parts of the country could be expected to benefit equally from a commercially attractive solar industry. The output of a PV system depends on the availability of sunlight, which of course varies significantly across the country. Because of transmission costs, most PV benefits would accrue where its electricity was produced.

Several distributive aspects of the solar energy program are likely to have had similar effects for all or most solar technologies. Sunlight, for example, is important to solar thermal and PV technologies, though less important for wind. Another common factor is the price of conventional and other energy sources, which also varies geographically. All solar technologies will be more attractive in areas where, all else equal, conventional energy sources are more expensive. Still another common factor is overlapping technology. For example, solar thermal and PV both use concentrators, so that advances in concentrator technology would benefit both. Moreover, the presence of technical commonalities could lead to geographic concentration of solar R&D across a spectrum of technologies. In fact, many firms and government labs worked simultaneously on several solar technologies. Thus the distributive benefits of photovoltaics, as well as ideological factors, constituted a basis for a solar coalition in Congress.

The demise of the energy crisis largely affected the value of short-term solar endeavors rather than PV, which was essentially a long-term R&D project. Nevertheless, support for the PV program declined as well. Furthermore, in the penurious 1980s, solar programs were forced to compete for funds among themselves, which clearly undermined the coalition. Finally, the shift in the PV program toward central system applications was at odds with the small-is-beautiful ideology of the solar coalition.

Econometric Analysis of Roll Call Voting

To test hypotheses about the politics of PV, I analyzed congressional roll call votes about photovoltaics using an econometric model. The votes analyzed (figure 11-2) include all roll call votes pertaining to photovoltaics from 1974 through 1988 except for votes for final passage of energy authorization or energy and water development appropriation bills and two lopsided votes, both from the House. Of the nine votes analyzed, eight affect solar authorizations or appropriations, and one affects only PV authorizations. Variable definitions are shown in figure 11-3 and sample statistics in table 11-12.[57]

57. Four variables were initially included in the analysis but were later dropped, and the equations reestimated. Authorizations committee membership, appropriations subcommittee membership, and authorizations subcommittee membership were uniformly insignificant. Natural gas prices were highly correlated with electricity prices, and less significant than electricity prices. The rationale for inclusion of natural gas prices had been that all but one of the votes pertain to solar energy programs in addition to solar electric, and solar programs such as heating and cooling substitute for other types of energy as well as for electricity.

ACA scores are from *Congressional Quarterly*, May 22, 1976; February 5, 1977; April 15, 1978; March 21, 1981. Electricity prices are from Department of Energy, Energy Information Administration, *State Energy Price and Expenditure Report 1970–1981*, DOE/

INTERPRETATION OF VOTES. The nine votes differed in content and expected support. First, though all three Senate amendments increase solar programs in general, the Gravel amendment does so to such an extent that supporters of moderate increases, such as proposed in the Glenn and Hart amendments, might find the increase too extreme. Because PV is an R&D program, too rapid expansion entails inefficiencies, including potentially sharply diminishing marginal returns and the danger of locking prematurely into technologies that are further developed at the expense of speculative options. Passage of the amendment could hurt the long-term prospects of the program.

Among the House votes, the Brown amendment reallocates funds among solar programs and consequently splits the solar coalition. Though solar heating proponents would favor the amendment, proponents of other pieces of the solar program would oppose it. Similarly, the Tsongas amendment addresses only PV. Insofar as a solar logroll depends on all parts of the program being contained in a single bill, coalitional support would be weaker for the Tsongas proposal than for the remaining amendments.[58] Furthermore, the Tsongas bill authorizes funds for procurement of PV systems rather than the federal R&D program, and consequently would lose the support of PV advocates who oppose the procurement strategy. The Fuqua amendment includes funds for fusion as well as solar programs and thus is distinguished by support and opposition from fusion proponents and opponents, coalitions that do not necessarily coincide with support and opposition to solar programs.

The interpretation of the Anderson and Brown amendments requires some explanation. The Jeffords amendment was to increase the fiscal

EIA-0376(81) (1984). Insolation estimates were derived from Jeff H. Smith, *Handbook of Solar Energy Data for South-Facing Surfaces in the United States*, vol. 1: *An Insolation, Array Shadowing, and Reflector Augmentation Model*, DOE/JPL-1012-25 (Pasadena, Calif.: Jet Propulsion Laboratory, 1980). PV spending figures were estimated using figures in the annual program summaries of the PV program, scaled up so that the sum over all districts and states equals the appropriation for that year. Since the program summarizes each contract, the associated spending, the name of the company, and the city, I assigned this spending to districts and states, using Charles B. Brownson, *Congressional Staff Directory 1977*, and Congressional Quarterly, *Congressional Districts in the 1970s*, 2d ed., 1974. Districts where a PV manufacturer was present were determined from Department of Energy, *National Photovoltaics Program: Electrical Power from Solar Cells*, prepared by the Jet Propulsion Laboratory (Pasadena, Calif., September 1980). Because of a lack of data on other years, this 1980 list is used for all years.

58. See also the discussion of coalitions and logrolls in chapter 10.

Figure 11-2. *Selected Roll Call Votes on Photovoltaics*

Senate:

Gravel (Democrat of Alaska) amendment to increase fiscal 1976 and transition quarter authorization for solar energy research and development by $63 million and $18 million, respectively. Rejected 34–59, July 31, 1975. *CQ* vote no. 366.

Glenn (Democrat of Ohio) amendment to increase appropriation for solar energy research and development in fiscal 1976 and the transition quarter by $46 million. Adopted 52–31, December 5, 1975. *CQ* vote no. 554.

Hart (Democrat of Colorado) amendment to increase fiscal 1977 appropriations for solar energy programs by $16.4 million. Adopted 54–41, June 23, 1976. *CQ* vote no. 312.

House:

Richmond (Democrat of New York) amendment to the McCormack (Democrat of Washington) amendment, to increase appropriation for solar energy research by $54.1 million for fiscal 1976 and $9.9 million for the transition quarter. Rejected 190–219, June 24, 1975. *CQ* vote no. 255. The McCormack amendment to increase appropriations for solar energy by $13 million in fiscal 1976 and $9.5 million for the transition quarter was subsequently passed by voice vote.

Anderson (Republican of Illinois) amendment to the Brown (Democrat of California) substitute amendment, to eliminate the $58 million increase for solar heating and cooling programs, thus cutting the 1977 increase for solar electric and other programs from $116 million to $58 million. Rejected 188–207, May 19, 1976. *CQ* vote no. 204.

Brown substitute amendment to the Jeffords (Republican of Vermont) amendment, to redistribute the $116 million increase for solar electric, ocean thermal, wind energy, biomass, and related programs to $58 million for these programs and $58 million for solar heating and cooling. Also deleted line item authorization for specific technologies and provisions to increase ERDA staffing. Adopted 265–127, May 19, 1976. *CQ* vote no. 205.

Jeffords amendment to increase fiscal 1977 authorization by $116 million for solar electric, ocean thermal, wind energy, biomass, and related items. Adopted 321–68 as amended by Brown amendment, May 19, 1976. *CQ* vote no. 206.

Tsongas (Democrat of Massachusetts) amendment to increase fiscal 1978 authorization by $28 million for federal purchase of solar photovoltaic systems and $10 million for technology development. Adopted 227–179, September 21, 1977. *CQ* vote no. 535.

Fuqua (Democrat of Florida) amendment to increase 1981 appropriations for energy supply, research, and development by $107 million, of which $49 million was for solar programs and the rest for fusion. Adopted 254–151, June 24, 1980. *CQ* vote no. 327.

Sources: *Congressional Quarterly Almanac*, various years; and *Congressional Record*, May 19, 1976, pp. H14410–27; June 24, 1980, pp. H16604–14; and September 21, 1977, p. H9765.

Figure 11-3. *Variable Definitions for Econometric Analysis*

Party: party affiliation of member (0 = Republican: 1 = Democrat).

Term: year term of senator ends (either 77, 79, or 81); variable not used in House regressions.

ACA: Americans for Constitutional Action score of member for year of vote, divided by 100.

App. dummy: member of Appropriations Committee, year of vote (0 = nonmember, 1 = member).

Elect. price: average electricity price in member's state, year of vote (1975 dollars per million BTU).[a]

Log of pop. den.: natural log of population density in state, year of vote, for Senate model (thousands of persons per square mile); natural log of 1970 population density in district for House model (persons per square mile).[b]

Insolation: estimated average annual insolation (sunlight) in state for Senate model; in district for House model (megawatt hours per meter).

PV spending: estimated real per capita federal PV expenditures in state in fiscal year being voted on, discounted at 10 percent (fiscal 1976 dollars per capita); for House model, total real federal PV expenditures (in millions of fiscal 1976 dollars, discounted at 10 percent) are used.[c]

PV manuf. dummy: PV manufacturer in state for Senate model; in district for House model.

a. Electricity price data are not available on a district basis.

b. Population estimates for intercensus years are not available for congressional districts. I use 1970 data because I could not find 1980 data per district (with the districts defined prior to the redistricting resulting from the 1980 census).

c. Total expenditures are used in House model since there is little variation in the population of congressional districts.

1977 authorization for solar electric, ocean thermal, wind, and biomass technologies by $116 million. The Brown amendment was a substitute amendment to the Jeffords amendment and redistributed this $116 million by giving half to solar heating and cooling and half to the technologies that were to receive the entire $116 million under the Jeffords amendment. The Anderson amendment to the Brown amendment allocated money to the same technologies as Jeffords but proposed to reduce the increase for all programs to $58 million. The sequence of votes was Anderson, then Brown, then Jeffords.

Assuming that no one is voting sophisticatedly, the interpretation of these is straightforward. The Anderson amendment (if Brown as amended is substituted for Jeffords and is adopted) results in an increase of $58 million to solar electric, wind, ocean thermal, and biomass, which is exactly the same result for these technologies if Anderson is rejected (and Brown is substituted for Jeffords and is adopted). Thus while photovoltaic

Table 11-12. *Mean Values of Variables Used in PV Regressions*[a]

Independent variable	Senate vote			House vote					
	Gravel	*Glenn*	*Hart*	*Richmond*	*Anderson*	*Brown*	*Jeffords*	*Tsongas*	*Fuqua*
Party	0.613	0.627	0.642	0.668	0.662	0.664	0.652	0.663	0.662
	(0.487)	(0.479)	(0.479)	(0.471)	(0.473)	(0.472)	(0.476)	(0.473)	(0.485)
Term	78.978	78.904	78.979	. . .	. . .	. . .	. . .	. . .	. . .
	(1.62)	(1.64)	(1.629)						
ACA	0.345	0.36	0.373	0.451	0.408	0.408	0.413	0.434	0.468
	(0.308)	(0.304)	(0.353)	(0.322)	(0.333)	(0.333)	(0.334)	(0.317)	(0.281)
App.	0.247	0.253	0.253	0.126	0.127	0.125	0.124	0.126	0.119
	(0.431)	(0.435)	(0.435)	(0.332)	(0.333)	(0.331)	(0.329)	(0.331)	(0.323)
Elect. price	8.616	8.628	8.641	9.314	9.424	9.44	9.408	9.775	10.579
	(2.653)	(2.751)	(2.52)	(2.642)	(2.519)	(2.481)	(2.513)	(2.457)	(2.638)
Log of pop. den.	−2.791	−2.773	−2.818	5.851	5.843	5.841	5.796	5.775	5.83
	(1.463)	(1.468)	(1.442)	(2.215)	(2.185)	(2.181)	(2.164)	(2.195)	(2.202)
Insolation	1.65	1.632	1.656	1.636	1.637	1.638	1.636	1.637	1.635
	(0.277)	(0.271)	(0.278)	(0.24)	(0.238)	(0.236)	(0.235)	(0.244)	(0.239)
PV spending	0.103	0.08	0.135	0.048	0.13	0.12	0.121	0.134	0.107
	(0.338)	(0.262)	(0.265)	(0.221)	(0.952)	(0.932)	(0.944)	(0.645)	(0.50)
PV manuf.	0.258	0.205	0.242	0.05	0.055	0.053	0.049	0.057	0.049
	(0.438)	(0.404)	(0.428)	(0.218)	(0.227)	(0.223)	(0.216)	(0.231)	(0.217)

a. The numbers in parentheses are standard deviations.

supporters have no preference per se, the rejection of Anderson (if Brown is substituted for Jeffords) leads to a $58 million increase in solar heating and cooling compared with no increase if Anderson is accepted. A no vote on Anderson is thus pro-solar. Once the Anderson amendment loses, the decision is on the Brown amendment. The Brown amendment increases the solar electric, wind, ocean thermal, and biomass authorization by $58 million; if Brown is defeated (and Jeffords is adopted), these technologies get $116 million. A no vote on Brown is thus pro-photovoltaic.

The rationale for sophisticated voting is that the probability of adoption of a proposal may be affected by the adoption or rejection of amendments to it. In this case, if one thought that a $116 million increase (Jeffords or Brown amendment) would be defeated, then PV proponents would vote for the Anderson amendment (a $58 million increase). Similarly, if increasing the authorization for heating and cooling was necessary to increase the authorization for PV (a logroll), then PV proponents should support the Brown amendment.

If expectations were consistent with the outcome of the votes, the second sophisticated strategy is plausible but not the first. The Jeffords amendment, as amended by Brown, passed, and hence the coalition as redefined by Brown was successful, whereas no evidence exists on the possible success of the original Jeffords amendment. However, because Jeffords passed 321 to 68, a large increase in solar spending had widespread support, so that the smaller increase defined by Anderson was clearly not all that could be obtained by solar proponents. Finally, the vote on the Jeffords bill was overwhelming, as is typical on final passage of acts rather than amendments. Consequently, tests for specific solar support are likely to be weaker for this bill, since it seems to represent a consensus in Congress.

REGRESSION RESULTS. All the reported equations are significant at the 0.001 confidence level, using the likelihood ratio test (table 11-13). The coding assumption of "no" votes on Anderson and Brown as pro-photovoltaic is supported by the fact that this assumption produces coefficients on the Americans for Constitutional Action (ACA) score, Appropriations Committee membership, electricity price, population density, and sunlight that have the hypothesized signs.

PARTY AND IDEOLOGY. Liberals are more likely to support solar programs, and, holding ideology constant, Republicans are more likely to support solar programs than Democrats are. Nevertheless, the program is identified with the Democratic party, for its members have, on average, much lower ACA scores. The party correction, rather, indicates that energy

Table 11-13. *Roll Call Regression Results*[a]

	Senate vote			House vote					
Independent variable	*Gravel, 1975*[b]	*Glenn, 1975*[c]	*Hart, 1976*[c]	*Richmond, 1975*[c]	*Anderson, 1976*[b]	*Brown, 1976*[b]	*Jeffords, 1976*[b]	*Tsongas, 1977*[b]	*Fuqua, 1980*[c]
Constant	−7.73	−7.68	−33.5	−.19	−.52	−2.36	1.24	2.60	5.05
	(.57)	(.50)	(2.40)	(.16)	(.40)	(1.89)	(.66)	(2.18)	(3.99)
Party	.81	−1.60	.08	−1.83	−.73	−.84	.32	−.74	−1.38
	(.98)	(2.00)	(.11)	(4.55)	(1.92)	(2.53)	(.82)	(2.01)	(3.16)
Term	.07	.05	.40	. . .	. . .	. . .	. . .	. . .	. . .
	(.39)	(.26)	(2.29)						
ACA	−3.43	−6.20	−4.59	−5.14	−5.41	−1.20	−5.11	−3.94	−5.31
	(2.23)	(3.96)	(3.67)	(7.52)	(7.98)	(2.22)	(6.17)	(6.22)	(6.17)
App.	−.65	−1.93	−1.09	−1.28	−.35	−.41	−.04	−.83	−2.24
	(1.02)	(2.70)	(1.68)	(3.33)	(.94)	(1.13)	(.09)	(2.41)	(5.66)
Elect. price	.30	.09	.04	.12	.20	.23	.17	.15	.12
	(2.12)	(.68)	(.30)	(2.36)	(3.43)	(3.99)	(1.82)	(2.67)	(2.13)
Log of pop. den.	−.65	−.19	−.31	−.09	−.14	−.09	.01	−.11	−.03
	(2.16)	(.80)	(1.31)	(1.44)	(1.90)	(1.44)	(.08)	(1.70)	(.42)
Insolation	−1.17	4.20	1.67	1.89	1.37	.63	1.01	−.52	−1.15
	(.85)	(2.62)	(1.31)	(3.47)	(2.27)	(1.14)	(1.21)	(.97)	(2.06)
PV spending	3.83	−2.18	3.33	−.21	−.05	−.20	−.17	.39	.27
	(1.06)	(1.13)	(1.82)	(.37)	(.25)	(.69)	(.76)	(1.53)	(1.02)
PV manuf.	−.82	1.87	−.69	.46	.21	.70	−.28	.66	.89
	(1.00)	(1.75)	(.81)	(.79)	(.35)	(1.40)	(.29)	(1.10)	(1.21)
Log likelihood	45.8	42.2	46.4	125.8	173.0	85.0	296.2	110.6	152.4
Nobs	93	83	95	419	402	399	388	406	405
Percent pro-PV	36.6	62.7	56.8	46.5	47.8	67.2	82.5	55.9	62.7

a. The numbers in parentheses are *t*-statistics.
b. Authorizations.
c. Appropriations.

policy is not unidimensional along the scale identified by the ACA. Relatively liberal Republicans, as defined by the ACA (for example, with scores around 50), are more likely to support solar programs than relatively conservative Democrats are (with equivalent ACA scores of about 50). This is precisely the opposite pattern to that found for support of the breeder reactor (see chapter 9) and suggests that solar and nuclear programs had different support constituencies, even after ideology is controlled for.

These trends are least significant in the Brown vote, which is consistent with the thesis that the Brown amendment split the solar coalition, and in the Gravel vote, which is consistent with a division of solar proponents over this vote owing to the large increase proposed.

DISTRIBUTIVE POLITICS. Program expenditures did not constitute a strong basis of support for these bills. The coefficient on program expenditures is at best marginally significant and is of the wrong sign in five of the nine votes. Alternative specifications of the spending variable yielded similar results. These findings are consistent with the other case studies. First, the absolute level of spending in all instances is low, and hence expenditures are unlikely to pass any threshold of importance for political support. Second, the pre-1977 votes predate the period of the largest program budgets. Retrospective importance, consequently, argues against finding any expenditure-based distributive support in these votes. Finally, the votes before 1977 were general solar votes rather than PV votes, which further dilutes specific PV effects. In the later two votes—Tsongas and Fuqua—the sign of the contract coefficient is positive and significant at the 90 percent confidence level in only the Fuqua vote.

The results for the presence of a PV manufacturer in one's district or state are similar to those for program expenditures. Given the long-term focus of the program and the extent to which federal R&D might limit the long-term advantage of being a current manufacturer, the absence of political appeal to manufacturers is not surprising.

Unlike program expenditures and the presence of PV manufacturing in a district, variables measuring ultimate program benefits are significant in explaining support. The electricity price variable is positive and significant in all House votes. Furthermore, the insolation (sunlight) variable is positive and significant for the Glenn and Hart Senate votes and for the early House votes, switching in sign for the Gravel, Tsongas, and Fuqua votes.

These results support the coalition arguments just discussed. In 1975 and 1976 solar programs drew support from districts with heightened energy concerns: those with high electricity prices. Gravel, by contrast, represented too big a jump in spending. The solar coalition had distributive backing, as shown in the insolation variable. However, PV alone, and specifically the procurement strategy (tested in the Tsongas vote), lacked backing from high-sunshine districts. By 1980 the drift of the solar programs undermined this initial coalition. Consequently, distributive support for PV depended on the solar coalition's support, even though its long-term prospects remained high, or perhaps became even better, as the PV program progressed.

The coefficient of population density is negative in all but one vote and significantly so in three votes. The reason low-density areas might benefit from PV is that they have more nongrid applications. Though nongrid applications are far smaller than grid-connected applications, they would become economic earlier.[59]

The distributive politics hypotheses receive a different test from the coefficients on terms of senators. Two aspects of distributive politics have different effects here. Distributive politics flowing from economic benefits of the technology suggests that the further off the election, the more likely there would be a yes vote, because senators viewed photovoltaics in particular (and solar programs in general) as having a long-term payoff. Those with short time horizons for economic benefits would favor other programs. But the time horizon for pork barrel is quite different. Those with elections coming up are more likely to vote for current spending programs. Thus the predicted sign for the coefficient of term depends on the relative strength of these two aspects of PV distributive politics.[60]

The fact that the significance levels for electricity and sunlight are higher than for photovoltaic spending suggests that program benefits may play more of a role than pork barrel in PV. In fact, in all three Senate votes, the further off a senator's election, the more likely he or she will vote for PV; however, this relationship is significant in only one equation.

COMMITTEE. Controlling for other factors, the Appropriations Committee in both the Senate and House was opposed to all these amendments, significantly so in half the cases. This result is consistent with a coalitional

59. Population density is also correlated with land prices, which affects the cost of PV in some applications. However, land prices are generally a small part of overall costs in central station applications, and in rooftop applications they are irrelevant. Thus, while the interpretation of the population density variable as a proxy for nongrid potential has its problems, it is defensible.

60. This assumes that the effect of party and ideology does not depend on the closeness of the senator's election. Statistical tests on this issue were inconclusive.

view of committees. All votes amended the bills that were reported out of committee (in some cases, the authorization committee, in others, Appropriations; however, the most consistently negative correlations are found for Appropriations bills). The extent of committee opposition to amending its own bill shows that durable compromises for energy budgets were reached within the committee structure. That most of these amendments were successful demonstrates the problems with formulating energy policy during the time investigated.

Evaluation of the Program

The evaluation of PV proceeds on two levels. First, I examine the evidence about whether the program achieved its economic goals; in particular, whether program choices were consistent with program results regarding the economics and commercial prospects for PV technology. This evaluation strays from the narrow goals initially defined for the program to more general social energy use considerations. Second, I examine evidence about the political requirements for an R&D program.

The PV Program and Economic Efficiency

The PV program grew rapidly in the 1970s and shrank dramatically in the 1980s. Whether the growth and subsequent cuts were justified is a complicated question. One place to begin is the shape of the budget path, given that the program was designed to pay off ten to twenty-five years after its inception. Growth of a program from $2.0 million in 1974 to $76.5 million in four years and $150 million in six years incurs waste. For example, William Ouchi, in his criticism of the PV program, claimed that program fragmentation caused contract awards to unqualified organizations and poor coordination of the research.[61] His point is consistent with the view that the program simply grew too fast.

The large cutbacks in the program were also probably in error for two reasons. First, while oil prices peaked in 1980–81, until 1984 they were higher in real terms than at any point during the 1970s. Although the decline in oil prices during the 1980s implies changes in the previous budget path for PV, the magnitude of the correction was far greater than justified by the change in the energy outlook. Second, although past

61. William G. Ouchi, *The M-Form Society: How American Teamwork Can Recapture the Competitive Edge* (Addison-Wesley, 1984).

expenditures are "sunk" and hence irrelevant in making decisions about future spending, rapid termination of projects probably resulted in not fully capturing the value of earlier research.

The program throughout was characterized by much uncertainty on both the supply side and the demand side. To cope with technical uncertainties, the program supported multiple approaches in a competitive environment. On the demand side, it devoted considerable analytical effort to understanding how PV might become competitive in different uses. Given its uncertainty, a decentralized, multifaceted program was a reasonable choice.

The program underwent several shifts in system configuration, specifically the issues of storage and of distributed versus central station PV systems. The shift in emphasis was consistent with economic considerations and was a reasonable response to the economic analysis performed on the program once it became a fairly important component of energy policy. On the analytical side, resolution of the second issue depended on the appropriate balancing of financial and tax considerations favoring residential use and cost factors favoring central station use. The cost differences between distributed and central station systems were not accurately assessed in the early 1970s. Some errors were obvious, such as the view that maintenance costs favored distributed use because homeowner maintenance costs were zero (they do it themselves). Another early error was the failure to perceive the large marketing and distribution costs of distributed systems, and complications such as utility interconnection and PV installer training that are exacerbated in distributed applications. The correction of these errors supports the shift in emphasis toward centralized systems. Such a shift probably would have been difficult had the program not been decentralized, for if it had not been, vested interests would have exerted political pressure not to change, leading to the technical inflexibility found in other cases.

The PV Program and Political Support

Political support for the PV program during the years of high budgets rested on proponents of nonnuclear energy. The econometric study of PV voting shows that the program was supported by legislators from districts with sunshine and high electricity costs, or with an ideological background that put them at odds with the federal nuclear strategy. These members represent constituents who have relatively high concern over energy prices, relative advantages in developing solar alternatives, and ideological biases against traditional federal energy endeavors.

No roll call votes on the PV program took place during its declining years; however, the progression of the program can be related to its initial basis of support and its evolution. As the program moved toward central applications, it ceased satisfying some of the goals of solar advocates. As federal funds for R&D declined, the solar coalition became attenuated because of competition among proponents of diverse technologies. After world oil prices declined, energy lost political salience, further diminishing available budgets and exacerbating the split in the solar coalition. In consequence, the program was vastly reduced, without regard to the substantial technical achievements to date.

Unlike the other technologies discussed in this book, PV had very good outcomes. Hence decisions related to outcome can be separated from decisions related to energy salience. The conclusion is that energy salience dominated the scene to the detriment of R&D development.

That energy salience dominated the PV situation is expected, since PV lacks politically salient distributive benefits. The history of PV development emphasizes an essential problem with federal R&D programs for technologies that are amenable to a decentralized, semicompetitive, fragmented strategy. This strategy can be credited with contributing to the technological success of the program; it can also be credited with contributing to its political failure. It achieved substantial cost reductions and technological progress, but it failed to create a political constituency that could outlive short-term saliency and survive long enough to develop an innovative, uncertain alternate energy technology.

Chapter 12

An Assessment of R&D Commercialization Programs

LINDA R. COHEN & ROGER G. NOLL

THE HISTORY of the federal R&D commercialization programs described in chapters 6 through 11 is hardly a success story. On the basis of retrospective benefit-cost analysis, only one program—NASA's activities in developing communications satellites—achieved its objectives and can be regarded as worth the effort. But that program was killed because it came into conflict with other more important political forces than the advancement of commercial technology. The photovoltaics program made significant progress, but it was dramatically scaled back for political reasons not related to its accomplishments. The remaining four programs were almost unqualified failures. The supersonic transport (SST) and Clinch River Breeder Reactor were killed before they had produced any benefits, and Clinch River, because of cost overruns, absorbed so much of the R&D budget for nuclear technology that it probably retarded overall technological progress. The space shuttle, though it became operational, lost its status as the primary vehicle for commercial launches after the *Challenger* disaster in 1986 because it costs too much and flies too infrequently. The synthetic fuels program produced one promising technology, the combined-cycle coal-gasification project (Cool Water), but billions were spent on other pilot and demonstration facilities that failed.

The case studies obviously justify skepticism about the wisdom of government programs that seek to bring new technologies to commercial practice. But the cases provide far more insight than that. They identify how and why government programs go wrong, and hence the problems in the federal decisionmaking process that need to be solved if performance is to be improved. Because federal R&D commercialization projects survived their most antagonistic presidency in the past half-century, that of Ronald Reagan, we suspect they are here to stay; therefore, identifying the source of their problems is potentially of great importance.

Management Pitfalls in R&D Programs

In chapters 2, 3, and 4 we provided the criteria for analyzing decisions about a government R&D commercialization program as it proceeds from initial exploratory research through final commercialization. In this section we compare the decisions that were made in the six case studies with the dictates of those criteria.

Initiation

Two general criteria should govern the initial decision to investigate a commercialization program: the technology in question should be susceptible to market failures in private R&D decisions—more than can be effectively compensated by general policies such as favorable tax treatment of R&D—and a reasonably promising technological path should be available for developing a new technology. All the cases examined in this book plausibly satisfied these criteria, with the possible exception of the SST.

Even without extensive commercialization, the three energy technologies would have been successful had they been sufficiently promising to cap the world price of oil or to reduce substantially American vulnerability to oil supply disruptions. And had they proved only somewhat more costly than established energy technologies, they might have been socially beneficial if they were environmentally superior. In addition, these technologies were designed for a sector of the economy that faces distorted incentives because it is heavily regulated. Also, the breeder reactor had national security ramifications.

Both of the space projects required a scale of demonstration (one satellite, one space shuttle) that was very large compared with the ultimate commercial scale of the market. Hence they faced risks that are especially difficult for the private sector to finance. Furthermore, they involved space launches, an activity that is almost the exclusive province of government, and technologies that have extensive applications to national defense.

The SST is the least plausible case for market failure. Because the airline industry was regulated, its incentives to adopt new aircraft were distorted, but even so, there is no evidence that technological progress had been insufficient. Moreover, although the United States would have experienced considerable economic disruption had it lost its commercial aircraft industry, there is no evidence that the leading private aircraft firms were reluctant to undertake new product innovations comparable in scale to the SST. During the 1960s these companies were developing short-haul and wide-body jet aircraft, usually without a comparable federal support program.

In all six cases, benefit-cost analysis confirmed the efficiency rationale for the program. For the most part, the studies undertaken by or for the supporting agencies were not wildly implausible. But in every instance the studies had two specific shortcomings. First, they assumed that the R&D program would achieve its objectives. Second, they compared the proposed program with a very narrow range of alternative paths of technological development. Both are consequences of the same phenomenon: technological optimism on the part of the government technologists who advocated the program. In addition, considering limited alternatives is a predictable response to the political requirements for a successful program.

A key element of all R&D projects is that their ultimate results are unpredictable. They may not succeed in meeting their original objectives, and they may lead to unforeseen innovations. Hence a valid prospective analysis of any R&D program requires taking into account the possibilities for failure (as well as unanticipated success). Doing so not only provides a more accurate assessment of potential costs and benefits but also ensures the development of a management plan for the program that makes economic sense. In fact, only in two cases—the SST and photovoltaics—did the management plan incorporate decisions about whether the program should continue as originally planned. (For the SST, these benchmarks were then ignored.)

Because the product of R&D is inherently uncertain, early, relatively inexpensive exploratory work should investigate alternative technological pathways, chosen on the basis of maximizing the expected payoff of the entire R&D effort. In a sense, all the programs studied here did investigate multiple alternatives in the early stages; however, in several cases too few alternatives were considered. For example, the early photovoltaics program focused on distributive systems because the technologists supporting the program hoped that solar technologies would "free" residential customers from dependence on large, monopolistic utilities, rather than because there was any analytic basis for believing that distributed systems were economically more attractive than central systems. The synthetic fuels program, until the energy panic following the 1979 oil crisis, focused on technologies using eastern coal, even though it was much easier and cheaper to make synthetic fuels from western coal. Thus the goal of reversing the decline of eastern coal states overrode the energy policy objectives of the program even after the 1973 crisis. The space shuttle program considered only minor variants of the craft that eventually was developed, and explicitly assumed that unmanned launch vehicles would not be developed further than they had been in the late 1960s. The communications satellite program was persistently limited to technological paths that, once launched, would be unthreatening to telecommunications common carriers. Both the SST and breeder reactor programs considered only alternatives that satisfied a highly binding technological constraint: a very high speed requirement for the SST and a high fuel conversion target for the breeder.

Technologists in sponsoring agencies were partly responsible for the narrow range of options considered. These limits sometimes reflected an overambitious vision of technology and sometimes policy concerns that went beyond near-term commercialization. Examples of the latter were the bias toward distributed solar energy systems and the National Aeronautics and Space Administration's (NASA) agenda in space exploration, which colored its choice of design for the space shuttle. But the narrowness of vision also had political origins outside the agencies, often representing the compromise that technologists made with interest groups and political leaders to garner their support. The satellite program, in particular, was persistently being redefined to stave off industry complaints about unfair encroachment on the private sector's domain. Likewise, the orientation of synfuels toward eastern coal had strong political support. President John F. Kennedy, who defined the objectives of the program, became the leading Democratic presidential candidate in 1960 after he won the West Virginia primary; he therefore counted eastern coal interests as among his prime constituency. Later two West Virginia politicians, Senator Jennings Randolph and Representative Harley Staggers, chaired congressional committees that had important responsibilities in energy policy (Public Works, and the Committee on Energy and Commerce, respectively).

The narrow scope of opportunities that agencies explore reflects the political necessity to obtain support for a program from many sources: first from the president, the agency technologists that assess the options, and Congress, and then from participants in the industry and sometimes other relevant organized interests such as environmentalists. A new technology is politically relevant only if it benefits the people who decide whether to pursue it. An alternativc can be politically irrelevant because it is uninteresting to the technologists who are developing the options to be considered, because its benefits and expenditures will accrue to a constituency that is not of interest to the president and a majority of each house of Congress, or because it imposes costs or other losses on constituencies represented by key politicians.

Commitment to Demonstration

The decision to progress from exploratory work toward commercial demonstration should be based on two conclusions. One, greater effort at a less expensive, more exploratory stage is unlikely to produce compensating benefits in cost and performance. Two, prospects for the technology are sufficiently promising to warrant the more expensive next stage. Our case studies support the conclusion that the decision to commit to demonstration rarely satisfies these criteria: the decision to build prototypes and demonstration facilities is usually made too hastily, and once a commitment is made to demonstration, a project is likely to become insufficiently flexible to take into account new information about its commercial prospects.

In all the cases except communications satellites, the government decided to build pilots, prototypes, or demonstrations despite concrete information that the technology was not ready for those projects. In the synfuels program, demonstrations were built after pilots had failed even to operate, let alone to meet performance goals. Photovoltaic demonstration facilities were constructed when the technology was about halfway to its ultimate performance goal. The first space shuttle was built after it was known that essential components for its commercial success were either technically infeasible (the space tug) or greatly delayed (the inertial upper stage). Construction of the SST prototype and at the Clinch River site began after it was recognized that serious technical problems had not been solved and that the prospects for commercial success were dim.

Once commitments to build large-scale facilities had been made, projects did not respond to new information, or did so only after a long delay. All energy projects became less attractive in the mid-1970s, when revisions in long-term-demand forecasts indicated that much less new electrical generation capacity would be needed during the 1980s and 1990s than had previously been expected. This information, however, had no effect on the scale of energy R&D programs. More recently, the failures of the space shuttle did not dissuade the president and Congress from building a new orbiter to replace the *Challenger*. The programs that were canceled after poor performance—the SST, Clinch River, and many synthetic fuels plants—still survived for several years after they were known to be unjustifiable.

Hasty decisionmaking and inflexibility result from convergence of two characteristics: technological optimism by advocates in the executive branch and impatience among political officials. Electoral politics causes politicians to favor programs that promise tangible results for the next election. Large, visible projects satisfy this political demand, not only because they stand as obvious signs of a return on expenditures but also because, unlike earlier research activities, they can deliver distributive benefits to constituents that pass the threshold of political saliency. As for technologists, even though they may be reluctant to test a technology prematurely, their optimism can cause them to make hasty commitments. Then, once the large visible projects are begun, politicians are prone to want to keep them going, regardless of performance, so as not to impose distributive costs through cancellation. Meanwhile technological optimists are prone to find sources of hope in a poorly performing project.

Long-Term Expenditure Path

The pattern of expenditures over the life of a commercialization program tends to depart from the optimum path described in chapter 3. In particular, the programs examined in this book did not adequately take into account that progress in earlier, exploratory research will be fairly insensitive to the rate of expenditure, and that the successful execution of a program requires sustained expenditures over a long period. Indeed, federal programs often have a boom-bust feature: an excessively rapid buildup of expenditures in the early stages, and then budgetary instability.

The best direct manifestations of the boom-bust phenomenon are in photovoltaics and synthetic fuels. In both programs, after the 1973 oil crisis the president and Congress seemed to compete over who could propose the larger increase in the budget for exploratory research on these technologies. In the case of photovoltaics, because so little was known about the technology, a considerable amount of fundamental research was needed in the early stages. Yet budgets were passed that exceeded the absorptive capacity of the research community, and Congress had difficulty staving off attempts to increase the budget even more rapidly than it actually did. In the case of synthetic fuels, much more was known, but the news was largely discouraging. Nevertheless, owing more to presidents than to Congress, there was a rapid budget buildup to support more developmental work on unpromising technologies. Then, a few years later, both programs suffered busts, because of the election of an ideologically unsympathetic president and the collapse of their support coalition in Congress.

Rapid budgetary increases for programs can also produce undesirable indirect effects. For the space shuttle and the Clinch River breeder, budget booms were financed in part by budget busts in other long-term research activities. The breeder caused massive cutbacks in other nuclear research programs, and the space shuttle caused cancellation or delay of several projects that, among other things, contributed to the demand for launches that was used to justify the shuttle.

Budgetary booms and busts are costly for all programs simply because to accommodate them requires sinking costs in resizing the organizations that will implement the program, both public and private. But for R&D the problem is especially severe, since the most important product of R&D—useful new knowledge—is easily lost during a bust and is not particularly susceptible to production speedup during a boom. Hence boom and bust greatly reduces the productivity of the expenditures invested in an R&D program.

The main cause of boom and bust is the fragility of the political coalitions that typically form around R&D commercialization programs. Risky, long-term programs are not likely to have a durable base of political support. Thus, in the first instance, they are not likely to be initiated. And if they are, they will probably move too quickly to solidify shaky support by generating immediate payoffs.

In general, R&D commercialization programs have two important, enduring sources of political opposition. Among Republicans, libertarian conservatives, though often supportive of basic research, are reluctant to undertake projects that overlap with private sector activities. Programs designed to bring to practice a new commercial technology tend to draw their opposition. Among Democrats, populist liberals, though also often supportive of basic research, are skeptical of large-scale, "hard" technical solutions to societal problems. They tend to oppose programs that promise contracts to big business and that they perceive will cause regressive redistribution of wealth and political power.

Because successful politicians usually have a strong pragmatic streak, these fundamental ideological predispositions do not entirely determine which programs they will support. In particular, the case studies reveal two pathways by which support for an R&D commercialization program can be garnered from political leaders who might otherwise oppose it. One is when R&D is tied to a salient national political issue, such as the space race with the Soviet Union during the 1960s or the energy crises of the 1970s. The other is distributive politics: contracts for constituents, or benefits accruing to a concentrated, organized constituency. Unfortunately, neither is a reliable basis for long-term political support for an R&D program.

Most of the R&D commercialization projects examined here saw the salient national political issue on which they were grounded recede from political importance before they were completed. An R&D program, because it is inherently long term, requires a durable political issue as the basis for its support, yet the space race and the energy crisis each endured for only about a decade. Once the issues had subsided, all that remained as an electoral basis for their continuation was distributive politics. Here R&D programs are not particularly attractive. Because the ultimate winning configuration of a technology is not predictable, neither are its distributive consequences. Moreover, the fundamental uncertainty of R&D causes many projects to fail in even the best-managed programs. And if the resulting new technology threatens an established industry, the distributive consequences can be negative. Hence a coalition put together at the beginning of a program may unravel because some parts of the program fail, or for some other reason its distributive support disappears.

Not only boom and bust but also haste and inflexibility are the result of the transitory nature of salient issues and the fragility of the initial supporting coalition. In essence, proponents of a program are driven to get to the stage of politically visible results and distributive benefits as quickly as possible, and then to try to retain as many projects as long as possible, to keep the support coalition for the whole program together. Indeed, this step may be necessary to maintain the momentum for a highly successful project, since even successful projects are likely to be politically irrelevant, or to have serious political liabilities, as stand-alone activities.

Policies that are a response to a transitory salient issue have still another problem. New societal problems normally do not fit neatly into the organizational structure of the government, whether the array of executive agencies or the committees of Congress. As a result, many bureaucrats and politicians may try to claim a piece of the responsibility for dealing with a new issue. Both space and energy had this characteristic: NASA, the Federal Aviation Administration, and the Defense Department claimed part of the turf for international competition in aerospace, and responsibility for the congressional response to the energy predicament was spread over most committees. The immediate consequence is excessive inclusiveness: the original coalition involves too many projects by too many agencies that are overseen by too many committees. Excessive inclusiveness, in turn, creates a weak institutional basis for long-term support for the program. Part of the reason for the boom phase of the budgetary cycle is excessive inclusiveness, and then part of the reason for the bust is that excessive inclusiveness brings failures and cancellations that erode the coalition supporting the more attractive policies.

Persistence in R&D Commercialization Programs

Among the paradoxes of R&D commercialization programs is that, despite the difficulty of starting them and their general instability, some important commercialization programs have always been pursued throughout the postwar era. Perhaps the best indicator of the durability of this class of programs is the portfolio actively backed by the Reagan administration. On entering office Ronald Reagan advocated a far more limited role for government support of R&D than his predecessors had, and he implemented this goal by proposing to eliminate numerous commercial R&D projects. Many of the proposals met with success, including the elimination of the Synthetic Fuels Corporation and the more applied activities in photovoltaics. But the Reagan administration was not single-minded in attempting to kill all commercialization projects. It tried unsuccessfully to keep alive the Clinch River breeder reactor demonstration and proposed several new programs of its own. The last budget proposal prepared by that administration, for fiscal 1990, contained several projects that support or would ultimately support commercial demonstrations and that continue some of the programs examined in the case studies.

NASA and several Defense Department agencies have programs investigating both supersonic commercial flight (the High Speed Civilian Transport program) and hypersonic flight for space and defense applications (the National Aerospace Plane, or NASP). During its last year in office, the Reagan administration spent $262 million on NASP and proposed a substantial increase for fiscal 1990.[1]

1. National Science Foundation, *Federal R&D Funding by Budget Function, Fiscal Years 1987–89* (Washington, 1988), pp. 14, 79; and National Science Foundation, *Federal R&D Funding by Budget Function, Fiscal Years 1988–90* (Washington, 1989), p. 14. Hereafter NSF(88) and NSF(89). For a more complete analysis of this program, see Linda R. Cohen, Susan A. Edelman, and Roger G. Noll, "The National Aerospace Plane: An

The space shuttle has many follow-on R&D programs. NASA is still trying to develop something like the space tug and other capabilities for launches from the cargo bay of the shuttle.[2] The orbital maneuvering vehicle, budgeted for $107 million in fiscal 1990, is now scheduled for first launch in 1994. About $100 million more is being spent on the development of upper-stage launch capabilities from the shuttle, including the completion of work on the inertial upper stage. Perhaps most revealing, NASA and the Department of Defense have a joint program, advanced launch systems, to develop new expendable launch vehicles—the technology that the shuttle was intended to make obsolete and that was ignored in NASA's benefit-cost analysis to justify the shuttle. Thus in 1991 NASA is still completing the development of shuttle capabilities that were expected to be ready a decade earlier and is engaged in a major effort to develop further the technology that the shuttle was supposed to replace.

The on-again, off-again history of the communications satellite program continued in the fiscal 1990 budget. The program that was canceled in 1974 (the "next logical step" in the advanced technological satellite program) was still the next logical step in 1990. The Reagan administration had agreed to initiate the advanced communication technology satellite program (ACTS) earlier in the decade, which was intended to launch a somewhat more elaborate version of the satellite project that had been canceled. But in the fiscal 1990 budget, the administration recommended that the program be canceled once again because ACTS is a program that "the Administration regards as a more suitable undertaking by the private sector."[3] Congress restored part of the budget cut, but the boom-bust cycle for communications satellite R&D was once again being repeated.

The Reagan commercial R&D program for nuclear power represented a follow-on not only to the breeder program but to the ubiquitous government presence in the entire nuclear power industry. The administration's goals for its program in nuclear fission were:

> the establishment of a predictable regulatory and institutional environment that enables nuclear power to compete freely in the market place; the support of industry efforts to revitalize the light-water reactor option in the United States; the exploration of the feasibility of advanced nuclear

American Technological Long-Shot, Japanese Style," in *American Economic Review*, vol. 81 (May 1991, *Papers and Proceedings, 1990*).

2. NSF(89), pp. 31–33.
3. NSF(89), p. 39.

> reactor systems; the support of international nuclear R&D projects; the strengthening of the U.S. position in international commerce; meeting long-term needs on uranium enrichment; and conducting development programs on compact space and terrestrial nuclear power systems.[4]

Whereas the traditional objectives relating to defense and nonproliferation can be inferred from some of these statements, the Reagan administration clearly went beyond those objectives. It wanted nuclear power to be important in the domestic energy industry and stood ready to help the industry develop the technology that would make this possible. And among the projects still under way is continued R&D on liquid metal reactors, which the Reagan administration believed could expect reduced budgets because of a revived "emphasis on government/private sector cost sharing initiatives."[5] Reagan's Department of Energy also had a much more ambitious long-term breeder program that was never initiated.[6]

Another main component of the nuclear effort was continued R&D on uranium enrichment. The Reagan administration's "long-term goal [was] to transfer the uranium enrichment enterprise to the private sector."[7] Nevertheless, it proposed more than $100 million in fiscal 1990 to develop new enrichment technology (the atomic vapor laser isotope separation program), with the objective being a demonstration program in the early 1990s.

Finally, the last few years of the Reagan administration saw a continued large commitment to the development of commercial fusion reactors. This work received approximately $300 million a year. The primary technical problem in fusion has been to create an environment with high enough temperature and pressure to sustain a fusion reaction. Up to now, no facility has managed to sustain a greater production of energy than is required to create fusion, so that the program has not yet reached the stage of constructing a pilot to demonstrate the technical feasibility of any commercial concept. Current work focuses on constructing test facilities that can be used to learn more about high-temperature fusion processes and that might lead to a facility that is a net energy producer. Consequently, the fusion program has not yet faced the early commercialization problem: the

4. NSF(89), p. 65.

5. NSF(89), p. 66

6. See Energy Research Advisory Board, *Review of the Proposed Strategic National Plan for Civilian Nuclear Reactor Development* (Department of Energy, 1986).

7. NSF(89), p. 68.

development of a relatively unpromising method for producing a net gain in energy that would require an enormous federal investment to "demonstrate." Technologists expect this test to occur sometime in the 1990s. At that time pressure will mount to build the fusion counterpart to Clinch River, probably at the cost of much of the ongoing development research in the program.

The Reagan administration's largest budgetary cutback was the elimination of virtually all the $100 billion appropriation for synthetic fuels development. Yet in its last year the Reagan administration signed on to a closely related new demonstration program, clean coal technology. The events instigating this about-face were, first, the Cool Water Coal Gasification Plant, the *only* successful synfuels demonstration project, which happened to be completed under Ronald Reagan's watch; and second, Reagan's international agreement with Canada about acid rain, which bought time for instituting a costly emissions control program for northeastern coal-fired generation facilities by promising to engage in extensive research on clean-burning coal technologies. Nevertheless, the scale of the Reagan clean coal program was awe inspiring. By 1988 the administration had approved twenty-nine demonstration projects, with a total expected cost of $2.3 billion. Moreover, it "anticipated that revised R&D funding for clean coal technology [would] increase significantly beyond the initial 1990 budget."[8] Even though $2.3 billion and climbing is peanuts compared with the $100 billion appropriated in the Carter era, it is substantially more than was ever actually spent on pilot, demonstration, and commercialization plants in the previous coal programs.

The Reagan administration steadfastly opposed all commercialization activities in solar energy and succeeded in greatly reducing the budget for all these programs in the 1980s. Congress has kept solar energy alive. In photovoltaics, Congress initiated one new demonstration program in the 1980s, the PVUSA project. But the administration did support some new initiatives in a somewhat related technology: semiconductors. First, it committed $100 million a year for five years to support SEMATECH, an organization for developing advanced manufacturing technologies for the semiconductor industry.[9] Second, it supported the development of the next generation of integrated circuit chips, and for fiscal 1990 it proposed a further increase in the Defense Department's MIMIC (microwave/

8. NSF(89), p. 62.
9. NSF(88), p. 14.

millimeter wave monolithic integrated circuits) program.[10] As with the aerospace plane, these programs have defense spillovers, because advanced military hardware uses the most advanced semiconductor technology. But they are also intended to reverse the decline (relative to Japan) of the U.S. commercial semiconductor industry.

As this record shows, commercial R&D projects remain alive and well, albeit trimmed down, at the beginning of the 1990s. Despite their political liabilities, and the poor historical record of successfully bringing such projects to conclusion, they persist. Why they do requires an explanation.

Certainly one reason the U.S. government tends to throw technology at perceived social problems is that the United States is very good, if not the best, at developing new technology. One of its great resources is its research enterprise in industry, government, and universities. It would be strange if the United States did not try to enlist these resources in solving its problems.

A second reason is that the present portfolio of projects is narrowly focused, as always, on two general areas: technologies related to defense and to energy. Indeed, only clean coal and nuclear fusion, among the programs advocated by the Reagan administration, do not have strong links with defense, and these are both part of the traditional conservative hard-technology agenda for energy policy. Moreover, some of the justification for both programs is environmental: clean coal promises to save one hard technology from environmental attacks, and fusion promises to be the ultimate benign application of the other hard technology.

Thus the Reagan administration's portfolio is not difficult to understand and is not inconsistent with the thesis that, in general, R&D commercialization programs are not particularly attractive politically. All the Reagan programs were responses to issues that were important to part of the president's primary constituency. Like presidents before him, Reagan did not have an overall strategy to apply American technological resources to improve the competitiveness of the U.S. economy or speed general economic growth, nor had he systematically examined the areas in which R&D market failures were most pronounced. Instead, his portfolio narrowly focused resources on R&D commercialization activities that solved a specific problem for the administration within its ideological constraints; but, as with those of previous presidents, his portfolio will be vulnerable

10. NSF(89), p. 14.

to an ideological swing in Congress and the presidency away from defense and hard energy technologies. The stage is therefore set for a future bust in the projects that boomed under Reagan's leadership.

Toward a Coherent R&D Commercialization Policy

The overriding lesson from the case studies is that the goal of economic efficiency—to cure market failures in privately sponsored commercial innovation—is so severely constrained by political forces that an effective, coherent national commercial R&D program has never been put in place. In brief, the problems are as follows.

— More fundamental exploratory research to create the technological foundation for a new commercial technology is largely irrelevant politically.

— Many R&D projects with potentially useful commercial applications have distributive liabilities because they threaten established firms.

— Program choices are likely to reflect excessive technical optimism and political compromises and therefore to be overambitious and overfocused.

— For projects requiring large investments in pilots and demonstrations, distributive politics can cause programs to be too inflexible and individual projects to be continued long after their justification has evaporated.

Because the early phases are inherently uninteresting politically, programs are rarely considered seriously unless they are tied to salient national political issues (defense, energy, environment). Quite often the saliency of the issue giving rise to a project disappears politically before a program is completed, so that the political coalition that started the program unravels, killing promising research projects prematurely. The purpose of this section is to examine ways to ameliorate some of these political problems.

Classification of R&D Projects

Federal R&D projects can usefully be conceptualized as falling into several distinct categories. First is research without a clear commercial objective, or at least without a potential future application that is sufficiently well defined for one to imagine the kinds of products or production processes to which it might apply. We call this type of research *fundamental R&D*, purposely avoiding the confusing term "basic research." It is undertaken to gain a deeper understanding of science and technology, without any particular application used to justify the work; however, the researcher and the sponsor may nonetheless believe it might eventually have useful applications, and use that belief as a general justification for its support. Most of the research supported by the science and engineering directorates of the National Science Foundation is in this category, as is most of the work supported by the National Institutes of Health, the space science program of the National Aeronautics and Space Administration, and the activities of the Department of Energy in basic energy sciences, such as experimental physics using the particle accelerators at various Energy Department national laboratories.

The second category of activities consists of R&D that is intended to contribute to the technological base of a particular industry. We call this type of activity *commercial R&D.* While some of this research is basic, it is nonetheless targeted at a specific set of technical problems affecting commercial technology. An example of such work is the research on air frames in NASA, including the agency's wind tunnel facilities and the metallurgy activities under way in NASA and the Defense Department to develop materials for the exterior shells of supersonic aircraft. Another example of commercial R&D is the research carried out in some university-business cooperative research facilities and in the Engineering Research Centers (ERC) supported by the NSF. The ERCs are multidisciplinary, problem-focused research centers that are required to obtain substantial financial support from industry to receive NSF grants.[11] For example, one of the first ERCs, located at Purdue University, was described by its director as follows: "The particular sector of industry that we address is that which deals with discrete product manufacturing. In terms of the Standard Industrial Classification codes, we focus on those that fall in the range from 34 through 39."[12]

The initial activities of SEMATECH also fell into this category. SEMATECH relied more on private initiatives and decentralized program choice than did our cases; however, the collaboration of otherwise competitive firms, arrangements made between them to share program results, the amount of federal financial support (50 percent, as was initially conceived for the breeder program), and the emphasis on generic manufacturing activities parallel the early phases of the commercial R&D programs examined here.

11. Industry support averaged about 55 percent of the budgets of the first group of centers that the NSF supported. See Nam P. Suh, "A New Experience: Lessons Learned by the National Science Foundation," in National Research Council, *The Engineering Research Centers: Leaders in Change* (Washington: National Academy Press, 1987), pp. 21–27.

12. James J. Solbert, "Engineering Research Center for Intelligent Manufacturing Systems," in National Research Council, *Engineering Research Centers,* p. 52.

The third category of activities are those designed to encourage the private sector to adopt specific new technologies. It includes construction of complete prototypes, demonstrations, and other forms of targeted subsidies of adoption. Examples are the final stages of the cases examined in this book, the clean-coal demonstration facilities that were begun at the end of the Reagan administration, and recent efforts by SEMATECH to construct a prototype manufacturing facility for semiconductors. Although much uncertainty remains to be resolved at the demonstration phase—in the programs examined here, excessive uncertainty—these activities can be thought of as commercialization efforts. To distinguish this category from the standard usage of the commercialization, we refer to it as *commercial applications*.

A fourth category of government activities, beyond the scope of this book, concerns efforts that directly address public concerns, such as defense and some of the research activities of the Environmental Protection Agency. While these occasionally spill over into the commercial sector, their motivation is usually different from that of our cases, so that they are only tangentially related to the current discussion.

Our proposal involves restructuring responsibilities in both Congress and the executive branch to separate institutionally the preliminary R&D phases from decisions on commercial applications. We do not regard our proposals as a panacea for the problems related to R&D policy, which we believe are in part fundamental to the political system, and our recommendation has some drawbacks that are discussed below. However, we believe on balance that the proposed reorganization addresses some of the difficulties encountered in the past and will lead to improved performance.

Restructuring in Congress and the Executive Branch

The relevant focus in Congress is its committee structure. Specifically, we propose that each body of Congress create committees that are solely responsible for comprehensive programs on federal support for commercial R&D. Moreover, new agencies should be established to be responsible for all commercial R&D.[13] These agencies and their congressional oversight committees would be responsible for technology development only to the

13. Because the structure of Congress is driven in large measure by the structure of the executive branch, a single congressional subcommittee in the House and the Senate for commercial R&D programs fits most naturally with the creation of a specialized agency that harbors such programs.

point where commercial applications projects would be considered. They would not regulate or otherwise influence the use of the new technologies. Commercial applications activities would remain the responsibility of operating agencies concerned with technology procurement or performance (for example, the Departments of Energy, Commerce, Transportation) and of the congressional committees that currently oversee these agencies. As we will discuss in detail, we are not proposing that all commercial R&D programs be pulled together into a single agency; diversity in sources is probably a strength in that it is likely to lead to a richer portfolio of exploratory research aimed at contributing to the technology base of an industry.

Two important distinctions between commercial R&D and commercial applications are the proximity of the work to incentives in the private sector and the cost of individual projects. Programs in the third stage of R&D are likely to have crossed a politically important budgetary threshold. In addition, concerns about the relative competitive positions of firms and structural aspects of the industry will probably figure into political decisions primarily at the last stage.

The budget of an individual project determines whether it is part of the technology pork barrel. The relevant budgeting threshold is when projects become politically relevant on expenditure grounds to a sufficient number of politicians, or to sufficiently important politicians (such as committee chairs or other legislators with important responsibilities in Congress). The threshold is sometimes reached in even fundamental R&D projects: the superconducting supercollider, for example, has been pursued for pork barrel reasons by politicians from Texas. Likewise, commercial R&D projects also on occasion reach the budget threshold for distributive significance. For example, the magnetic fusion program has maintained budgets in excess of several hundred million dollars annually throughout the penurious 1980s. This project has an extraordinarily long time horizon, for its commercial potential awaits significant technical breakthroughs not expected for decades, if ever, but its political support has not been undermined by its distant and uncertain payoffs. In general, that is highly unusual for fundamental R&D or long-term commercial R&D. The threshold of distributive budgetary significance is normally not reached until programs enter the demonstration phase.

The present government organization combines commercial applications projects with, at one end, fundamental and commercial R&D, and at the other, regulatory and other promotional activities. This structure exacerbates the problems caused by the distributive nature of large commercial applications projects. First, members of Congress belong to few committees, and their opportunities for constituency service and distributive politics center on the responsibilities of the committees to which they belong. Because the political payoff to R&D is distant and uncertain, legislators are prone to pay too much attention to distributive benefits from near-commercial pork barrel projects relative to more fundamental R&D activities. Second, once a commercial project has been attempted, if its commercial R&D phase is integrated with demonstration, operation, and promotional responsibilities, the opportunity is created to pursue programs that induce or even require the industry to adopt the technology, even if it is unwarranted. Doing so converts a technical and economic failure into a (short-term) political success. Examples of such actions abound in the histories of the space shuttle and commercial nuclear power.[14]

Agencies and committees with responsibilities for fundamental R&D are prone to pursue commercial applications projects at least in part for their technical interest, thereby giving insufficient weight to commercialization possibilities. Moreover, the narrow, less practical orientation of agencies with major responsibilities for fundamental R&D breeds technological optimism. Both factors contribute to the pervasiveness of attempts at premature commercial application, based not on attempts to shore up R&D failures but rather on incorrect assessments of the likely commercial benefits and risks. Finally, when applications projects in agencies that also support fundamental and commercial R&D go sour, the result is to finance cost overruns by reducing other research activities.

Creating a specialized commercial R&D agency and oversight committees in each house of Congress ameliorates each of these problems. First, it creates a group of legislators who are reliant on commercial R&D for achieving their political objectives and therefore less able to trade off long-term uncertain commercial R&D projects for politically more attractive large-scale demonstration projects. Of course, that is a two-edged sword, for with political stability comes the likelihood that pork barrel will play a

14. The Price-Anderson Act and policies on the fuel cycle exemplify inducement through subsidy, and the threat to let the Atomic Energy Commission compete with electric utilities if the latter did not pursue the commercialization of light-water reactors was tantamount to a forced adoption of the technology. See George T. Mazuzan and J. Samuel Walker, *Controlling the Atom: The Beginning of Nuclear Regulation, 1946–62* (University of California Press, 1984).

larger role in these programs, owing to the very absence of viable alternatives for distributive politics among members of the committees. Second, a specialized commercial R&D committee would not have the opportunity to force commercial adoption of their failures without the consent of other committees that have cooperating responsibilities. Similarly, the technological optimism of narrowly scientific agencies and their overseers could infect commercial R&D applications projects only if they succeeded in leaping institutional barriers. Again, that is a less likely eventuality in a specialized commercialization organization, because its natural constituency is developers and users of commercial technology rather than scientists. Likewise, commercial applications failures would not be financed out of fundamental and commercial R&D.

A primary motivation, and a key political rationale, for all the research activities examined in this book is to enhance industrial activities through technological innovation. But the proximity of federal activities to private industry is a critical determinant of the political constraints for R&D programs. Without distributive benefits, commercial applications projects are particularly sensitive to boom-bust cycles, as summarized above. When combined with fundamental and commercial R&D activities, the latter tend to be subject to the same fate as the larger demonstrations and commercialization activities. Furthermore, when commercial applications programs have distributive benefits but suffer from cost overruns, frequently the primary casualty is commercial R&D. Thus the conjunction of the two activities exacerbates financial ups and downs in commercial R&D, where sustained long-term efforts tend to be a prerequisite for efficient performance.

Activities in the first and, to a lesser extent, second stages of research are typically nonthreatening to established firms. Alternatively, the adoption or availability of identified products can change the competitive structure of an industry. For example, satellite development in the 1960s and early 1970s had immediate competitive implications for the telecommunications and broadcasting industries. Recently technology developed for cellular telephones—where commercialization, not surprisingly, has been a completely private effort—has the potential for a major restructuring of telecommunications. Democratic political systems incorporate industry preferences into decisions; hence such efforts face extraordinary hurdles as government initiatives.

The case studies examined here identify two additional reasons why research may cease prematurely when industry becomes closely involved in attempted commercialization. First, the reluctance of ideologically conservative legislators to become involved in private sector activities increases with increased private participation. Because commercial R&D is currently bundled with commercial applications, strengthening the coalition against the latter tends to undermine commercial research budgets as well. Second, saliency declines for federal efforts as the immediate problem appears resolved. The case study of satellites demonstrates this point; another particularly unfortunate example is the history of light-water reactor development in the United States.[15]

Congressional committees are widely recognized as the mechanism that forges enduring support coalitions for programs, and our analysis provides some confirmation for the importance of committees in R&D projects. An especially important lesson can be deduced from the rapid rise and fall of energy projects after the 1973 and 1979 energy crises. When energy became a salient issue in the 1970s, Congress did not have a convenient committee structure with which to deal with it. Existing committees fought for jurisdiction, and responsibilities became fragmented. As a result, a strong base of support for energy generally, and commercial energy R&D in particular, was never constructed by the subcommittees. This lack of support contributed to a spending competition in 1975–80 that probably spent far too much on energy-related commercial R&D and on activities intended to enhance private adoption of new energy technologies, followed by a rapid contraction as the subgroups of energy interests failed to coalesce around a smaller yet politically viable program.

One can view this history in two ways. First, the absence of a coherent congressional structure caused boom and bust rather than a balanced, stable R&D program, thereby dooming the enterprise to failure. Or, second, only the fragmentation of Congress prevented a largely irrational set of projects from hanging on indefinitely, much as Clinch River hung on despite the Carter administration's attempt to kill it. Although we lean toward the second interpretation in this case, the positive conclusion is that the extent of committee fragmentation does affect a program's sensitivity to the political fads of the moment, both pro and con.

Sensitivity to political ups and downs contributes to the problems identified in commercial applications projects; in particular, the tendency to initiate large-scale spending prematurely and to rush through projects to deliver quick results that will keep the program going. Furthermore,

15. See Mazuzan and Walker, *Controlling the Atom.*

skepticism about whether a program can be maintained to completion contributes to a deemphasis of the commercialization goals of a program relative to short-term pork barrel considerations. When the economic goals assume secondary importance, commercial applications programs are more likely to become distorted, as the breeder reactor program gave way to the Clinch River demonstration even though the early results of the program and changes in underlying economic conditions decreased the probability of ultimate commercial success. Thus political vicissitudes are at the root of both premature cancellations and unwarranted continuations.

Separating commercial R&D from demonstrations and promotions has clear drawbacks. The two activities frequently require joint costs and shared expertise. More generally, the synergy between commercially oriented research and commercialization may be so great that it needs to be taken into account at all stages of R&D. Furthermore, the viability of the separation proposal depends in part on whether commercial R&D is sufficiently large to justify a specialized agency and oversight committee. Obviously, expenditures on such programs were cut substantially in the 1980s; however, they remain significant. To address these issues we turn to the current portfolio of R&D projects in space and energy.

The National Aeronautics and Space Administration

NASA has a dazzling array of projects that either are now supported or are ready for support as soon as space becomes politically more attractive, including space science (American participation in the race to establish a base on the moon and to return samples from Mars) as well as commercial projects (manufacturing on the space station and the aerospace plane). In addition, NASA has been identified as the lead agency for the "mission to planet earth," a proposed program involving many satellites to investigate and monitor terrestrial climate.[16] Disentangling NASA's commercial applications projects from its space science and defense activities is problematic. Greater effort in space science and defense probably increases the attractiveness of commercial R&D in space by offering additional sources of

16. While technological features of the Mission to Planet Earth program—satellite platform development and space launches—build on traditional NASA activities, the environmental focus is a considerable departure. The environmental application places it partially in the category of research for public consumption rather than of commercial R&D. The discussion of commercial application of Defense Department projects may apply generally to issues involved in the program; however, its lack of precedent regrettably places it outside the scope of this book.

potential benefits from the fixed costs of R&D. But commercial application is almost certain to lag behind the development of technical capabilities for space exploration and space-based defense. To begin commercial applications programs before the technical capabilities of the other projects are reasonably well established is especially premature given that technological predictions of capabilities will probably be optimistic. Moreover, NASA may well be a poor choice for housing responsibilities to develop potential commercial applications.

The Department of Defense and NASA are in some ways analogous. Both produce pure public goods for national consumption: international security and space exploration. The programs at both agencies require substantial research effort, in part because of international competition and in part because their achievement demands capabilities beyond the reach of current technology. At the same time, both programs often expand commercial possibilities in ways that are unrelated to the public good they produce. As illustrated by the studies in Nelson's compendium on federal R&D programs,[17] in a wide variety of industries defense R&D has laid the foundation for important commercial advances. Nevertheless, the Defense Department is not required to pick R&D projects and paths of technical explorations on the basis of potential commercial spinoffs. The job of the department is to focus on projects that are desirable solely from the standpoint of national security. It does not even have a systematic process of identifying and communicating possible commercial potential into its selection of programs.

The key issue is whether the Defense model makes sense for NASA. Although the Defense Department has been criticized for its own version of technological optimism regarding the contribution of sophisticated technology to national security, the narrow focus of its responsibilities prevents its technological optimism from easily spilling over into commercial projects. Applied to NASA, the analogous structure would be an agency focused solely on space science and exploration, with commercial R&D and applications programs (if any) located elsewhere.

We suspect that such an arrangement would have led to a far different history for the development of the space shuttle. An agency focusing purely on the exploration of space and space science would have designed a research vehicle; the commercialization agency may or may not have investigated modifications and improvements to be used for commercial

17. Richard R. Nelson, ed., *Government and Technological Progress* (Pergamon Press, 1982).

launches; and even if commercial R&D had taken place, the Department of Transportation might or might not have ordered its own orbiters for commercial demonstration. In any case, if commercial applications were housed separately in, say, the Transportation Department, no agency would have had an incentive to commit in the early 1970s to exclusive reliance on the shuttle for commercial launches in the 1980s.

The counterargument to such an arrangement is that both functions had to be served for the shuttle to be desirable. Even though this argument has some technical validity, it must be stacked up against the theoretical and empirical support for the argument that the absence of such a separation increases the chance for overoptimistic and poorly managed R&D programs. The economic and human disasters delivered by the space shuttle were only proximately caused by bad decisions by NASA officials; they were fundamentally caused by the political necessity to commit to commercial development prematurely, and to live up to some semblance of the initial launch schedule despite technological bad news from 1978 on. NASA made poor decisions from 1978 through 1986 largely because its budget constraints and performance requirements were inconsistent with technological feasibility. An early separation of technological development from commercial application would have focused the costs of technological optimism on NASA's ultimate organizational purpose, which is exploration and science.

The Department of Energy

The Energy Department is still active in developing nuclear power and clean coal technology, but a renewed energy crisis is unlikely to focus on these technologies for increased effort. The payoff to nuclear R&D is too distant, and clean coal deals only with coal-fired electricity plants and thus does not contend with the problems of liquid fuels. To the extent that late-1980s energy budgets should be viewed as buying information to guide technology choices during the next energy crisis, the budget allocations were too narrowly focused.

The Defense Department model is less clearly applicable to energy R&D because of the peculiar policy significance of nuclear power. Nuclear weapons research is located in the Energy Department because of the desire to keep control of nuclear weapons out of the hands of the military. The impetus for a federal role in civilian uses of nuclear power was fueled in part by its connection to defense. One reason the United States invested heavily in commercial nuclear power was to gain leverage in controlling its use and to prevent the proliferation of nuclear weapons. Initially the commercial development of nuclear power was organizationally colocated with defense-related nuclear R&D because both made use of nuclear reactors. Eventually the creation of, first, the Energy Research and Development Administration, and then the Department of Energy, was partly motivated by the hope that an all-purpose energy research organization would make more informed trade-offs among the array of energy technologies that might be developed. Of course, though each link of the argument makes sense, the result is an agency (and a parallel structure in Congress) that is also capable of logrolling across industries and technologies and that is prone to eschew real trade-offs among them to retain a winning coalition of support. The difficulty with applying the logic of separation between generic research and commercial projects to the Energy Department is the perception of a social value in retaining a viable American nuclear power industry. If an American presence in the industry is necessary for the country to have maximal influence in controlling weapons proliferation, then a continued connection between defense and support for commercial R&D makes sense. Indeed, federal subsidies to keep the failing nuclear power industry alive would also be justified.

Nonetheless, the argument is less persuasive for continuing the Energy Department's role in both generic research and commercialization projects for the other technologies. The absence of balance between nuclear and other energy technologies is good evidence against the validity of combining work on all technologies under one roof and suggests that little would be lost by a different organizational form. An attractive alternative would be to place generic research in the comprehensive commercial R&D agency and to create a new organization, similar to the Synthetic Fuels Corporation, for demonstration and commercialization projects in all nonnuclear energy technologies. Although such projects are now relatively few, Congress's PVUSA and Reagan's clean coal demonstrations show that they are a durable part of energy policy.

A similar argument can be made for the further separation of generic research on nonnuclear technologies and all nuclear R&D, which could be accomplished by creating an exclusively nuclear agency, from weaponry to commercial reactors. Separation would prevent the rationale for a governmental presence in civilian uses of nuclear energy from infecting decisions about other energy technologies and energy policy more generally. The Department of Energy (or an Energy Administration in the

Department of the Interior)[18] would focus on energy regulation, energy policy analysis, and generic research on nonnuclear energy technologies.

Summary

The preceding analysis is closely related to more general points that have been made for decades in other economic analyses of federal R&D.[19] Successful commercial demonstration programs have one overriding characteristic: the basic technology supporting them must be fairly well understood before a commitment is made to commercial demonstration. Walter Baer and others made several additional observations in their 1976 report: successful programs are integrated with research, they are not very expensive and not very visible politically, and they have a built-in tolerance for parallel activities and schedule slippages.[20] Nancy Rose later added the important insight that commercialization objectives should not be combined with other political objectives in which success is measured more by the production of a working technology than by satisfying the demands of a market test.[21] She also reiterated Nelson, Peck, and Kalachek's arguments concerning the differences between programs to develop basic technical know-how (generic R&D) and those that have specific commercial applications.

The lessons from our work are broadly consistent with these views. Expanding the range of technological opportunities available for industry through the promotion of generic research avoids some of the problems exhibited by commercial demonstration programs. Specifically, it can impede the rush to commercial demonstration. Moreover, by concentrating effort on the early, relatively small tail of the time distribution of expenditures for introducing new technologies, generic centers are more likely than commercial demonstrations to avoid the late-stage problems associated with distributive politics. Finally, they are unlikely to be threatening to incumbent firms in an industry if the knowledge is openly produced and

18. Interior has responsibility for managing and assessing important energy resources.

19. See especially Walter S. Baer, Leland L. Johnson, and Edward W. Merrow, "Analysis of Federally Funded Demonstration Projects: Executive Summary," R-1925-DOC (Santa Monica, Calif.: Rand Corp., 1976); and Richard R. Nelson, Merton J. Peck, and Edward D. Kalachek, *Technology, Economic Growth and Public Policy* (Brookings, 1967).

20. Baer and others, "Analysis," pp. 10–11.

21. Nancy L. Rose, "The Government's Role in the Commercialization of New Technologies: Lessons for Space Policy," Working Paper 1811-86, Sloan School of Management, Massachusetts Institute of Technology, 1986.

shared, if the industry is itself doing little or no generic research, and if the decisions about demonstration projects are separately made in another political forum.

The proposal to support generic research, however, has a serious problem: it is not clear that it responds to the political forces that give rise to R&D commercialization projects. Specifically, in times of crisis the presence of blue-sky work is not likely to be a satisfactory response to the political imperative to act. Indeed, there is a danger that the presence of generic research centers for a spectrum of industries will constitute a source of ideas, not to mention proponents, for great new technologies that the private sector is overlooking and that with only a few hundred million federal dollars could be brought to commercial practice. An unsettling precedent is the synthetic fuels program of the 1970s, which was shaped by the small, unsuccessful generic research program in the Office of Coal Research during the 1960s.

Regardless of the ultimate merits of either generic programs or commercialization projects, the lesson seems to be to keep them institutionally separated. Doing so creates independent sources of technical knowledge and reduces the chance of a confused melding of technical and economic objectives. It also separates the activities that are most susceptible to the vicissitudes of pork barrel influences and of crisis politics from the fundamental objective of expanding the technological base of industry, which is most in need of managerial flexibility and long-term policy stability. Finally, it maximizes the chance that commercialization decisions can be separated from ex ante decisions about the lines of research to pursue in generic centers and made more explicitly dependent on the ex post products of research to expand the technological base.

Conclusion

The analysis in this book leads to the conclusion that the commercial aerospace projects between Sputnik and the space shuttle, and the energy projects of the 1970s and early 1980, were exceptional. Nevertheless, the question remains what specific conditions could cause another round of exceptions. If connection to a salient national issue is necessary, three candidates emerge: international competitiveness, space competition with the Russians, and energy disruptions in the Middle East.

During the 1980s the United States suffered large deficits in its balance of trade in manufactured and primary products and, concomitantly, declining production and employment in some key traditional industries. Because the phenomenon was common to many industries and most regions of the country, it attained national political significance. Although the causes and potential cures of the trade deficit and economic stagnation in several key industries are a complex issue, increased spending on commercial R&D is widely touted as an attractive political response.

Nonetheless, the status of international trade will probably not cause a major commitment of federal funds to commercial R&D. First, such a commitment would be an uncertain and very long-term response to a problem that is salient because of its immediate effects. Should political leaders have to act on the trade issue, they are unlikely to find that long-term R&D will relieve the political pressure placed on them by people whose present jobs and businesses are in trouble. Second, the saliency of the international competitiveness issue is not likely to last as long as the time horizon for R&D commercialization projects. If so, the support coalition behind such a program would be likely to disintegrate soon after the program began, much as happened to energy R&D in the early 1980s. Third, many R&D projects are likely to fail, a fact political leaders will know from the beginning. Because expected efficiency does affect decisions, and because many in Congress on both left and right will be philosophically predisposed against these projects, resistance to an R&D-based industrial policy will be high.

As for the two other candidates for a salient issue to motivate R&D commercialization activities, a reprise of Soviet successes in space and another world energy crisis are both likely during the 1990s. The Soviet space program was substantially revivified in the 1980s, and several impressive projects with great potential visibility are scheduled, culminating in the return of samples from the surface of Mars about the year 2000. Meanwhile, the world remains vulnerable to disruptions in oil supplies from the Middle East. The final consequences of the attack by Iraq on Kuwait, another conflict among oil-producing states, or an internal revolution in Saudi Arabia could recreate conditions resembling those of 1973 and 1979.

Because both Soviet successes in space and another oil crisis are likely, a sensible policy would be to decide now whether a commitment to commercial R&D for space and energy constituted part of a good response to these circumstances and, if so, to begin now a careful, unpanicked set of R&D programs in those areas. Probably that will not happen; despite its long-term attributes, R&D seems most salable as a reaction to a crisis. The problem seems to us to be fundamental; we have not been able to fashion any realistic way to avoid it. Here we propose only to advertise the existence of the problem and hope for the best. Specifically, even opponents of R&D commercialization projects should consider whether a research-oriented program in space and energy technologies with possible commercial potential is not good insurance for warding off to some degree a wasteful, ill-directed response to likely future crises.

We are not sanguine about the prospects that this or any other recommendation about structuring the process by which decisions are made will dramatically raise the batting average of R&D commercialization projects. The recommendations do not address many of the managerial problems we have identified, including the mismatch of market failure rationales for projects with the structure of political incentives. Nevertheless, the knowledge that this is an imperfect world should not cause one to lose the will to improve matters where one can. Two firm lessons can be drawn from the nation's experiences with commercialization projects: it is desirable to maintain a base program designed to expand commercially relevant technological knowledge, and it is desirable to separate these programs institutionally—in Congress and in the federal bureaucracy—from basic scientific research and from operational responsibilities.

Index